T0360553

Cambridge Studies in Biological and Evolutionary Anthropology 56

Between Biology and Culture

Humans adapt to their environment through a unique amalgamation of culture and biology. Both are intrinsic to our existence and constitute the dual aspect of human nature. This book addresses topics and themes exploring this close interrelationship by presenting principles and applications of scientific approaches to human remains. Their appreciation within a human ecological context, incorporating conditions of the natural environment as well as cultural, social and political circumstances of the past, provides the framework for the detection and interpretation of our biocultural identity. Written for academic researchers and students alike, *Between Biology and Culture* assembles chapters that encompass topics from taphonomy to individual life histories, from seasonality to food, from well-being to disease, from genetics to mobility, and from body theory to forensic individualization. In doing so, the contributions probe the potential of skeletal analysis to look beyond the face value of observations and to detect the biological outcomes of cultural strategies encoded in human remains.

HOLGER SCHUTKOWSKI is a Reader in Biological Anthropology in the Division of Archaeological, Geographical and Environmental Sciences at the University of Bradford, UK.

Cambridge Studies in Biological and Evolutionary Anthropology

Series editors

Also available in the series

40 *Shaping Primate Evolution* Fred Anapol, Rebecca L. German & Nina G. Jablonski (eds.) 0 521 81107 4

41 *Macaque Societies – A Model for the Study of Social Organization* Bernard Thierry, Mewa Singh & Werner Kaumanns (eds.) 0 521 81847 8

42 *Simulating Human Origins and Evolution* Ken Wessen 0 521 84399 5

43 *Bioarchaeology of Southeast Asia* Marc Oxenham & Nancy Tayles (eds.) 0 521 82580 6

44 *Seasonality in Primates* Diane K. Brockman & Carel P. van Schaik 0 521 82069 3

45 *Human Biology of Afro-Caribbean Populations* Lorena Madrigal 0 521 81931 8

46 *Primate and Human Evolution* Susan Cachel 0 521 82942 9

47 *The First Boat People* Steve Webb 0 521 85656 6

48 *Feeding Ecology in Apes and Other Primates* Gottfried Hohmann, Martha Robbins & Christophe Boesch (eds.) 0 521 85837 2

49 *Measuring Stress in Humans: A Practical Guide for the Field* Gillian Ice & Gary James (eds.) 0 521 84479 7

50 *The Bioarchaeology of Children: Perspectives from Biological and Forensic Anthropology* Mary Lewis 0 521 83602 6

51 *Monkeys of the Taï Forest* W. Scott McGraw, Klaus Zuberbühler & Ronald Noë (eds.) 0 521 81633 5

52 *Health Change in the Asia-Pacific Region: Biocultural and Epidemiological Approaches* Ryutaro Ohtsuka & Stanley J. Ulijaszek (eds.) 978 0 521 83792 7

53 *Technique and Application in Dental Anthropology* Joel D. Irish & Greg C. Nelson (eds.) 978 0 521 870 610

54 *Western Diseases: An Evolutionary Perspective* Tessa M. Pollard 978 0 521 61737 6

55 *Spider Monkeys: The Biology, Behavior and Ecology of the Genus Ateles* Christina J. Campbell 978 0 521 86750 4

Between Biology and Culture

Edited by

Holger Schutkowski

University of Bradford

Shaftesbury Road, Cambridge CB2 8EA, United Kingdom

One Liberty Plaza, 20th Floor, New York, NY 10006, USA

477 Williamstown Road, Port Melbourne, VIC 3207, Australia

314–321, 3rd Floor, Plot 3, Splendor Forum, Jasola District Centre, New Delhi – 110025, India

103 Penang Road, #05–06/07, Visioncrest Commercial, Singapore 238467

Cambridge University Press is part of Cambridge University Press & Assessment, a department of the University of Cambridge.

We share the University's mission to contribute to society through the pursuit of education, learning and research at the highest international levels of excellence.

www.cambridge.org
Information on this title: www.cambridge.org/9780521859363

First published 2008
First paperback edition 2012

A catalogue record for this publication is available from the British Library

Library of Congress Cataloging-in-Publication data
Between biology and culture / edited by Holger Schutkowski.
p. cm. – (Cambridge studies in biological and evolutionary anthropology)
Includes index.
ISBN 978-0-521-85936-3
1. Physical anthropology. 2. Ethnology. 3. Human biology. I. Schutkowski, Holger.
GN60.B48 2009
599.9 – dc22 2008030111

ISBN 978-0-521-85936-3 Hardback
ISBN 978-1-107-41065-7 Paperback

Contents

Contributors

Dr Ian Barnes
School of Biological Sciences, Royal Holloway University of London,
Egham, Surrey, TW20 0EX, UK

Professor Sue Black OBE
Anatomy and Forensic Anthropology, School of Life Sciences, MSI/WTB
Complex, University of Dundee, Dow Street, Dundee, DD1 5EH, UK

Professor Don Brothwell
Department of Archaeology, University of York, The King's Manor,
York Y01 7EP, UK

Mr Michael Buckley
BioArCh, Department of Biology (S Block), University of York, PO Box 373,
York YO10 5YW, UK

Dr Enrico Cappellini
BioArCh, Department of Biology (S Block), University of York, PO Box 373,
York YO10 5YW, UK

Professor Matthew J. Collins
BioArCh, Department of Archaeology, University of York, The King's
Manor, York Y01 7EP, UK

Professor M. Thomas P. Gilbert
Department of Biology, University of Copenhagen, Universitesparken 15,
DK-2100 Copenhagen Ø, Denmark

Dr Rebecca C. Griffin
School of Dental Sciences, Edwards Building, University of Liverpool,
Daulby Street, Liverpool L69 3GN, UK

Dr Louise T. Humphrey
The Natural History Museum, Department of Palaeontology, Cromwell Road, London, SW7 5BD, UK

Dr Hannah Koon
BioArCh, Department of Archaeology, University of York, The King's Manor, York Y01 7EP, UK

Dr Kirsi Lorentz
School of Historical Studies, Newcastle University, Armstrong Building, Newcastle NE1 7RU, UK

Professor Gabriele A. Macho
Division of Archaeological, Geographical and Environmental Sciences, School of Life Sciences, University of Bradford, Bradford BD7 1DP, UK

Professor Donald J. Ortner
Smithsonian Institution, National Museum of Natural History, Department of Anthropology, Washington DC 20560, USA

Dr Kirsty E. H. Penkman
BioArCh, Department of Chemistry, University of York, Heslington, York YO10 5DD, UK

Professor T. Douglas Price
University of Wisconsin-Madison, Laboratory for Archaeological Chemistry, Observatory Drive, Madison WI 53706, USA

Dr Holger Schutkowski
Division of Archaeological, Geographical and Environmental Sciences, School of Life Sciences, University of Bradford, Bradford BD7 1DP, UK

Professor Beth Shapiro
Department of Biology, The Pennsylvania State University, 208 Mueller Laboratory, University Park, PA 16802, USA

Professor Richard H. Steckel
Department of Economics, Ohio State University, 410 Arps Hall, 1945 N High Street, Columbus OH 43210, USA

Foreword

Between Biology and Culture arose out of the inaugural symposium of the Biological Anthropology Research Centre (BARC) at the University of Bradford. The Research Centre was established in 2003 and the symposium, entitled 'Biological Anthropology at the Interface of Science and Humanities', was held on 24 May of the same year.

The University of Bradford has a long tradition of international-level, science-based research in the analysis of human skeletal remains, both ancient and modern. The BARC was formerly the Calvin Wells Laboratory and houses the largest collection of human skeletal remains in an archaeology department in the UK, including material that dates from the Neolithic to the nineteenth century. Consistent with the legacy and the current strengths of the new centre, this volume emphasizes how human skeletal remains can be used to both infer human lifestyle and living conditions in the past and investigate aspects of human identity in the context of the cultural and natural environment. The theme running throughout the volume is that humans, unlike other animals, adapt to their environment through both biology and culture. It is the task of modern biological anthropologists to disentangle the effects of these two variables on human adaptation in the past.

Between Biology and Culture showcases modern, largely British, research that meets this challenge and also points the way towards promising avenues for future research. The papers in the volume can be divided roughly into two general areas. The first stressing fundamental aspects of skeletal analysis and the second focusing on the application of cutting edge techniques to analyse the interface between biology and culture. Examples in the first category include papers on determining identity of human remains in forensic anthropology, assessing human well-being from skeletal material, analysing disease ecology, relating seasonality and climate change to the hominin fossil record, and reconstructing past human diets. In the second category, contributions demonstrate how different approaches to isotope analysis can be used to gather dietary information from calcified tissues and determine the age of weaning of individuals, as well as human migration and mobility patterns. DNA analysis has heralded a revolution in our understanding of humans in the past and contributions presented here both review the use of genetics in the understanding of phylogenetic, kinship and individual genetic relationships, and highlight taphonomic considerations that must be taken into consideration in the application of modern genetic approaches.

What makes this volume all the more impressive is the relatively difficult funding environment in which UK-based biological anthropologists work today. It is down to the strong motivation and ambitious vision of Holger Schutkowski and his colleagues at Bradford that the BARC both exists and is a vibrant centre for research into human skeletal biology. It is also down to the creativity of UK-based biological anthropologists that quality research continues to be carried out on what scientists in other fields and in other countries might consider a shoe-string.

Leslie C. Aiello

Acknowledgements

The idea for this collection of essays arose from a symposium, funded by the Wellcome Trust, and held to mark the inauguration of the Biological Anthropology Research Centre (BARC) at Bradford in 2003, in which most of the authors assembled here participated. It has been an interesting journey since, not always straightforward, but nonetheless highly rewarding and pleasurable. I am most grateful to all authors who, despite busy schedules and a multitude of commitments, agreed to contribute, and I would like to extend my special thanks for their co-operation and intellectual company during the gestation and growth of this book. I am also indebted to those colleagues who most thankfully gave their time for peer-review and thus greatly helped improve the final product. My great appreciation goes to Cambridge University Press, first Tracy Sanderson and then Dominic Lewis and Alison Evans, who have been most supportive and forthcoming throughout. Thanks are due to Clare Lendrem for excellent copy-editing and to Eleanor Collins for smooth production of the volume. Finally, I would like to express my gratitude to present and former members of the BARC: Anthea Boylston, Jo Buckberry, Christopher Knüsel, Alan Ogden, Robert Pastor and Darlene Weston, without whom there would have been little to inaugurate.

1 *Introduction*

HOLGER SCHUTKOWSKI

Petrarch's famous ascent of Mount Ventoux in April 1336 is often said to be the first recorded example of contemplative nature experience as a value of its own, a sharp departure from the perception of nature in the context of utilitarian appropriation. Whether or not this interpretation is true, Petrarch's own account of this day (1999, p. 11) is contemplative in quite a different manner; self-reflective, pondering human virtues and vices, and realising weaknesses and imperfections in his own life course. We may be tempted to view this as part of a quest for meaning and identity, and as such as an allegory for being able to look beyond the face value of observations and facts. In this respect, then, it would resemble what the contributions in this volume try to convey by probing the potential of skeletal analysis.

It is true, what human skeletal remains reveal to us in the first instance is factual, informative, and often highly conducive to answering scientific questions. But their meaning does and must go beyond that, and is probably best described in a biocultural interpretive framework. Even though the chapters in this book represent a somewhat diverse array of essays, there are two broad themes common to the approaches put forward here; (i) the detection of human lifestyle and living conditions, providing an understanding of the cultural setting in which populations lived; (ii) the investigation of aspects of human geographical, genetic or social identity in the context of the cultural and natural environment.

This is what the title *Between Biology and Culture* attempts to invoke. Humans, unlike other animals, adapt to their environment through both culture and biology. In the course of evolution, the cultural component has assumed increasing importance and has become the signifying feature of humans. Through the invention and permanent modification of cultural traits humans have found a unique way to respond to and cope with the environment and its constraints. But however sophisticated their material representations, social organisations, subsistence modes or belief systems, these attempts eventually have repercussions on their biology, their chances for survival, opportunities for

Between Biology and Culture, ed. Holger Schutkowski. Published by Cambridge University Press.

reproduction and prospects of well-being. Human identity is essentially shaped by the biological outcomes of cultural strategies. Hence, humans sit firmly, yet not always comfortably, between their biology and culture.

To disentangle the interplay and mutual dependence of human biology and culture in past populations requires a holistic effort. The approach outlined here has its starting point in the skeletal evidence, largely represented by human remains from archaeological assemblages, which constitutes the most direct and immediate access to the study of past human biology. A wide range of advanced analytical techniques are now available to provide the data that facilitate interpretations of biocultural adaptations and identities.

Human identity is multifaceted, despite a common notion that it is always social and therefore, by inference, cultural (Jenkins, 2004, p. 4). It is in fact expressed through the interplay of biological and cultural representation in a broad sense, and would thus have to be seen as more inclusive to reflect these two sides of human nature. The majority of chapters in this volume actually address issues of identity, albeit not always explicitly, but certainly in accord with fundamental elements of its concept, i.e. similarity, difference, classification or association (Jenkins, 2004, p. 4). All these elements are connected with a process of identification, which in analytical terms would refer to the statement of links, traits, correlations and evidence sufficient to eventually establish causal relationships and meaning, which constitute and help ascribe identities.

Human activities and behaviour are characterised by cultural mediation of biological needs and expressions. Culture is intrinsic to human nature; we literally cannot live without it. Consequently, human presence in any ecological setting leaves a cultural imprint; the natural environment is shaped by cultural activity. Adaptation therefore is biocultural, and it occurs against local conditions as much as its outcomes reflect local diversity (Gamble, 2007, p. 277). Human biocultural adjustment to prevailing and fluctuating surroundings thus inevitably creates a multitude of identities, variations of those fundamental categories of place, resource and behaviour. Translated into ecological terms these identities reflect flows of matter, energy and information, the control, steering and manipulation of which is a hallmark feature of the human ecological niche (Schutkowski, 2006). The ability of humans to create and respond to the cultured environment is a cornerstone of the dual human nature. Whether biology or culture takes precedence is almost a moot point, as the two are so intimately intertwined to the point of being inseparable.

The consequence of human dependence on culture kits for survival and the maintenance of life support systems is that information gleaned from the skeletal record always has a cultural connotation, which is eventually reflected in the biochemical, morphological or genetic signatures. For example, a dietary signal ascertained from isotope analysis can indicate what food components contributed to an individual's or a population's diet. While at face value an observation may establish the consumption of C_3 plant protein with some input from marine resources, this information at the same time provides us with a better understanding of the subsistence regime and strategies of resource use viewed against the environmental setting of a certain locality. Arguably, this is what we really want to find out. Likewise, the detection of differences in dietary habits between groups of people within a society, whether socially or biologically defined, reveals more than just group-level distinction in food consumption patterns, but rather the socio-cultural framework of availability and access to resources. Governance of material and energy flows thus becomes measurable in biological or molecular terms and, through classification and association, allows aspects of identity to become evident.

This is even more obvious when human remains reveal ancestry and relatedness through genetic information as the most direct and immediate signifier of identity. The term genetic fingerprint speaks for itself to denote uniqueness at an individual level. But similar or even shared traits at group and population level are the molecular equivalent of a biologically founded identity, which is at the basis of a strong sense of belonging to genealogical entities. Kinship is essential for the formation of social and behavioural identities, from the day-to-day conduct within a community to marriage patterns, inheritance and mythology (e.g. Parkin and Stone, 2003). Being able to start to reveal patterns of kinship in the remote and not so distant past provides a crucial tool for hinging human biology and culture at a very fundamental level, and thus permits us to ascribe cultural meaning to genetic evidence. One may even want to mention appearance here, including morphological shape, accepting that there is distinctiveness behind variation, both culturally induced and biologically evolved.

Often, this coincides with place, geographical location or habitat. Human remains are a biocultural archive, which holds information that can be taken as a reflection of geographical identity. The cycling of chemical elements in the biosphere creates distinctive patterns visible in, for example, isotopic signatures that distinguish local from foreign, in exceptional cases even unveil traces of origin. Beyond the face value of these signatures there are clues that reveal mobility, migration and the formation of communities. But they also provide access to more elusive behavioural patterns, such as the formation of new identities by immigrants into existing, established communities, when the molecular information is combined with artefactual archaeological evidence. On the other hand, social identities can be strong and so persistent that communities would adhere to the known and familiar even under the stress of dramatically changing environmental conditions, as the example of the Norse communities in Greenland demonstrates (McGovern *et al.*, 1988). Identity may thus serve as a socio-cultural buffer that maintains resilience until ecological disruption becomes overwhelming and destabilising.

It would seem, therefore, that the analysis of the tangible evidence of human remains and their interpretation against ambient environmental or ecological circumstances allow us to go beyond this level and explore what is at the heart of human nature – a biocultural identity.

The collection of essays assembled in this volume therefore aims to explore approaches that allow this agency to be discovered. The scope ranges from taphonomic aspects to individual life histories, from seasonality to food, from well-being to disease, from genetics to mobility, from body theory to forensic individualisation. No specific reference will be made to the acquisition of basic anthropological information, such as age or sex from the skeleton. These aspects have recently been discussed elsewhere in the context of social archaeology (e.g. Gowland and Knüsel, 2006). After a brief historical review (Chapter 2) the biocultural theme is developed first through the consideration of fundamental aspects; the establishment of identity in forensic investigations (Chapter 3), approaches to the measurement of human well-being in past and present populations (Chapter 4), the ecology of disease (Chapter 5), the importance of seasonality and climate change for the interpretation of the fossil hominid record (Chapter 6), and connotations of food and the reconstruction of past diets (Chapter 7). This is followed by a suite of essays that focus on the use of certain analytical techniques to address the biology/culture interface; the detection of early life history events taking weaning as an example (Chapter 9), the elucidation of phylogenetic, kinship and individual genetic relationships (Chapter 10) and the meaning of mobility and migration in the past (Chapter 11), preceded by an appreciation of basic taphonomic considerations and their impact on the survival of molecular and chemical information in archaeological human remains (Chapter 8). Body theory and the materiality of human remains (Chapter 12) throw a bridge back to the initial identity theme.

Chapter 2 by Don Brothwell is a concise personal account on the study of human populations over the last 150 years or so. Written by one of the doyens of international biological anthropology, the chapter offers an increasingly rare opportunity for an educating journey through topics and time, outlining developments and illustrating milestones. It is, not surprisingly, concerned with the nature of human variation and with the many attempts to systematise the complexity and interrelatedness of aspects that shape human nature in order to gain a better understanding of what biologically constitutes the uniqueness of our species. Contributions often came from scholars whose academic affiliation was outside anthropology. While this could be perceived as benevolent amateur preoccupations with a discipline that is struggling to find its focus, it was, in fact, advantageous and tremendously helpful in developing the genuine multi-valence approach of biological anthropology. The chapter identifies certain areas as having been drivers towards the development of a comprehensive scientific remit, from evolutionary aspects to adaptability and population genetics, at the same time alluding to major themes discussed in the chapters to follow. In particular, human palaeontology still captivates large audiences in the attempt to satisfy a widespread curiosity about our ancestry in deep time. No matter whether one tends to be a lumper or a splitter, the role of environmental change in shaping the evolution of large brains, and morphological and behavioural variation in general, remains a key issue, and the importance of ecological considerations in clarifying this cannot be underestimated. In a similar vein, this also holds for the appreciation of Holocene biometric and non-metric human skeletal variation. Essentially, at the heart of this notion is the question of how the nature-nurture dichotomy can be reconciled by integrating the properties of the natural and social environments in our interpretation of growth patterns and demographic profiles or the remarkable human capacity for adaptability. Brothwell's conclusion that the study of human remains and, in particular, the deciphering of their chemical and molecular information, will continue to grow and ramify (Katzenberg and Saunders, 2008), anticipates one of the threads running through this volume.

Quite unlike connotations in the popular trivialisation of crime scene work in the media, forensic anthropology has a deeply humanitarian and ethical role; the biological identification of human remains with a view to establishing identity and thus providing not only closure to an ongoing inquest, but making an important contribution to the healing and grieving process. In Chapter 3, Sue Black describes the context and the practical implications of forensic anthropology and its service to justice. Subject to courtroom scrutiny, the sometimes arduous task of extracting biological information from often degraded, commingled or otherwise compromised remains seeks to attach statistical probability to 'sameness' of remains under study and a putative target individual, in order to assign beyond reasonable doubt the congruence of information needed to match biological profile and real persona. Verification of methods and procedures used to ascertain this information is crucial, and the combination of antemortem records and postmortem data in achieving identification must be guided by scientific rigour. Case studies relating to investigations of homicides and war crimes, as well as to aiding the identification of victims from natural mass disasters, demonstrate what a powerful tool forensic anthropological analysis, with its multiple strands of evidence gathering, has become. But it shows first and foremost how instrumental it is in helping to find certainty and consolation for family and friends, and in establishing identity for the deceased.

To take anthropological data beyond individual characterisation and to collate them at population level or even in meta-analyses of multiple assemblages offers the possibility for the detection of long-term trends and patterns in the standard of living of both biological and socio-cultural groups within past societies. Whilst measures to assess quality of life have a long tradition in economics using goods and services as proxy currencies, the use of (skeletal) biological information has only recently been recognised as a promising method to identify the biological standard of living, health and, by implication, well-being. In Chapter 4, Richard Steckel provides a detailed account of the potential of this cross-over approach by comparing traditional material-led analyses with the obvious advantages provided by direct observation of archaeological human remains. One major conclusion, borne out by the Western Hemisphere Project (Steckel and Rose, 2002), is that the mass-statistical analysis of certain skeletal indicators, assembled in a health index, against social and ecological variables does produce meaningful correlations and evidence of biocultural adaptation. These results are inevitably broad-brushed, but this is inherent to this type of analysis. It is even more difficult, however, to match trends derived from the skeletal record with a widespread concept in the social sciences: that of happiness. But it might be worthwhile to explore if and how this translation could be tackled. The chapter ends with a fascinating outlook on the potential role of nanotechnology in the assessment of health and well-being in modern times, leaving the reader with a sense of disappointment over the naturally limited scope of skeletal data in the face of prospects to measure brain activity and biochemical markers as a reflection of well-being.

No doubt, the detection of overall associations in global studies between terrain, vegetation or settlement patterns and morbidity is an encouraging start. But there remains the need for a better understanding of the ecological factors, both natural and cultural, that govern the ability of humans to survive under the widest possible range of environmental conditions, yet at the potential risk of increased susceptibility to disease. Whatever the occurrence and frequency of skeletal lesions are actually able to reveal about health in the past (see Wood *et al.*, 1992), the incorporation of ecological concepts currently applied to questions in public health and disease control into palaeopathological interpretation seems to hold promise. By focusing on certain aspects of human disease ecology in Chapter 5, Donald Ortner and Holger Schutkowski explore the effects of the interplay of biological and cultural factors in the co-evolution of human hosts and pathogens, while at the same time considering the reflection of deficient natural environments in skeletal alterations. The latter provide an inexhaustible reservoir of instructive case studies, of which some are presented here, for example fluorosis and scurvy. Major subsistence transitions and mobility appear to have had the greatest impact on human morbidity. Fundamental changes, such as the adoption of agriculture and sedentism, would have resulted in the re-adjustment of material and energy flows in a given habitat and provided the means for novel socio-cultural expression as well as population growth. However, they would have also been accompanied by challenges or even threats due to zoonotic or vector-borne diseases. Expansion into new habitats, be it through pioneering or conquest, provided new pathways for infections and spread of diseases, often with devastating consequences. Despite such difficulties biocultural adaptation supports at all levels the survival and maintenance of individuals and populations.

In deeper time, evolutionary change in the hominin line is frequently associated with environmental variability as a proxy for circumstances that allow selective advantages to become manifest. Analysis of mineralisation patterns archived in dental enamel provides a tool that utilises indelible recording structures for the reconstruction of seasonality, here as an indication of rainfall, and the effect of short-term oscillations in environmental conditions on evolutionary pathways. In Chapter 6, Gabriele Macho suggests that it is the unpredictable nature of such short-term variability rather than global climatic fluctuations, which would have likely created a scenario where large brains, and by implication more behavioural flexibility, provided the capacity for innovation and thus for better and more sustained chances for differential reproduction and survival. In contrast, longer-term variance in climatic conditions seemed to have little effect on, for example, the evolution of Australopithecines, while the combination of overall change and short-term unpredictability of environmental conditions, i.e. the constant challenge of coping with changing circumstances on different scales, would have facilitated the evolutionary emergence of large-brained species of the genus *Homo*. The explicit incorporation of ecological/environmental information to interpret climate-sensitive micromorphological features from both fossil hominids and associated faunal remains not only contributes to reconstructing scenarios of climate change but to establishing the environmental evolutionary framework of our own remote ancestry.

Arguably, much of our evolutionary history is connected with the necessity to discover, explore and use food resources from a variety of very different habitats. Environmental variation at variable scales and the selection pressure towards bigger brains and a flexible behavioural repertoire would have greatly helped the inquisitiveness required to exploit the advantages of a truly omnivorous and non-specialist diet. Since the invention and mastery of controlled fire, at the latest, humans have developed a capacity to take food consumption beyond the mere satisfaction of basic needs. How far back exactly this can be traced will probably remain unknown forever, but it would appear reasonable to assume that today's pronounced desire to use food and diet as a signifier of cultural identity represents the sophisticated echo of a tradition that is connected with the very beginnings of humanity. Humans transfer diet into meaning and in Chapter 7 Holger Schutkowski investigates how food as cultured biology can be analysed in the skeletal record. Chronic or sustained episodes of nutritional stress, often culturally induced, can be detected as osseous pathological alterations but would rather reflect the negative side of a deficient condition. Isotopic analysis, however, has generated data that allow direct dietary information to be gleaned from calcified tissues. This has led to the detection of major and more subtle dietary differences on the level of entire groups or populations. Being able to identify such patterns then facilitates their interpretation against ecological basic conditions and socio-cultural variation to reflect both adaptation through human/environment interaction and self-expression.

Any attempt to extract such biomolecular information from archaeological samples rests on the assumption that we have a basic understanding of those processes that govern the preservation of this information, or at least, given the complexity of it, understand enough to ascertain whether the data generated are reliable and not taphonomically compromised. In Chapter 8, Matthew Collins and colleagues have contributed an erudite lesson on the potentials as well as the numerous pitfalls associated with this approach. By taking ancient proteins as an example, they guide us through the different stages of decay that contribute to deterioration but also differential preservation of this promising group of biomolecules. Some of the deterioration that can be detected is culturally induced, for example the kind of heat treatment we refer to as cooking, which causes proteins to be structurally altered and in this modification become indicative of the thermal history of a sample. Another form of chemical modification in proteins, racemisation, the gradual transformation of one enantiomer into its mirror sterical configuration, has become an important dating tool. Protein residues in prehistoric pottery are able to take the identification of dietary components beyond the bulk analysis, while the detection of protein sequences offers a possibility to differentiate faunal samples at the genus level and thus contribute to the understanding of resource exploitation patterns. The list of potential applications is long and the prospects exciting.

The combination of isotopic analysis and the recording of macroscopic mineralisation defects in dental enamel offer a unique opportunity to discover events in early stages of life history, as Louise Humphrey demonstrates in Chapter 9. One of those stages is the weaning period, when breastfeeding is gradually replaced by the introduction of solid foodstuffs. Whilst in dietary terms this simply constitutes the transition from a carnivorous to an omnivorous human being, the physiological and often also psychological effects can be traumatic enough to leave visible traces, known as linear enamel hypoplasias. The increasing intake of protein sources other than breast milk alters the isotopic signal recovered from dental tissues formed at various points in early life and thus helps define the transition period more exactly. A detailed knowledge of the timing of weaning in past populations supports the interpretation of mortality patterns associated with those periods when children are particularly susceptible to disease and immunologically compromised. High or low mortality rates around the time of weaning also reflect the buffer capacity of individuals and populations and are therefore a powerful indicator of living conditions and coping mechanisms under different socio-ecological regimes, in other words a proxy measure for successful biocultural adaptation. Perhaps even more importantly, the identification of patterns in the timing of weaning has an evolutionary ecological dimension (Voland, 1998). Not only is an earlier onset of weaning associated with a more sedentary as opposed to a foraging nomadic lifestyle, it also allows inferences to be made about birth intervals and thus offers a potential route into elucidating reproductive behaviour and parental strategies in the past.

Since the invention of the polymerase chain reaction (Mullis and Faloona, 1987) the study of ancient and modern DNA has received a boost commensurate with the significant advances it has facilitated for the detection of genetic relationships between taxa, populations and individuals, not only, but particularly, related to our hominin ancestry and anatomically modern humans. In Chapter 10 Beth Shapiro and colleagues review a number of high-profile applications where DNA has helped to clarify or shed new light on conventional views. With regard to our own phylogenetic relations there seems to be consistent evidence now that significant evolutionary differences exist between Neanderthals and modern humans, even though the absolute extent of this is currently under investigation. The reconstruction of gene flows plays an important role here. Migration events, which have shaped the population, and thereby genetic, history of countries and geographical areas can now be underpinned by DNA data and provide a much better understanding of genetic roots and relationships. The Anglo-Saxon colonisation of Britain is a case in point for frequent large-scale genetic admixtures throughout Europe which, despite the distinct cultural differences we are seeing today, should remind us of continued shared and common ancestry (Cavalli-Sforza *et al.*, 1994). The analysis of DNA equally contributes to the elucidation of natural selection and human variability, for example in those traits that are linked to major biocultural developments (lactose tolerance) or adaptations that were once beneficial but prove deleterious today (thrifty genes) under changed environmental and cultural conditions.

Sparked by the seminal study on Bell Beaker people ten years ago (Grupe *et al.*, 1997) strontium, and later lead, oxygen and sulphur, isotope analysis has become an established tool for attempts to provenance materials and people. Comparable to studies using DNA, it is now being widely applied to investigate migration and mobility in the past, and contribute answers to the paradigmatic question of whether it is ideas or people that travel (or both) and spread biocultural innovation in their course. In Chapter 11 Douglas Price develops this idea, providing examples mostly from the Americas, by demonstrating how foreign individuals can be identified in skeletal assemblages and whether or not a likely place of origin can be assigned. While isotopic ratios are usually very useful in revealing those samples that do not match local isotopy, it is much harder to tell where these immigrants actually came from. Very detailed knowledge and data of the surrounding geology, soils and water are necessary to make any informed suggestions, as ongoing discussions over the provenance of the 'Amesbury Archer' demonstrate. Three prominent examples, from Grasshopper Pueblo, Teotihuacán and Copán are successful in both respects; they help understanding of the supra-regional importance of a proto-urban pueblo site, as well as the social fabric and funerary archaeology of ancient cities. The case study of the Iceman should be taken as a memento that different approaches and different results may in fact support the same conclusion. Provenance analysis from the chemical composition of calcified tissues provides geographical identity and reflects a life history, not just the circumstantial place of burial.

The bioarchaeology of human remains does encompass all the themes introduced here, but has its poignant representation in an attempt to interpret the body as an expression of socio-cultural concepts and circumstances. Archaeological human remains allow for the reconstruction of the living body, because skeletal traits that denote shape, size and alteration of form due to, for example, disease or cultural practice reflect events and episodes of individual life histories. The living body is mindful and active, it is engendered and material and hence much more than La Mettrie's notion of 'man a machine'. It is in fact a product of both biological and cultural influences, and the dynamic of these influences translates into skeletal appearance – an obvious example would be culturally prescribed cranial deformation. As Kirsi Lorentz advocates, in Chapter 12, there are both the need and the means for a more comprehensive approach that uses different, complementary avenues to take the meaning of skeletal remains beyond the bone.

The continued advancement of diagnostic methods, both morphological and molecular, helps establish the biological identity of human skeletal remains in unprecedented detail and accuracy. Yet, only its appreciation within a human ecological context, i.e. by incorporating conditions of the natural environment as well as cultural, social and political circumstances of the past, provides the framework required for the detection and interpretation of what is at the heart of human dual nature, our biocultural identity. Be it individual life histories, ascertained through osteobiography, or signifiers at group or population level, human remains hold the keys to understanding our place between biology and culture.

References

Cavalli-Sforza, L. L., Mendozzi, P. and Piazza, A. (1994). *The History and Geography of Human Genes*. Princeton: Princeton University Press.

Gamble, C. (2007). *Origins and Revolutions. Human Identity in Earliest Prehistory*. Cambridge: Cambridge University Press.

Gowland, R. and Knüsel, C. (eds.) (2006). *Social Archaeology of Funerary Remains*. Oxford: Oxbow.

Grupe, G., Price, T. D., Schröter, P. *et al.* (1997). Mobility of Bell Beaker people revealed by strontium isotope ratios of tooth and bone: a study of southern Bavarian skeletal remains. *Applied Geochemistry*, **12**, 517–25.

Jenkins, R. (2004). *Social Identity*. 2nd edn. London: Routledge.

Katzenberg, M. A. and Saunders, S. R. (eds.) (2008). *Biological Anthropology of the Human Skeleton*. 2nd edn. New York: Wiley-Liss.

McGovern, T. H., Bigelow, G. F., Amorosi, T. and Russell, D. (1988). Northern islands, human error, and environmental degradation: a preliminary model for social and ecological change in the medieval North Atlantic. *Human Ecology*, **16**, 45–105.

Mullis, K. B. and Faloona, F. (1987). Specific synthesis of DNA in vitro via a polymerase-catalysed chain reaction. *Methods in Enzymology*, **155**, 335–50.

Parkin, R. and Stone, L. (eds.) (2003). *Kinship and Family: An Anthropological Reader*. Oxford: Blackwell.

Petrarca, F. (1999). *Selections from the "Canzoniere" and other works*, transl. M. Musa. Oxford: Oxford University Press.

Schutkowski, H. (2006). *Human Ecology. Biocultural Adaptations in Human Communities*. Berlin: Springer.

Steckel, R. H. and Rose, J. C. (eds.) (2002). *The Backbone of History: Health and Nutrition in the Western Hemisphere*. New York: Cambridge University Press.

Voland, E. (1998). Evolutionary ecology of human reproduction. *Annual Review of Anthropology*, **27**, 347–74.

Wood, J. W., Milner, G. R., Harpending, H. C. and Weiss, K. M. (1992). The osteological paradox: problems of inferring prehistoric health from skeletal series. *Current Anthropology*, **33**, 343–70.

2 *Historic dimensions to the biological study of human populations*

DON BROTHWELL

There are probably well over 50 000 references in biological anthropology which could be considered in a detailed evaluation of the history of the subject. Any reduction of this vast literature is likely to present some bias, although there are clearly major themes which should not be excluded. My concern in this brief review is to provide as balanced account of the development of the subject as possible. Slotkin (1965) has provided some evidence that we can extend the history back well before 1800, and indeed Johann Blumenbach (1752–1840) is usually regarded as the 'father' of the subject. Nethertheless, I will maintain that the nineteenth century was one of special relevance, with a significant expansion of interest in our species at a scientific level.

Because of the extensive literature, there has clearly been a need to be very selective and, when appropriate, books rather than single papers are quoted. The subject has had its share of controversies, most being resolved now. While biological variation in human evolution is accepted now as basically the result of Darwinian natural selection, the place of cultural factors in our evolution is still one of the topics demanding further debate and exploration. What then should be included in a review of biological anthropology? Clearly we are concerned with the evolutionary history of our species, and to a lesser degree the primates as a broader perspective. The nature of human variation, at a molecular, structural, growth, demographic, epidemiological and behavioural level, provides the basic data of the subject. Contributors to this field have often not had academic status as anthropologists, and the bibliography to this review provides ample evidence of researchers who are associated with departments of genetics, biochemistry, medicine, zoology, anatomy, palaeontology and demography. This has been to the great advantage of the subject, has ensured the intrusion of ideas and expertise from other sciences, and has assisted in the maturation of the subject. Humans are the most complex, the most unique, the most planet-threatening species, and there remains an urgent need to understand this – at times – frightening animal.

Between Biology and Culture, ed. Holger Schutkowski. Published by Cambridge University Press.

The nineteenth century

In the two centuries during which biological anthropology has mainly been developing, I would guess that the majority of my colleagues would see the most progress as being during the twentieth century. But I confess to being far from convinced of this. Of course at an academic level, and in terms of numbers of publications, it became firmly established last century, but for the excitement of exploring new territory, studying human groups in the nineteenth century may well have been far more thrilling.

William Lawrence, a London surgeon, published in 1819 his 'Lectures on Comparative Anatomy, Physiology, Zoology and the Natural History of Man', and interest in this subject was such that the book ran into at least seven editions. A year after this edition, Samuel Morton (1839) initiated ancient skeletal studies in America, and at his death in 1851, his collection amounted to parts of 900 individuals. Clearly he was concerned with populations and their variation, not type specimens. Others followed, and it is interesting that Jeffries Wyman in 1871 was stating 'brain measurements cannot be assumed as an indication of the intellectual position of races any more than individuals' (Buikstra, personal communication).

In Europe, Prichard (1843) had continued a broad interest in human variation, and while regional differences especially intrigued these early workers (Nott and Gliddon, 1854; Zeising, 1854), skeletal variation, pigmentation, hair form and even 'acclimatisation' were to some extent discussed. This last topic was particularly discussed by Nott (1857) who realised that climate and the impact of disease could influence the nature of human populations. Such early investigators also realised that cultural factors could impinge on morphology, and Gosse (1855) provides one of the early studies of head deformation.

With the appearance of Darwin's (1859) 'Origin of Species', studies on human groups to some extent changed gear. Incidentally, the secondary title to this book was 'The Preservation of Favoured Races in the Struggle for Life'. The term 'races' was used in preference to 'varieties' and applied to all the species he considered; I mention this because 'race' has become such an abused and controversial term in the past fifty years.

In the same year as Darwin's masterpiece, Paul Broca (Schiller, 1979) began to publish on Europeans. Shortly after, Lyell (1863) attempted to consolidate evolutionary theory with a wide range of geological and palaeontological observations, and it is some indication of public interest that the book ran into three editions within the year. Would such a thing happen today? The same year also saw C. Loring Brace (1863) write on Old World 'races' with a comment that the variation was influenced by linguistic and geographic factors. Further writing of this kind, as well as methodology, included Huxley (1863; 1866), Topinard

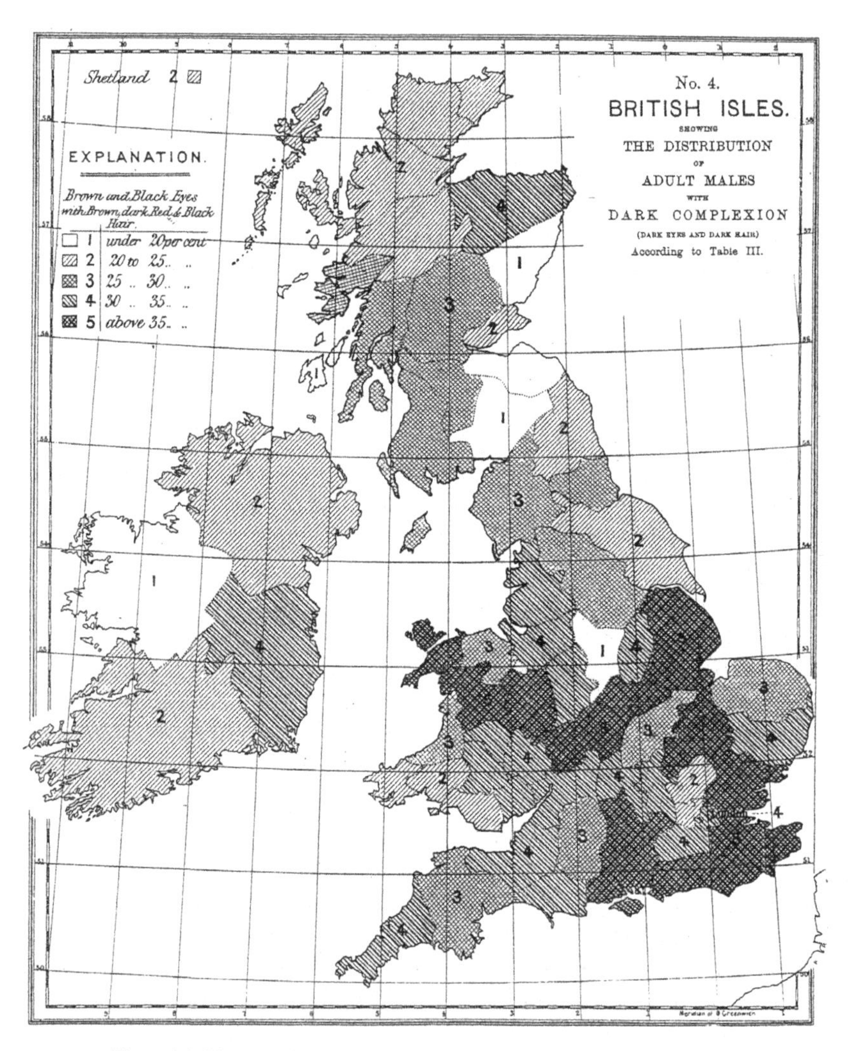

Figure 2.1 Pigmentation map of Britain, from Roberts, 1878.

(1876), Bowditch (1877), Roberts (1878), Broca (1879), Quatrefages (1883) and Beddoe (1885), see Figures 2.1 and 2.2.

Methodology and pigmentation were also demanding more attention (Schmidt, 1888). These decades also saw the Societé d'Anthropologie de Paris (1859) founded, as well as the Deutsche Anthropologische Gesellschaft (1866) and Royal Anthropological Institute (1870) and journals. Such organisations allowed a greater interaction of the early anthropologists and those of similar interests, and it is important to remember that fossil hominids were by then being

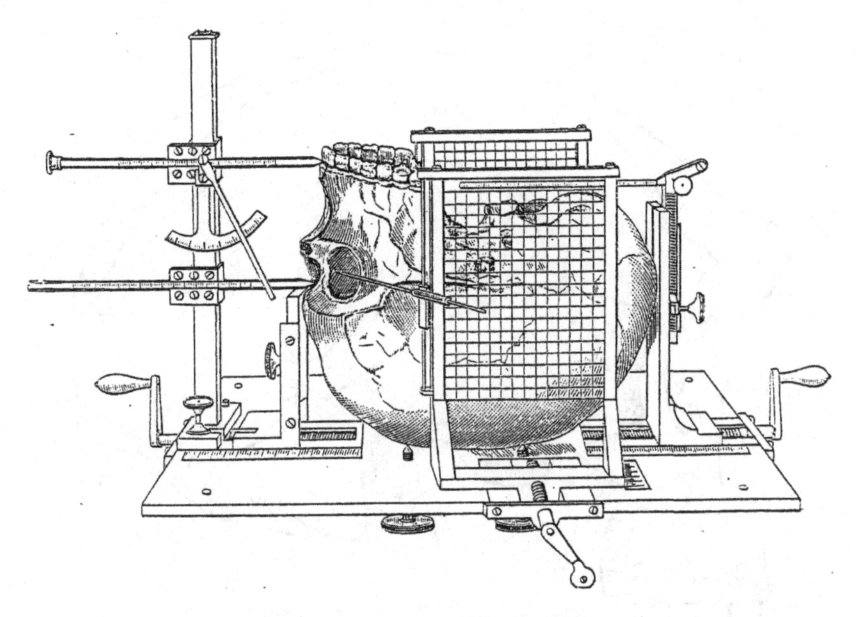

Figure 2.2 Early craniometric apparatus (Schmidt, 1888).

described (Figures 2.3 and 2.4), especially Neanderthal and Upper Palaeolithic remains (Oakley, 1964). Although this was a period when human population genetics was undeveloped, interest in variation in relation to inheritance was debated, and Galton (1883), the British mathematician whose life work was the investigation of human variation and heritability, provided considerable stimulus in this respect, including the investigation of twins. The term 'eugenics' (good birth) was first used in 1883, and was associated in particular with Galton. As in the case of animal and plant breeding, he wondered how improvements might similarly be made in human populations. His studies included the heritability of intelligence, a research topic which has caused much controversy over the years. By 1907, eugenic societies had started to be established, but the subject continued to attract much debate, and there were rightly fears about possible sterilisation laws. Today, the term tends to be avoided, but of course social medicine is active in advising families on the dangers of perpetuating seriously damaging genes, e.g. Huntington's chorea. The long-term result of such medical advice may be that certain detrimental genes will be slowly lost from populations, although mutation will replenish such genes to some extent.

Disease, both ancient and modern, was attracting various specialists and observers. Statistical reports on morbidity and mortality clearly indicated regional differences (War Office, 1853, for instance). Specific diseases were being noted in relation to the past, and Whitney (1883) records what he

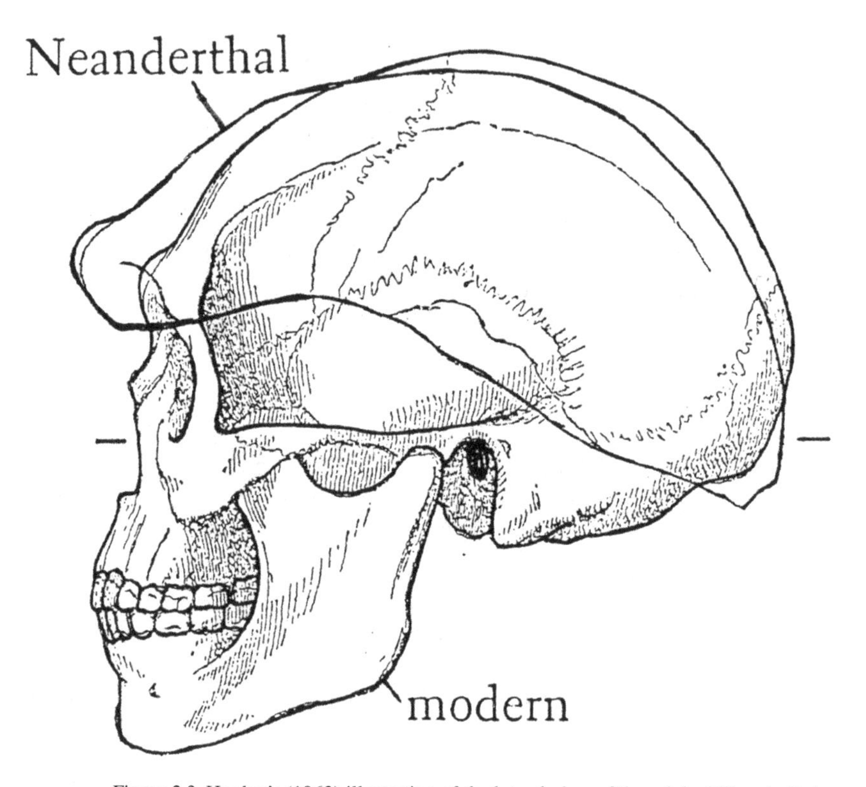

Figure 2.3 Huxley's (1863) illustration of the lateral view of the original Neanderthal calotte superimposed over a modern skull.

believes to be syphilis in pre-Columbian America. Bland Sutton (1890) moreover attempted to view disease within a broader evolutionary background, and stated that 'disease is controlled by the same laws which regulate biological processes in general'.

In the final year of the nineteenth century, Ripley (1899) consolidated the common interest in evaluating 'races', which was to continue well into the next century. And Haeckel (1899) was to include *Pithecanthropus alalus* (speechless) into an evolutionary tree, together with *Homo stupidus*. The latter species has been totally ignored by the scientific community, although it seems to me to have far greater justification than the more arrogant *Homo sapiens*, which our history over the last century clearly proves to be ridiculous.

The twentieth century

While the greatly expanding literature of this century demands a sub-division of topics for discussion, some introduction would seem necessary. In some

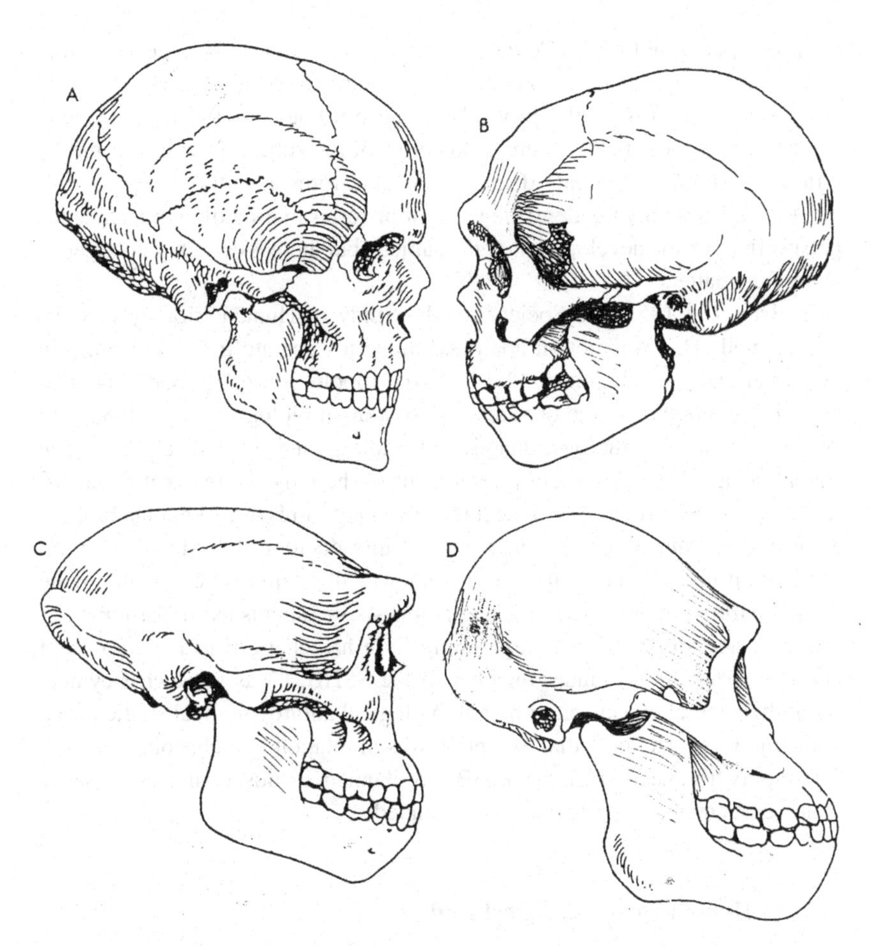

Figure 2.4 Fossil species known by the end of the nineteenth century: (A) Upper Palaeolithic, (B) Neanderthal, (C) *Homo erectus*, together with a more recent Australopithecine find.

respects, research and production of literature continued through into the twentieth century. Laboratory guides or handbooks have periodically appeared (Duckworth, 1904; Martin, 1928; Montagu, 1960; Brothwell, 1963a; Weiner and Lourie, 1969). Most have been especially good in traditional recording in field or laboratory, but the extent of investigations, especially at a laboratory level, has now meant that we have gone beyond general guides. How, for instance, does one include the interpretation of CT Scans, or procedures for isotope analysis, or the techniques of dental microwear analysis? These issues are addressed comprehensively in more recent volumes (e.g. Katzenberg and Saunders, 2008).

It is probable that we have also 'outgrown' detailed histories of the subject. Brief reviews were appearing early in the century (Haddon, 1910), as well as more specialised ones (Comas, 1969; Spencer, 1982). A large and more definitive work (Spencer, 1997) failed to include most currently active teachers and researchers in the field, probably because it could no longer adequately represent everyone. Is the future of histories of the subject to be answered by a range of sub-divisions into dental, molecular, forensic anthropology and so on? I would certainly not deny the value of histories in providing much needed perspectives on the development of the subject, but there is certainly no urgency about this matter.

In Britain in 1957, the Society for the Study of Human Biology (SSHB) was formed. This was not an apartheid move to separate from colleagues in the American Association of Physical Anthropologists or European equivalents, but to emphasise the broader base of human biology, with its links into anatomy, genetics, ethology, adaptive physiology and other disciplines. The annual symposia by this society demonstrated the range of research interests, and some of these volumes will be referred to later. A résumé of human biology as seen in Britain is outlined in a general introductory text (Harrison *et al.*, 1964). A primary hope of this endeavour was that universities would create such departments, or at least change emphasis in established departments of anatomy, and so on. In fact surprisingly little has changed in Britain except in one respect. As a founder member of the SSHB, I was somewhat cynical about the general expansion of human biological anthropological studies at an academic level. The potential exception was in relation to archaeology, and in the roughly fifty years since the SSHB was planned, in most countries the study of ancient remains has significantly expanded.

Human palaeontological studies

Without doubt, the investigation of fossil and sub-fossil skeletons and populations has been a major impetus in defining and developing biological anthropology. It was clear early on that Upper Palaeolithic and Neanderthal groups were physically distinctive and displayed tool kit differences. With the discovery of *Pithecanthropus erectus* (Dubois, 1894) there was no doubt that there were at least three forms in the evolutionary tree leading to our species. The discovery of a contrasting mandible from Heidelberg (Schoetensack, 1908) seemed to provide another fossil species, if not genus. These various fossils, the sites, faunas and possible antiquity, were reviewed by Keith (1915; 1931), one of the outstanding anatomists of the time. Sadly, he failed to understand the incompatibility of the mandible (Figure 2.5) and partial cranium from Piltdown, and eventually these and the associated fauna were shown to be fakes (Weiner, 1955; Spencer, 1990). With the discovery of an immature Australopithecine skull in South Africa (Dart, 1925), the main building blocks of an evolutionary

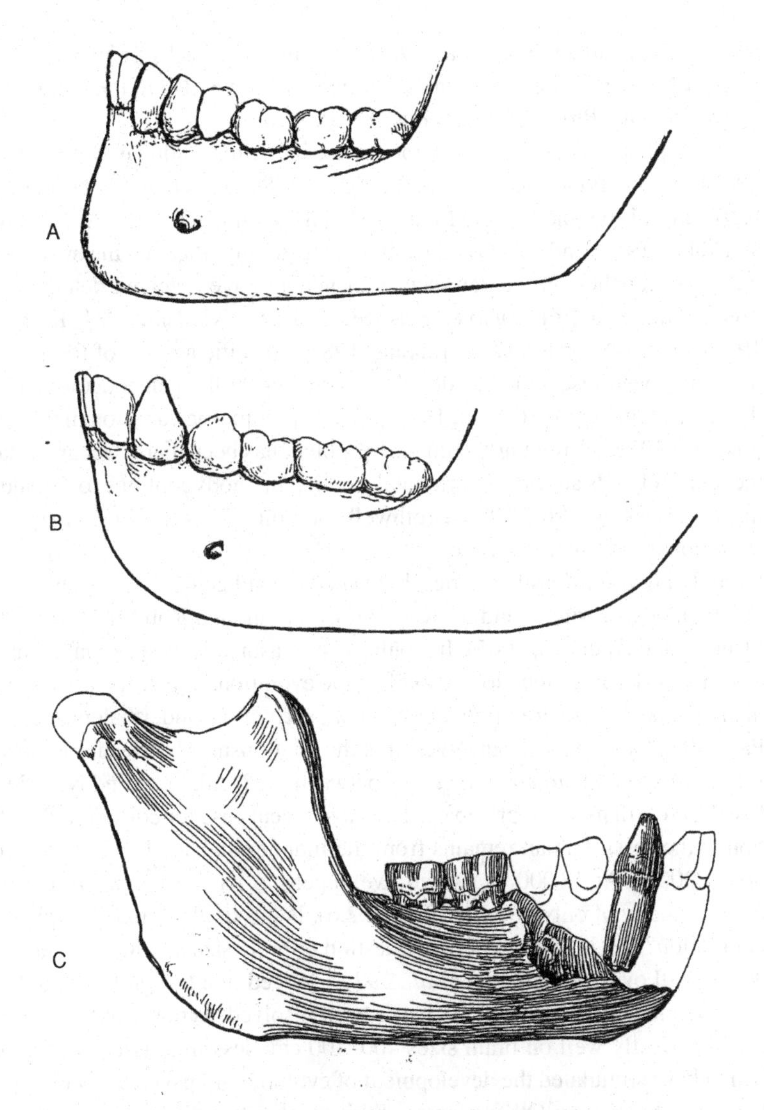

Figure 2.5 The Piltdown mandible (C) (Woodward, 1913), compared with Topinard's (1876) views of a modern human (A) and chimpanzee (B) mandible.

sequence appeared to be in place. Added to these fossils by 1856 were discoveries of Miocene hominoids, and other genera were added to the list over time (*Dryopithecus*, *Proconsul*, *Limnopithecus*).

Subsequent discoveries at various sites, including Olduvai Gorge (Tobias, 1967), Koobi Fora (Leakey and Leakey, 1978), Ternifine (Arambourg and Hoffstetter, 1963) the western Mediterranean (de Lumley, 1973), Amud (Suzuki and Takai, 1970) and various other sites, have greatly filled out the evolutionary tree and branches. Particular species of genera have received detailed attention, as in the case of *Homo erectus* (Sigmon and Cybulski, 1981; Rightmire, 1990) and the Neanderthals (Trinkaus, 1983). Specific aspects of fossil groups have also been researched in detail, as for instance the palaeodemography of the Australopithecines (Mann, 1975), and morphometric variation in this group (Oxnard, 1975). A regularly recurring problem has been that of securely dating the hominid finds and sites. Fortunately, dating methods continue to expand and improve (Oakley, 1969; 1980; Brothwell and Pollard, 2001), thus impacting on evolutionary evaluation (Aitken *et al.*, 1993).

Publications on fossil varieties, both specific and general texts, have continued to appear (Stringer and McKie, 1998; Barham and Robson-Brown, 2001; Stringer and Andrews, 2005). It would seem reasonable to state that many discoveries and studies tend to consolidate the evolutionary picture as established many years ago (Australopithecines, *Homo erectus*, Neanderthalers and Upper Palaeolithic ancestors of ourselves), but the relationships between, for instance, *Homo habilis* and *Homo ergaster*, or between the Heidelberg and Neanderthal fossils, are still matters for debate. The most recent case for considerable reflection is the fossil human remains from the Indonesian island of Flores (Brown *et al.*, 2004). At 18 000 years ago, we appear to have a pygmy form with a brain capacity of only 380 cm^3 (Figure 2.6). Is the single brainbox typical of a population? If it is, then the whole question of brain size and IQ will need to be debated all over again (Hahn *et al.*, 1979). Indeed, the reason for large brains of some modern individuals has never been resolved, as many in the world get along perfectly well on brain sizes 400–500 cm^3 less. It is issues of this kind which have stimulated the development of evolutionary psychology or psychological anthropology. Other aspects of this much-expanding subject include the nature of language, abnormal mental states and ‘ethnopsychology’ (Schwartz *et al.*, 1992; Ingham, 1996). Although there are now numerous palaeoecological studies in relation to fossil hominids (Howell and Bourlière, 1964; Tuttle, 1975; Coppens *et al.*, 1976; Jolly, 1978; Behrensmeyer and Hill, 1980; Vrba *et al.*, 1995), it is really not possible to call on environmental change and its challenges to explain brain increase and variation. To me, it seems equally difficult to argue in the early Pleistocene for behavioural or proto-cultural changes as strong-enough selection pressures.

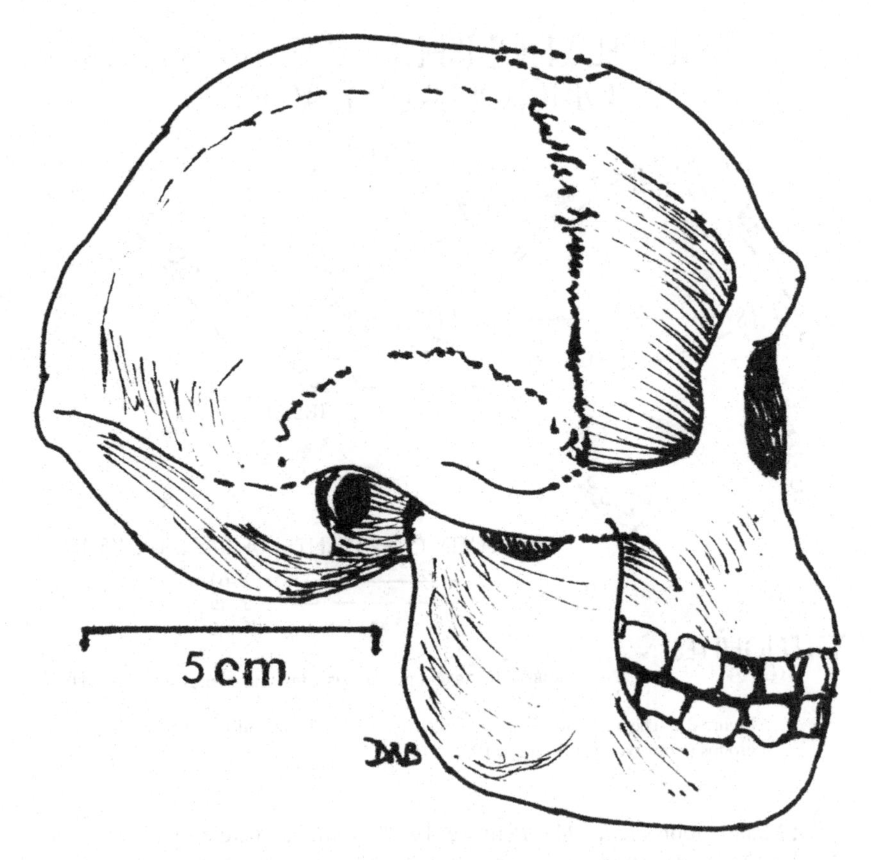

Figure 2.6 The small-brained Flores skull, dated to circa 18 000 years.

Biometric and non-metric studies of Holocene populations

While the theme of race continued to occupy some biological anthropologists this century (Coon, 1939; 1966; Morant, 1939; Demerec, 1950; Garn, 1965; Bernhard and Kandler, 1974; Molnar, 1975), there were growing doubts about the value of such classifications (Montagu, 1952; Livingstone, 1962). Race was also caught up in the debate about the long-term regional evolution of human groups, the 'multiregionalism' which Coon (1962) favoured and Wolpoff and Caspari (1997) perpetuated.

Because skeletal material is excavated regularly, and particularly spans the period from about 4000 BC to 1500 AD, there are many studies of these remains, and it has been a constant research challenge (Hooton, 1930; Larsen, 2002). At an osteometric level, there have been refinements and alternative methods of statistical analysis (Figure 2.7) offered over the years (Pearson, 1926; Fisher,

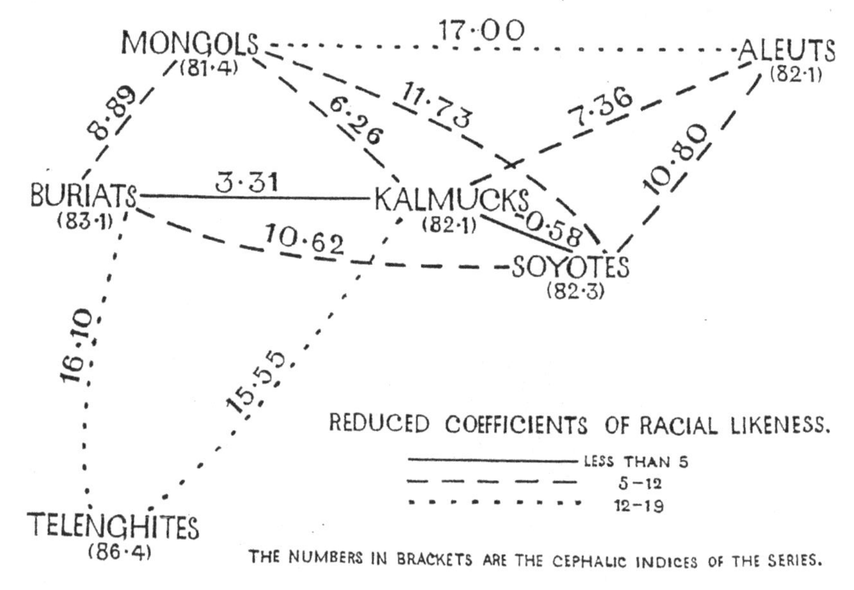

Figure 2.7 The application of Pearson's CRL to the classification of some Asiatic groups (from Woo and Morant, 1932).

1936; Mahalanobis *et al.* 1949; Penrose, 1954; Weiner and Huizinga, 1972; van Vark and Howells, 1984). The extensive work by Howells (1995) has demonstrated the continued value of such studies. Somewhat in parallel with metric studies has been the development of surveys of non-metric skeletal variation (Le Double, 1903; Berry and Berry, 1967; Berry, 1974; Finnegan, 1978), although there is still a need to understand the genetic background to such skeletal variation. At a personal level, I have somewhat moved ground over the past fifty years, and now suspect that environmental factors affecting growth and development probably are more important in determining non-metric traits than I had originally considered.

The human dentition has also come under closer study over the years, both metric and non-metric variation (Campbell, 1925; Hellman, 1928; Dahlberg, 1945; Pedersen, 1949; Moorrees, 1957; Garn, 1959; Brothwell, 1963b; Dahlberg, 1971; Kurtén, 1982; Hillson, 1996; Scott and Turner, 1997). While the studies considered regional variation, microevolution, growth and even genetic aspects, new lines of research continue to be explored (Ryan, 1979; Covert and Kay 1981; Dobney and Brothwell, 1986). With the recent development of micro-CT scanning, we may well see the development of a new range of studies on the dentition, both fossil and recent.

Growth and demography

Linked to both skeletal and dental development are questions of age and sex, but studies on living populations have also shown an increasing concern with demographic parameters of anthropological relevance (Brass, 1971; Harrison and Boyce, 1972; Zubrow, 1977). Since Malthus (1798), population studies have greatly diversified, no more so than in the field of palaeodemography and historical demography (Glass and Eversley, 1965; Acsadi and Nemeskéri, 1970; Swedlund and Armelagos, 1976; Hassan, 1981; Hoppa and Vaupel, 2002). But there are still major problems, especially with early skeletal populations. The evaluation of sexual dimorphism, in my experience, is not always as simple as textbooks encourage us to believe. Moreover, the accuracy of ageing adults is usually to within ten years, and in the case of those of 50+, the range of the estimate can be much more. Clearly, there is also a need to evaluate world variation in growth, and understand the impact of the environment, natural and social, on growth rates and variation (Boas, 1912; Shapiro and Hulse, 1940; Eveleth and Tanner, 1976; Hoppa and FitzGerald, 1999).

Since the time of Thomas Dwight (1843–1911), a 'father figure' of forensic anthropology, there have been close links with those researching on ageing and sexing, as well as the question of ethnic identity. So there is a tendency to find considerable overlap in the interests of forensic anthropology (Krogman, 1962; Gustafson, 1966; Stewart, 1979; Byers, 2002; Haglund and Sorg, 2002), and bioarchaeology, but this is to the advantage of both.

Human adaptability and ecology

Studies on the nature of human adaptability began to gather momentum about four decades ago (Baker and Weiner, 1966; Bresler, 1966), and during the following decade there was significant activity in this field (Damon, 1975; Cohen, 1977; Collins and Weiner, 1977; Hardesty, 1977; Harrison, 1977; Baker, 1978). These studies covered a wide range of research under the general heading of human adaptability, and include adaptations to food stress, high altitudes and temperature extremes. While adaptation at that time was initially more concerned with living or recent groups, its relevance to the past has become more discussed in more recent decades (Minnis, 1985; Eckhardt, 2000). However, at a skeletal, rather than a modern physiological or theoretical level, it is far from easy to isolate traits which can be regarded as markers of an adaption to, say, high altitudes or extreme heat. Clearly this is an area deserving further research.

From blood groups to molecular anthropology

It was becoming clear by the early years of the twentieth century, that human blood group reactions were variable and their genetics far from simple (Nuttall, 1904). Within two decades, there were a variety of studies concerned with the extent of blood group types, relationships with disease, and variation within human populations (Buchanan and Higley, 1921; Landsteiner and Levine, 1929; Steffan, 1932; Boyd, 1939; 1950; Race and Sanger, 1950; Mourant, 1954). The gene frequency maps which were being constructed showed major clines and some puzzling variation (Figure 2.8). But for those of us who were students in the 1950s, they demonstrated clearly that clear-cut racial differences between tribes or other regional groups did not exist, and instead there were gene-frequency gradients from one local group into another. Boyd's (1950) 'Genetics and the Races of Man' in particular became an influential text, and he included what

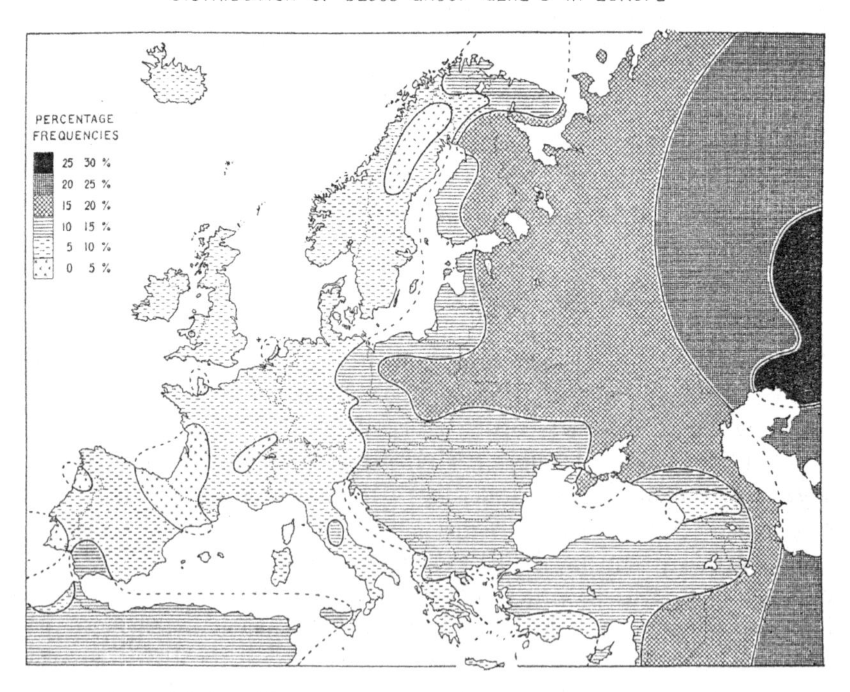

Figure 2.8 One of the gene maps produced in Mourant's (1954) classic study.

was then known of the biology of pigmentation, of taste thresholds to phenylthiocarbamide, mid-digital hair, colour blindness, and other variations. This biological cocktail of variables, together with information on growth, dermatoglyphics, anthropometry and demography, became the standard methodology to apply to field studies on specific populations (Salzano and Freire-Maia, 1970; Littlewood, 1972; Friedlaender, 1975; Harper and Sunderland, 1986).

Serology and human biochemical genetics continued to contribute significantly to an understanding of human microevolution. The abnormal haemoglobin which produced sickle-cell anaemia (Pauling *et al.*, 1949) was found to protect against malaria (Allison, 1954). Such findings closely indicated that polymorphisms could have important adaptive value (Ramot, 1974). Other serum proteins (e.g. haptoglobins) were found (Harris, 1959), but of more importance perhaps are the human lymphocyte antigens, proteins on the surface of white blood cells, which also display much polymorphism (Brock and Mayo, 1978).

'Molecular anthropology' is said to have come of age at a Wenner Gren meeting (Washburn, 1963), and has developed rapidly since then (Goodman, 1976; Cavalli-Sforza *et al.*, 1994; Boyce and Mascie-Taylor, 1996). A good example of how this field is having an impact on the population history of a region is given by Crawford's (1998) detailed evaluation of the 'Origins of Native Americans'. But such precise molecular information does not, however, eliminate controversies regarding the number of founding lineages involved. While some would seem to favour mitochondrial DNA in considering variation and the spread of modern populations (Cann, 1985; see also Chapter 10), others look towards nuclear DNA. Kidd and colleagues (1991), and others, have expressed support for the view that nuclear DNA is likely to provide better evidence of evolutionary relationships.

As in the 'old' genetics, studies on DNA variation in populations do not exclude the need to consider founder effect, drift and mutation in human groups. There is another dimension which is probably more relevant to a consideration of humans in contrast to other species, and that is 'relaxed selection' (Post, 1971). Human society, by its complexity and unusual altruistic and protective measures, cuts away the harshness of Darwinian selection, at least to some degree. There can therefore be survival in the gene pool of variants which would not have survived in much earlier societies. For instance, albinism or colour blindness may have been severely disadvantaged in some earlier environments.

Disease and epidemiology

While interest in mummies extended back into the nineteenth century, as also did medical historical mapping (McGlashan, 1972), palaeopathology and palaeoepidemiology became subjects of considerable academic interest in the early decades of the twentieth century (Elliot Smith and Wood-Jones, 1910; Ruffer, 1921; Pales, 1930). This work lost some impetus with the coming of the Second World War, but has since seen very considerable research in these fields (Brothwell and Sandison, 1967; Steinbock, 1976; Buikstra, 1981; Aufderheide and Rodrígues-Martín, 1998; Ortner, 2003). This work was concerned not only with the accurate diagnosis of pathology and establishing its antiquity, but in viewing the evidence in population terms. Indeed, there now appears to be a trend towards major reviews of specific diseases or aspects (Arnott *et al.*, 2003; Roberts and Buikstra, 2003; Powell and Cook, 2005).

In parallel with such studies has been a growing interest in changes in the pattern of disease, both in tribal societies and advanced western societies (Rothschild, 1981; Trowell and Burkitt, 1981). There has also been growing interest in well-preserved ancient bodies, and the extent to which biological information can be derived from soft tissues (Hansen *et al.*, 1991; Spindler *et al.*, 1996; Lynnerup *et al.*, 2003; Raven and Taconis, 2005). Also, of course, anthropologists in general are now conscious of the fact that health and disease issues affect all communities and influence the daily lives and beliefs of communities (Alland, 1970; Grollig and Haley, 1976; Moore *et al.*, 1980). In this general field of medical anthropology, the most neglected but difficult aspect is that of mental illness. It is of course difficult enough to evaluate mental instability or more advanced psychopathological states in populations today, and therefore it has been viewed as somewhat hazardous to extend back such studies in time. Nethertheless, progress is being made and it is at least established as a research field within anthropology (Mezzich and Berganza, 1984; Bailey, 1987; Fábrega, 2002). The related field of sociobiology has also considered mental illness (Wenegrat, 1984), and indeed the extent to which it occurs in younger individuals raises the question of the reproductive fitness of these members of a community.

Sociobiology began with Wilson's (1975) classic text, and gained support in biology and anthropology (Barash, 1979; Reynolds *et al.*, 1987), as well as justified criticism (Sahlins, 1976). Interest in evolutionary aspects of human behaviour and culture has a longer history (Schmidt, 1936; Washburn, 1961; Montagu, 1968; 1973; Chapple, 1970), but the main thrust of research has been in the past two decades (Standen and Foley, 1989; Foley, 1991; Oliver *et al.*, 1994; Runciman *et al.*, 1998; Dunbar *et al.*, 1999; Richerson and Boyd, 2005). How the evidence from skeletal biology can provide evidence for behavioural interpretation has also been reviewed in some detail (Larsen, 1997). The evolution of language has particularly held the attention of numerous researchers (Dunbar, 1996; Noble and Davidson, 1996; Carruthers and Chamberlain, 2000; Knight *et al.*, 2000). Some of this debate is frankly puzzling, in the sense that numerous species of primate and other mammals communicate by body or sound signals, so that its elaboration in large-brained humans is not so surprising. Once the value of differential sounds is understood, proliferation of such 'language' inevitably grows.

Discussion and conclusions

This brief review of historic aspects of biological anthropology will, I hope, provide a reasonably balanced assessment of the subject from its nineteenth century beginnings through to modern times. Perhaps I could begin by saying that work in this field seems to be relatively healthy, and in a sample of 73 individuals who have contributed in some way to the field over the past two centuries, nearly 36 % died between their 80th and 89th year. Also, there is no evidence that we display increased suicide rates in comparison with the population at large, unlike the medical and veterinary profession in general. There are still plenty of problems to stimulate our interest, and perhaps in the next few decades, the numbers of researchers in different areas of biological anthropology will change. From the evidence of over 800 articles appearing in the 'American Journal of Physical Anthropology' and 'Human Biology' between 1960 and 1969 (Salzano, 1973), 74 % covered only palaeoanthropology, osteology, population genetics, growth, anthropometry and anthroposcopy. Other categories, including comparative anatomy, dermatoglyphics, primate behaviour, physiological variation, demography and comparative pathology, were all included in the remaining 26 %.

Predicting changes in emphasis is not easy, but it seems very likely that skeletal studies will remain strong, and I would suggest that increases in this research field will be particularly in relation to chemical variation, including molecular differences (Price, 1989). Both inorganic and organic chemical analysis promise to provide answers as divergent as detecting human migration and dietary variation (the latter through both isotopic ratios and elemental concentration) as well as determining the time of weaning (Larsen, 1997; Eckhardt, 2000), to the ritual application of anointing or other activities for preparing the ancient dead (Aufderheide, 2003).

Population genetics, including of past populations, will probably also remain strong, with growth research remaining steady. In contrast, I suggest that anthropometric field work will decline. Behavioural studies seem already to be on the increase, and I would predict that further research will be productive on abnormal states of mind. For me, brain size remains the most puzzling feature of our evolution, and there is still a need to explain it convincingly. Darwin was impressed by human brain size, which had significantly increased by *Homo erectus* times, over two million years ago. The beginnings of language, early stone tool use and social modifications are flimsy arguments for such changes. Is there in fact a simpler and biologically more sustainable answer? Could the driving force for cerebral change be the earlier adaptation for bipedalism and resulting pelvic remodelling? Then, as a result of dietary changes, with increased meat eating and especially fatty acid intake was there then a greater foetal stimulation of brain growth? Pelvic remodelling was advanced by the emergence of *Homo* and could have enabled the survival, without obstetric problems, of larger-brained babies, with perhaps greater creative potential. Having said all that, there is still the additional question of why some of us have such large brains when smaller ones would do? It was recorded many years ago (Berry and Büchner, 1913) that brain capacity in a large sample of Australian criminals ranged from 1191 cm^3 to 1771 cm^3, a difference of 580 cm^3. Other evidence of small brains in normal individuals suggests that perhaps there is a 'Rubicon' of about 1000 cm^3 which enables individuals to sustain a normal life in advanced societies, but there is an urgent need for a clarification of this.

We also need to give far more consideration to mental health abnormality in evolutionary terms. It has been long known that schizophrenia is relatively common in various world populations, and neuroses even more so. What is the evolutionary significance of this? This question is beginning to be discussed (Fábrega, 2002) but demands more consideration. Were there psychotic Australopithecines, or are abnormal mental states the result of later adaptive responses gone wrong? Does increasing brain size and complexity carry with it the seeds of mental malfunctioning – just as an erect posture can be linked to back stress and pain?

Three final points. Firstly, in my lifetime in the field of anthropology, there has been a very great reduction in the number of unmodified tribal communities throughout the world. There is still a need to consider their biology, but in the future we must take into account the influences of technologically more advanced societies. Also, as urbanism takes in more and more of the world's population, adaptations or otherwise to high density situations will need study. Secondly, as it is becoming more difficult to study the biology of individuals in detail, there is a need for better collaboration with biological and veterinary colleagues who may be exploring aspects of other mammals which potentially have relevance for ourselves. For instance, if copper deficiency affects the skeleton of the calf (Suttle and Angus, 1978), then can a similar situation arise in humans? Similarly, if inbreeding in rabbits can result in skeletal variations and malformations (Chai, 1970), what changes can occur in humans in the same inbred situations?

Finally, the problem of collections, especially of skeletal remains. A century ago, there was general concern about the collection and curation of such material

(Hrdlicmčka, 1900). Now, there is the prospect of much skeletal material being returned for reburial. There is an urgent need for education, negotiation and compromise with local communities, so that the material remains available for future study, even if protected and curated by local groups (Brothwell, 2004). This is by no means linked just to tribal communities, but affects the survival of material even in Britain. So availability of material in the future, as well as changing research emphasis, will provide new challenges and problems for the future biological anthropologist. But one thing is certain; there will still be plenty of questions to ponder and explore.

Acknowledgements

I am grateful to Holger Schutkowski, for this opportunity to attempt my own brief interpretation of the history of biological anthropology. I am also most grateful to Jane Buikstra for educating me, especially in the history of the subject in the New World.

References

Acsadi, G. and Nemeskéri, J. (1970). *History of Human Life Span and Mortality*. Budapest: Akadémiai Kiado.

Aitken, M., Stringer, C. and Mellars, P. (eds.) (1993). *The Origin of Modern Humans and the Impact of Chronometric Dating*. Princeton: Princeton University Press.

Alland, A. (1970). *Adaption in Cultural Evolution: An Approach to Medical Anthropology*. New York: Columbia University Press.

Allison, A. C. (1954). Protection afforded by sickle-cell trait against subtertian malarial infection. *British Medical Journal*, **1**, 290–4.

Arambourg, C. and Hoffstetter, R. (1963). *Le Gisement de Ternifine*. Archives de L'Institut de Paléontologie Humaine, Mémoire 32. Paris: Masson.

Arnott, R., Finger, S. and Smith, C. (2003). *Trepanation: History, Discovery, Theory*. Lisse: Swets and Zeitlinger.

Aufderheide, A. C. (2003) *The Scientific Study of Mummies*. Cambridge: Cambridge University Press.

Aufderheide, A. C. and Rodríguez-Martín, C. (1998). *The Cambridge Encyclopedia of Human Paleopathology*. Cambridge: Cambridge University Press.

Bailey, K. G. (1987). *Human Paleopsychology: Applications to Aggression and Pathological Processes*. Hillsdale: Erlbaum.

Baker, P. T. (1978). *The Biology of High Altitude Peoples*. Cambridge: Cambridge University Press.

Baker, P. T. and Weiner, J. S. (eds.) (1966). *The Biology of Human Adaptability*. Oxford: Clarendon.

Barash, D. (1979). *Sociobiology: The Whisperings Within*. Glasgow: Fontana.

Barham, L. and Robson-Brown, K. (eds.) (2001). *Human Roots: Africa and Asia in the Middle Pleistocene*. Bristol: Western Academic Press.
Beddoe, J. (1885). *The Races of Britain*. Bristol: Arrowsmith.
Behrensmeyer, A. K. and Hill, A. P. (eds.) (1980). *Fossils in the Making, Vertebrate Taphonomy and Paleoecology*. Chicago: Chicago University Press.
Bernhard, W. and Kandler, A. (eds.) (1974). *Bevölkerungsbiologie*. Stuttgart: Fischer.
Berry, A. (1974). The use of non-metrical variations of the cranium in the study of Scandinavian population movements. *American Journal of Physical Anthropology*, **40**, 345–58.
Berry, A. C. and Berry, R. J. (1967). Epigenetic variation in the human cranium. *Journal of Anatomy*, **101**, 361–79.
Berry, R. J. and Büchner, L. W. (1913). The correlation of size of head and intelligence as estimated from the cubic capacity of brain of 33 Melbourne criminals hanged for murder. *Proceedings of the Royal Society of Victoria*, **25**, 254–67.
Bland Sutton, J. (1908). *Evolution and Disease*. London: Scott.
Boas, F. (1912). *Changes in Bodily Form of the Descendants of Immigrants*. New York: Columbia University Press.
Bowditch, H. P. (1877). *The Growth of Children*. Eighth Annual Report of the Board of Health of Massachusetts, Boston.
Boyce, A. J. and Mascie-Taylor, C. G. N. (eds.) (1996). *Molecular Biology and Human Diversity*. Cambridge: Cambridge University Press.
Boyd, W. C. (1939). Blood groups of American Indians. *American Journal of Physical Anthropology*, **25**, 215–35.
Boyd, W. C. (1950). *Genetics and the Races of Man*. Oxford: Blackwell.
Brace, C. L. (1863). *The Races of the Old World: A Manual of Ethnology*. London: Murray.
Brass, W. (ed.) (1971). *Biological Aspects of Demography*. London: Taylor and Francis.
Bresler, J. B. (1966). *Human Ecology: Collected Readings*. Reading: Addison-Wesley.
Broca, P. (1879). Instructions relatives a l'étude anthropologique du système dentaire. *Bulletin de la Société d' Anthropologie de Paris*, **2**, 128–65.
Brock, D. J. and Mayo, O. (eds.) (1978). *The Biochemical Genetics of Man*. London: Academic.
Brothwell, D. R. (1963a). *Digging up Bones*. London: British Museum (Natural History).
Brothwell, D. R. (ed.) (1963b). *Dental Anthropology*. London: Pergamon Press.
Brothwell, D. R. (2004). Bring out your dead: people, pots and politics. *Antiquity*, **78**, 414–8.
Brothwell, D. R. and Pollard, A. M. (eds.) (2001). *Handbook of Archaeological Sciences*. Chichester: Wiley.
Brothwell, D. R. and Sandison, A. (1967). *Diseases in Antiquity*. Springfield: Thomas.
Brown, P., Sutikna, T., Morwood, M. *et al.* (2004). A new small-bodied hominin from the late Pleistocene of Flores, Indonesia. *Nature*, **431**, 1055–61.
Buchanan, J. A. and Higley, E. (1921). The relationship of blood groups to disease. *British Journal of Experimental Pathology*, **2**, 247–55.
Buikstra, J. E. (ed.) (1981). *Prehistoric Tuberculosis in the Americas*. Illinois: Evanston.
Byers, S. N. (2002). *Introduction to Forensic Anthropology*. London: Allyn and Bacon.

Campbell, T. D. (1925). *Dentition and Palate of the Australian Aboriginal*. Adelaide: Hassell Press.
Cann, R. I. (1985). Mitochondrial DNA variation and the spread of modern populations. In *Out of Africa*, eds. E. Szathmary and R. Kirk. Canberra: Australian National University, pp. 113–22.
Carruthers, P. and Chamberlain, A. (eds.) (2000). *Evolution and the Human Mind: Modularity, Language and Meta-cognition*. Cambridge: Cambridge University Press.
Cavalli-Sforza, L., Menozzi, P. and Piazza, A. (1994). *The History and Geography of Human Genes*. Princeton: Princeton University Press.
Chai, C. K. (1970). Effect of inbreeding in rabbits; skeletal variations and malformation. *Journal of Heredity*, **61**, 2–8.
Chapple, E. D. (1970). *Culture and Biological Man: Explorations in Behavioural Anthropology*. New York: Holt, Rinehart and Winston.
Cohen, M. N. (1977). *The Food Crisis in Prehistory: Overpopulation and the Origins of Agriculture*. London: Yale University Press.
Collins, K. J. and Weiner, J. S. (1977). *Human Adapatability: A History and Compendium of Research*. London: Taylor and Francis.
Comas, J. (1969). *Historia Sumaria de la Asociación Americana de Antropólogos Físicos (1928–1968)*. Mexico: Instituto Nacional de Antropología e Historia.
Coon, C. S. (1939). *The Races of Europe*. New York: Macmillan.
Coon, C. S. (1962). *The Origin of Races*. London: Cape.
Coon, C. S. (1966). *The Living Races of Man*. London: Cape.
Coppens, Y., Howell, F. C., Isaac, G. and Leakey, R. E. (eds.) (1976). *Earliest Man and Environments in the Lake Rudolf Basin*. Chicago: University of Chicago Press.
Covert, H. and Kay, R. (1981). Dental microwear and diet: implications for determining the feeding behaviour of extinct primates, with a comment on the dietary patterns of Sivapithecus. *American Journal of Physical Anthropology*, **55**, 331–6.
Crawford, M. H. (1998). *The Origins of Native Americans: Evidence from Anthropological Genetics*. Cambridge: Cambridge University Press.
Dahlberg, A. A. (1945). The changing dentition of man. *Journal of the American Dental Association*, **32**, 676–90.
Dahlberg, A. A. (ed.) (1971). *Dental Morphology and Evolution*. Chicago: University of Chicago Press.
Damon, A. (ed.) (1975). *Physiological Anthropology*. New York: Oxford University Press.
Dart, R. (1925). Australopithecus africanus: the man-ape of South Africa. *Nature*, **115**, 195–9
Darwin, C. (1859). *The Origin of Species by Means of Natural Selection*. London: Murray.
Demerec, M. (ed.) (1950). *Origin and Evolution of Man*. Cold Spring Harbor Symposia on Quantitative Biology, XV, New York.
Dobney, K. and Brothwell, D. (1986). Dental calculus; its relevance to ancient diet and oral ecology. In *Teeth and Anthropology*, eds. E. Cruwys and R. A. Foley. Oxford: BAR International Series No. 291, pp. 55–82.
Le Double, A. F. (1903). *Traité des Variations des Os du Crâne de L' Homme*. Paris: Vigot.

Dubois, E. (1894). *Pithecanthropus erectus, eine menschenähnliche Übergangsform aus Java*. Batavia.
Duckworth, W. L. H. (1904). *Morphology and Anthropology: A Handbook for Students*. Cambridge: Cambridge University Press.
Dunbar, R. (1996). *Grooming, Gossip and the Evolution of Language*. Cambridge: Harvard University Press.
Dunbar, R., Knight, C. and Power, C. (eds.) (1999). *The Evolution of Culture*. Edinburgh: Edinburgh University Press.
Eckhardt, R. B. (2000). *Human Paleobiology*. Cambridge: Cambridge University Press.
Elliot Smith, G. and Wood-Jones, F. (1910). Report on the human remains. *Archaeological Survey of Nubia*. Vol. 11 (1907–1908), 375pp. Cairo: Egyptian Survey Department.
Eveleth, P. B. and Tanner, J. M. (1976). *Worldwide Variation in Human Growth*. Cambridge: Cambridge University Press.
Fábrega, H. (2002). *Origins of Psychopathology: The Phylogenetic and Cultural Basis of Mental Illness*. New Brunswick: Rutgers University Press.
Finnegan, M. (1978). Non-metric variation of the infracranial skeleton. *Journal of Anatomy*, **125**, 23–37.
Fisher, R. A. (1936). The 'coefficient of racial likeness' and the future of craniometry. *Journal of the Royal Anthropological Institute*, **66**, 57–63.
Foley, R. A. (ed.) (1991). *The Origins of Human Behaviour*. London: Unwin Hyman.
Friedlaender, J. S. (1975). *Patterns of Human Variation: The Demography, Genetics and Phenetics of Bougainville Islanders*. Cambridge: Harvard University Press.
Galton, F. (1883). *Inquiries into Human Faculty and its Development*. London: Macmillan.
Garn, S. M. (1959). *The Dentition of the Growing Child*. Cambridge: Harvard University Press.
Garn, S. M. (1965). *Human Races*. Springfield: Thomas.
Glass, D. V. and Eversley, D. E. (eds.) (1965). *Population in History, Essays in Historical Demography*. London: Arnold.
Goodman, M. (ed.) (1976). *Molecular Anthropology*. New York: Plenum.
Gosse, L. A. (1855). Essai sur les déformations artificielles du crâne. *Annales d'Hygiène Publique*, **3**, 317–93; **4**, 5–83.
Grollig, F. and Haley, H. B. (eds.) (1976). *Medical Anthropology*. The Hague: Mouton.
Gustafson, G. (1966). *Forensic Odontology*. New York: Elsevier.
Haddon, A. C. (1910). *History of Anthropology*. London: Watts.
Haeckel, E. (1899). *Über unsere gegenwärtige Kenntniss vom Ursprung des Menschen*. Berlin: Strauss.
Haglund, W. D. and Sorg, M. H. (eds.) (2002). *Advances in Forensic Taphonomy*. Boca Raton: RC Publications.
Hahn, M., Jensen, C. and Dudek, B. (eds.) (1979). *Development and Evolution of Brain Size*. London: Academic Press.
Hansen, J. P., Meldgaard, J. and Nordqvist, J. (eds.) (1991). *The Greenland Mummies*. London: British Museum Publications.
Hardesty, D. L. (1977). *Ecological Anthropology*. London: Wiley.

Harper, P. and Sunderland, E. (eds.) (1986). *Genetic and Population Studies in Wales*. Cardiff: University of Wales Press.

Harris, H. (1959). *Human Biochemical Genetics*. Cambridge: Cambridge University Press.

Harrison, G. A. (ed.) (1977). *Population Structure and Human Variation*. Cambridge: Cambridge University Press.

Harrison, G. A. and Boyce, A. J. (1972). *The Structure of Human Populations*. Oxford: Clarendon.

Harrison, G. A., Weiner, J. S., Tanner, J. M. and Barnicot, N. A. (1964). *Human Biology*. London: Oxford University Press.

Hassan, F. A. (1981). *Demographic Archaeology*. London: Academic.

Hellman, M. (1928). Racial characters in human dentition. *Proceedings of the American Philosophical Society*, **67**, 157–74.

Hillson, S. (1996). *Dental Anthropology*. Cambridge: Cambridge University Press.

Hooton, E. A. (1930). *The Indians of Pecos Pueblo: A Study of the Skeletal Remains*. New Haven: Yale University Press.

Hoppa, R. D. and FitzGerald, C. M. (eds.) (1999). *Human Growth in the Past*. Cambridge: Cambridge University Press.

Hoppa, R. and Vaupel, J. (eds.) (2002). *Paleodemography: Age Distribution from Skeletal Samples*. Cambridge: Cambridge University Press.

Howell, C. and Bourlière, F. (eds.) (1964). *African Ecology and Human Evolution*. London: Methuen.

Howells, W. W. (1995). *Who's Who in Skulls: Ethnic Identification of Crania from Measurements*. Papers of the Peabody Museum of Archaeology and Ethnology, 82. Cambridge: Harvard University Press.

Hrdlička, A. (1900). Arrangement and preservation of large collections of human bones for purposes of investigation. *The American Naturalist*, **34**, 9–15.

Huxley, T. H. (1863). *Man's Place in Nature and Other Essays* (collected, 1906). London: Dent.

Huxley, T. H. (1866). Notes upon human remains from Keiss. In *Prehistoric Remains of Caithness*, ed. S. Laing. London: Williams and Norgate, pp. 83–161.

Ingham, J. M. (1996). *Psychological Anthropology Reconsidered*. Cambridge: Cambridge University Press.

Jolly, C. (ed.) (1978). *Early Hominids of Africa*. London: Duckworth.

Katzenberg, M. A. and Saunders, S. R. (eds.) (2008). *Anthropology of the Human Skeleton*. 2nd edn. New York: Wiley-Liss.

Keith, A. (1915). *The Antiquity of Man*. London: Williams and Norgate.

Keith, A. (1931). *New Discoveries Relating to the Antiquity of Man*. London: Williams and Norgate.

Kidd, J., Black, F., Weiss, K., Balazs, I. and Kidd, K. (1991). Studies of three Amerindian populations using nuclear DNA polymorphisms. *Human Biology*, **63**, 775–94.

Knight, C., Studdert-Kennedy, M. and Hurford, J. (eds.) (2000). *The Evolutionary Emergence of Language*. Cambridge: Cambridge University Press.

Krogman, W. M. (1962). *The Human Skeleton in Forensic Medicine*. Springfield: Thomas.

Kurtén, B. (ed.) (1982). *Teeth: Form, Function and Evolution*. New York: Columbia University Press.

Landsteiner, K. and Levine, P. (1929). On the racial distribution of some agglutinable structures of human blood. *Journal of Immunology*, **16**, 123–31.

Larsen, C. S. (1997). *Bioarchaeology, Interpreting Behaviour from the Human Skeleton*. Cambridge: Cambridge University Press.

Larsen, C. S. (2002). *Bioarchaeology of the Late Prehistoric Guale: South End Mound 1, St Catherine Island, Georgia*. Anthropological Papers, **84**. New York: American Museum of Natural History.

Lawrence, W. (1819). *Lectures on Comparative Anatomy, Physiology, Zoology and the Natural History of Man*. London: Taylor.

Leakey, M. and Leakey, R. (eds.) (1978). *Koobai Fora Research Project, The Fossil Hominids and an Introduction to Their Context, 1968–1974*. Oxford: Clarendon.

Littlewood, R. A. (1972). *Physical Anthropology of the Eastern Highlands of New Guinea*. Seattle: University of Washington Press.

Livingstone, F. B. (1962). On the non-existence of human races. *Current Anthropology*, **3**, 279–81.

de Lumley, M. A. (1973). *Anténéandertaliens et Néandertaliens du Basin Méditerranéen Occidental Européen*. Études Quaternières, Memoire 2. Marseille: Université de Provence.

Lyell, C. (1863). *The Antiquity of Man*. London: Dent.

Lynnerup, N., Andreasen, C. and Berglund, J. (eds.) (2003). *Mummies in a New Millenium*. Copenhagen: Danish Polar Center.

Mahalanobis, P., Majumdar, D. and Rao, C. (1949). Anthropometric survey of the United Provinces, 1941. *Sankhya*, **9**, 89–324.

Malthus, T. R. (1798). *An Essay on the Principle of Population*. London: Lock.

Mann, A. E. (1975). Paleodemographic aspects of the South African Australopithecines. *Publications in Anthropology, No. 1*. Philadelphia: University of Pennsylvania.

Martin, R. (1928). *Lehrbuch der Anthropologie*. Jena: Fisher.

McGlashan, N. D. (1972). Medical geography: an introduction. In *Medical Geography: Techniques and Field Studies*, ed. N. D. McGlashan, London: Methuen, pp. 3–15.

Mezzich, J. and Berganza, C. (eds.) (1984). *Culture and Psychopathology*. New York: Columbia University Press.

Minnis, P. E. (1985). *Social Adaptation to Food Stress. A Prehistoric Southwestern Example*. Chicago: University of Chicago Press.

Molnar, S. (1975). *Races, Types and Ethnic Group: The Problem of Human Variation*. New Jersey: Prentice-Hall.

Montagu, M. A. (1952). *Man's Most Dangerous Myth: The Fallacy of Race*. New York: Harper.

Montagu, M. A. (1960). *A Handbook of Anthropometry*. Springfield: Thomas.

Montagu, M. A. (ed.) (1968). *Culture: Man's Adaptive Dimension*. New York: Oxford University Press.

Montagu, M. A. (ed.) (1973). *Man and Aggression*. London: Oxford University Press.

Moore, L., Van Arsdale, P., Glittenburg, J. and Aldrich, R. (1980). *The Biocultural Basis of Health: Expanding Views of Medical Anthropology*. London: Mosby.

Moorrees, C. F. (1957). *The Aleut Dentition*. Cambridge: Harvard University Press.

Morant, G. M. (1939). *The Races of Central Europe*. London: Allen & Unwin.
Morton, S. G. (1839). *Crania Americana*. Philadelphia: Dobson.
Mourant, A. E. (1954). *The Distribution of the Human Blood Groups*. Oxford: Blackwell.
Noble, W. and Davidson, I. (1996). *Human Evolution, Language and Mind*. Cambridge: Cambridge University Press.
Nott, J. C. (1857). Acclimation; or, the comparative influence of climate, endemic and epidemic diseases, on the races of man. In *Indigenous Races of the Earth*, eds. A. Maury, F. Pulszky, J. Meigs, J. Nott and G. Gliddon. London: Trübner, pp. 353–401.
Nott, J. C. and Gliddon, G. (eds.) (1854). *Types of Mankind*. Philadelphia: Lippincott.
Nuttall, G. H. (1904). *Blood Immunity and Blood Relationship*. Cambridge: Cambridge University Press.
Oakley, K. P. (1964). The problem of man's antiquity: an historical survey. *Bulletin of the British Museum (Natural History), Geology*, **9**, 85–155.
Oakley, K. P. (1969). *Frameworks for Dating Fossil Man*. London: Weidenfield and Nicolson.
Oakley, K. P. (1980). Relative dating of the fossil hominids of Europe. *Bulletin of the British Museum (Natural History), Geology*, **34**, 1–63.
Oliver, S., Sikes, Ń. and Stewart, K. (eds.) (1994). *Early Hominid Behavioural Ecology*. London: Academic Press.
Ortner, D. J. (2003). *Identification of Pathological Conditions in Human Skeletal Remains*. London: Academic Press.
Oxnard, C. (1975). *Uniqueness and Diversity in the Human Evolution*. Chicago: University of Chicago Press.
Pales, L. (1930). *Paléopathologie et Pathologie Comparative*. Paris: Masson.
Pauling, L., Itano, H., Singer, S. and Wells, I. (1949). Sickle-cell anaemia, a molecular disease. *Science*, **110**, 543–8.
Pearson, K. (1926). On the coefficient of racial likeness. *Biometrika*, **18**, 105–17.
Pedersen, P. O. (1949). *The East Greenland Eskimo Dentition*. Copenhagen: Reitzels.
Penrose, L. S. (1954). Distance, shape and size. *Annals of Eugenics*, **18**, 337–43.
Post, R. H. (1971). Possible causes of relaxed selection in civilized populations. *Human Genetics*, **13**, 253–84.
Powell, M. and Cook, D. (eds.) (2005). *The Myth of Syphilis, The Natural History of Treponematosis in North America*. Gainesville: University Press of Florida.
Price, T. D. (ed.) (1989). *The Chemistry of Prehistoric Human Bone*. Cambridge: Cambridge University Press.
Prichard, J. C. (1843). *The Natural History of Man*. London: Bailliere.
Quatrefages, A. D. (1883). *The Human Species*. London: Paul, Trench and Trübner.
Race, R. R. and Sanger, R. (1950). *Blood Groups in Man*. Oxford: Blackwell.
Ramot, B. (ed.) (1974). *Genetic Polymorphisms and Diseases in Man*. New York: Academic Press.
Raven, M. J. and Taconis, W. K. (2005). *Egyptian Mummies: Radiological Atlas of the Collections in the National Museum of Antiquities at Leiden*. Turnhout: Brepols.
Reynolds, V., Falger, V. S. and Vine, I. (eds.) (1987). *The Sociobiology of Ethnocentrism*. London: Croom Helm.

Richerson, P. J. and Boyd, R. (2005). *Not by Genes Alone: How Culture Transformed Human Evolution*. Chicago: Chicago University Press.
Rightmire, G. P. (1990). *The Evolution of Homo Erectus*. Cambridge: Cambridge University Press.
Ripley, W. Z. (1899). *The Races of Europe: A Sociological Study*. London: Paul, Trench and Trübner.
Roberts, C. (1878). *A Manual of Anthropometry*. London: Churchill.
Roberts, C. A. and Buikstra, J. E. (2003). *The Bioarchaeology of Tuberculosis*. Gainesville: University of Florida Press.
Rothschild, H. R. (ed.) (1981). *Biocultural Aspects of Disease*. London: Academic.
Ruffer, M. A. (1921). *Studies in the Palaeopathology of Egypt*. Chicago: University of Chicago Press.
Runciman, W., Smith, J. M. and Dunbar, R. (eds.) (1998). *Evolution of Social Behaviour Patterns in Primates and Man*. Oxford: Oxford University Press.
Ryan, A. S. (1979). Scanning electron microscopy of tooth wear of Australopithecus afarensis. *American Journal of Physical Anthropology*, **50**, 478.
Sahlins, M. (1976). *The Use and Abuse of Biology*. Ann Arbor: University of Michigan Press.
Salzano, F. M. (1973). Physical anthropology: retrospect and prospect. In *Physical Anthropology and its Extending Horizons*, eds. A. Basu, A. Ghosh, S. Biswas and R. Ghosh. Bombay: Orient Longman.
Salzano, F. M. and Freire-Maia, N. (1970). *Problems in Human Biology: A Study of Brazilian Populations*. Detroit: Wayne State University Press.
Schiller, F. (1979). *Paul Broca: Founder of French Anthropology, Explorer of the Brain*. Berkeley: University of California Press.
Schmidt, E. (1888). *Anthropologische Methoden*. Leipzig: Von Veit.
Schmidt, R. R. (1936). *The Dawn of the Human Mind: A Study of Palaeolithic Man*. London: Sidgwick and Jackson.
Schoetensack, O. (1908). *Der Unterkiefer des Homo heidelbergensis aus den Sanden von Mauer bei Heidelberg*. Leipzig: Wilhelm Engelmann.
Schwartz, T., White, G. M. and Lutz, C. A. (eds.) (1992). *New Directions in Psychological Anthropology*. Cambridge: Cambridge University Press.
Scott, G. R. and Turner, C. G. (1997). *The Anthropology of Modern Human Teeth*. Cambridge: Cambridge University Press.
Shapiro, H. L. and Hulse, F. (1940). *Migration and Environment*. London: Oxford University Press.
Sigmon, B. A. and Cybulski, J. S. (eds.) (1981). *Homo Erectus: Papers in Honor of Davidson Black*. Toronto: University of Toronto Press.
Slotkin, J. S. (ed.) (1965). *Readings in Early Anthropology*. Viking Fund Publications No. 40. New York: Viking Fund.
Spencer, F. (ed.) (1982). *A History of American Physical Anthropology, 1930–1980*. New York: Academic.
Spencer, F. (1990). *Piltdown, a Scientific Forgery*. London: Oxford University Press.
Spencer, F. (1997). *History of Physical Anthropology*. New York: Garland.
Spindler, K., Wilfing, H., Rastbichler-Zissernig, E. *et al.* (eds.) (1996). *Human Mummies*. New York: Springer.

Standen, V. and Foley, R. (1989). *Comparative Socioecology: The Behavioural Ecology of Humans and Other Mammals*. London: Blackwell Scientific.

Steffan, P. (1932). *Handbuch der Blutgruppenkunde*. Munich: J. F. Lehmann.

Steinbock, R. T. (1976). *Paleopathological Diagnosis and Interpretation*. Springfield: Thomas.

Stewart, T. D. (1979). *Essentials of Forensic Anthropology*. Springfield: Thomas.

Stringer, C. and Andrews, P. (2005). *The Complete World of Human Evolution*. London: Thames and Hudson.

Stringer, C. and McKie, R. (1998). *African Exodus: The Origins of Modern Humanity*. London: Pimlico.

Suttle, N. F. and Angus, K. W. (1978). Effects of experimental copper deficiency on the skeleton of the calf. *Journal of Comparative Pathology*, **88**, 137–48.

Suzuki, H. and Takai, F. (eds.) (1970). *The Amud Man and His Cave Site*. Tokyo: University of Tokyo Press.

Swedlund, A. C. and Armelagos, G. J. (1976). *Demographic Anthropology*. Dubuque: Brown.

Tobias, P. V. (1967). *Olduvai Gorge. The Cranium of Australopithecus (Zinjanthropus) Boisei*. Cambridge: Cambridge University Press.

Topinard, P. (1876). *L'Anthropologie*. Paris: Reinwald.

Trinkaus, E. (1983). *The Shanidar Neandertals*. London: Academic.

Trowell, H. and Burkitt, D. (eds.) (1981). *Western Diseases: Their Emergence and Prevention*. London: Arnold.

Tuttle, R. H. (ed.) (1975). *Palaeoanthropology: Morphology and Palaeoecology*. The Hague: Mouton.

Van Vark, G. N. and Howells, W. W. (eds.) (1984). *Multivariate Statistical Methods in Physical Anthropology*. Dordrecht: Reidel.

Vrba, E., Denton, G., Partridge, T. and Buckle, L. (1995). *Paleoclimate and Evolution with Emphasis on Human Origins*. New Haven: Yale University Press.

War Office Returns (1853). *Statistical Reports on the Sickness, Mortality and Invaliding Among the Troops in the United Kingdom, the Mediterranean, and British America*. London: Clowes.

Washburn, S. L. (ed.) (1961). *Social Life of Early Man*. Chicago: Aldine.

Washburn, S. L. (ed.) (1963). *Classification and Human Evolution*. London: Methuen.

Weiner, J. S. (1955). *The Piltdown Forgery*. London: Oxford University Press.

Weiner, J. S. and Huizinga, J. (eds.) (1972). *The Assessment of Population Affinities in Man*. Oxford: Clarendon.

Weiner, J. S. and Lourie, J. A. (1969). *Human Biology: A Guide to Field Methods*. IBP Handbook No. 9. Philadelphia: Davis.

Wenegrat, B. (1984). *Sociobiology and Mental Disorder: A New View*. Menlo Park: Addison-Wesley.

Whitney, W. F. (1883). On the existence of syphilis in America before the discovery of Columbus. *Boston Medical and Surgical Journal*, **108**, 365–6.

Wilson, E. O. (1975). *Sociobiology: The New Synthesis*. Cambridge: Harvard University Press.

Wolpoff, M. and Caspari, R. (1997). *Race and Human Evolution: A Fatal Attraction*. Boulder: Westview Press.

Woo, T. L. and Morant, G. M. (1932). A preliminary classification of Asiatic races based on cranial measurements. *Biometrika*, **24**, 108–34.

Woodward, A. S. (1913). The Piltdown Man. *Geological Magazine*, **10**, 433–4.

Zeising, A. (1854). *Neue Lehre von den Proportionen des menschlichen Körpers*. Leipzig: Weigel.

Zubrow, E. (ed.) (1977). *Demographic Anthropology*. Albuquerque: University of New Mexico Press.

3 *Forensic anthropology serving justice*

SUE BLACK

Until relatively recently, forensic anthropology in the UK has tended to be regarded as one of those specialties that might be of some 'occasional assistance' but was not routinely employed in either homicide or mass disaster investigations. The tide of opinion began to change when the value of anthropological expertise came to the fore in the identification of the victims of war crimes and violations of human rights in South America and then in Rwanda (Ferllini, 1999). Close on the heels of these horrendous situations came the atrocities witnessed in East Timor, Sierra Leone and of course the former Yugoslavia (Glenny, 2000; Komar, 2003; Baraybar and Gasior, 2004; Steadman and Haglund, 2005). Now, following the Bali bombing, 9/11 terrorist attack, Iraq war, Asian Tsunami, London bombings, terrorist attacks on Sharm el Sheikh and hurricane Katrina, the position of forensic anthropology has been secured within war crimes investigations and global Disaster Victim Identification (DVI) response requirements.

Anthropology has a significant role to play at both a national and an international level, assisting the judicial investigation of homicide, mass disaster incidents, crimes against humanity, war crimes and genocide (Steadman and Haglund, 2005). This chapter will examine briefly three very different cases to illustrate the part that forensic anthropology can play within a multidisciplinary forensic team. It should be appreciated though that every deployment is unique, requiring different skill sets as the situation dictates and demands, and so the scenarios selected are not an exhaustive representation of the involvement of forensic anthropology in judicial and humanitarian missions but serve rather to illustrate the flexibility and potential utility of the subject.

The American Board of Forensic Anthropology defines the discipline as follows –

> *Forensic anthropology is the application of the science of physical anthropology to the legal process. The identification of skeletal, badly decomposed, or otherwise unidentified human remains is important for both legal and humanitarian reasons. Forensic anthropologists apply standard*

Between Biology and Culture, ed. Holger Schutkowski. Published by Cambridge University Press.

> *scientific techniques developed in physical anthropology to identify human remains, and to assist in the detection of crime. Forensic anthropologists frequently work in conjunction with forensic pathologists, odontologists, and homicide investigators to identify a decedent, discover evidence of foul play, and/or the postmortem interval.*
>
> (ABFA, 2005).

At the centre of this definition are the words 'identification', 'identify', and therefore by natural extension and implication 'identity'. Establishing and confirming the identity of the deceased is central to the role of forensic anthropology, culminating in the ultimate goal of assigning a personal name to the individual. Society holds tremendous store by the name that we choose to live with and yet, as the embodiment of our true identity, it is such a transient and fugacious concept. Our name is a transferable commodity that can be changed at will for both legal and illegal purposes. Yet, in his semi-autobiographical account, Goethe expounded that '*A man's name is not like a cloak that merely hangs around him, that may be loosened and tightened at will. It is a perfectly fitting garment. It grows over him like his very skin. One cannot scrape and scratch at it without injuring the man himself*' (cited in Zabeeh, 1968). Unfortunately that 'snug coat' can indeed be altered at will and it forms a poor and highly unreliable verifier of our identity.

Identity is an integral part of our modern global society and being able to prove our identity is fundamental to many daily domestic interactions and routines, yet almost every verification method is to some extent vulnerable to abuse or open to misinterpretation (Nanavati *et al.*, 2002). The word 'identity' derives from the Latin '*idem*,' meaning 'the same', yet the irrefutable equality of 'same' is something that can never be proven absolutely. All of our scientific techniques for verification rely on a statistical probability of 'sameness' and it is that small possibility of mismatch or error that gives lawyers the opportunity to demand that we prove identity 'beyond reasonable doubt'.

Therefore by implication, a name cannot be an indicator of identity but rather it is the catalyst whereby a set of data may be established and amassed that will permit comparison with a second data set to facilitate the conclusion and therefore prove, or disprove, beyond reasonable doubt that both sets of information originated from the same individual of a given name. Indeed, few of us will ever own a truly unique name and so we must share this most personal of labels with others who, very obviously, cannot possibly be of the same identity. For example, the 2001 UK electoral register shows that 12 547 persons were registered under the name 'John Smith', 35 as 'Anthony Blair' and amazingly 20 as 'Elvis Presley'. So whilst it may be essential to assign a name to a deceased, great caution must be placed on the value and validity of that information for the judicial and humanitarian purposes of verification.

Most of us are blissfully unaware that our identity has a commercial value, and indeed identity theft and identity fraud are amongst the fastest growing crimes in the UK, with a rise of over 45 % in the past year (Cabinet Office, 2002; Collins, 2006). It is highly probable that every person reading this article will be a victim of identity theft (perhaps more than once) in their lifetime and it is estimated that it may take over 400 hours of your time and easily £8000 of your money for you to prove your innocence. Identity theft is virtually the only crime where the victim is presumed guilty until they can prove their own innocence. A truly exceptional account of identity theft, recounted by a senior officer from Devon and Cornwall Constabulary (Sincock, 2006), is worthy of note to illustrate the importance of a name in assigning correct identity to the deceased.

However without a name for the deceased, the investigative authorities find it virtually impossible to reach a successful conclusion in a homicide investigation. Similarly it is essential for both the legislative and humanitarian issues surrounding an international investigation into war conflicts (ICRC, 2004) that, where possible, the identity of the victim be confirmed not only to expedite judicial matters of estate for the survivors and dependents of the deceased but indeed to secure prosecutions whilst also upholding the human rights of the accused. Of equal importance is the role that identification of the deceased plays in the healing and grieving processes of the survivors and families (Walter, 2005). Repatriation of the remains of a loved one that facilitates the observance of religious and cultural rites has a significant impact on the psychological acceptance of the death (PAHO, 2004). Therefore whilst the natural instinct in the aftermath of a mass disaster may be to consider the deceased 'in due course', the importance of early attention to mortuarial matters is vital for the initiation of healing and the culmination of acceptance at personal, national and even international levels (Edson *et al.*, 2004).

The speed by which identity can be confirmed is strongly correlated with the time elapsed since death (i.e. the time death interval – TDI). The more recent the event then the more likely it is that confirmation will be forthcoming without delay and as time passes and decomposition and disruption or dispersal of the remains occurs, so the chances of a successful identification generally decrease (Haglund and Sorg, 2001). Therefore time is the natural enemy of identification and this was very evident following both the Asian Tsunami and hurricane Katrina where visual identification of the deceased ceased to be of value within days of the event.

Following the Interpol DVI guide (2002), the identity of a deceased can be established through both circumstantial and physical evidence. Forensic anthropology plays no significant role in identifications that are based on **circumstantial evidence** such as visual recognition of the deceased by friends or family members or perhaps through the recovery of personal effects. Although these provide valuable corroborative evidence of identity, neither is acceptable in the absolute confirmation of identity which requires objective and verifiable criteria.

Although identification is normally achieved through the matching of antemortem and postmortem data, in its absence some authorities (Morgan *et al.*, 2006) may be willing to rely solely on the visual recognition of remains, although this lacks scientific rigour and has been shown to be an unreliable means of identification. The face in death can often bear little resemblance to the living face (Wilkinson, 2004). Equally a family member, traumatised by the event will find it difficult to recognise their loved one due to various factors of distortion, lack of muscle tone and alteration in colouration, and indeed their own potential lack of objectivity given the stressful nature of the situation. Therefore it is generally accepted that confirmation of identity cannot rely solely on visual recognition but requires a combination of criteria that are scientifically verifiable and will withstand the rigours of courtroom scrutiny.

Identification from associated personal effects is also to be viewed with some caution (Thompson and Puxley, 2006). Descriptions of clothing, jewellery and pocket or bag contents can be extremely useful in the confirmation of identity, provided reliable antemortem association can be established. It must be borne in mind that in a disaster scenario it is relatively commonplace for loose items to become lodged with another body and so unless they can unequivocally be associated with the body, i.e. a ring on a finger or a watch on a wrist, then attribution of these artefacts of identity must be considered with some caution. An added complication is that data loggers from different countries will have different terminology for items of clothing and this can lead to confusion in the identification process, e.g. 'pants' in the UK refers to underwear whereas in the US they will mean trousers.

Forensic anthropology is of greatest value in the analysis of indicators of **physical identity** in terms of both external and internal corporeal information. External features of identification include determination of sex from visible external genitalia, a broad estimation of the age of the deceased from physical appearance, muscle and fat volume estimations of body build, colour of skin, hair colour, eye colour, physical or developmental anomalies etc. Specific features of personal identification are also noted and these may include scars, birth marks, moles, tattoos, piercings etc. Fingerprints are a specific external feature and these are generally recorded by an attendant police officer and do not fall under the remit of the anthropologist (Kahana *et al.*, 2001).

Physical evidence of internal identity is revealed during the autopsy investigation, when non-invasive approaches are insufficient to secure identity (Shepherd, 2003). The investigation, which is usually led by a pathologist, may include the retention of samples for further analysis, e.g. blood, tissue biopsy, stomach contents, vitreous humour (fluid from inside eye) etc. There may be medical findings which will assist with identification, e.g. previous fractures, disease, surgery etc., and the role of the anthropologist will be to assist both the pathologist and the odontologist.

Dental evidence is often of particular importance in the field of identification due to the potential availability of good antemortem data for comparison and the survivability of teeth. In the 2004 Asian Tsunami, forensic odontology was responsible for the majority of the identifications of foreign nationals within the first six months of the operation, although the process was not without difficulties (Kieser *et al.*, 2006).

The removal of bone and tooth samples for DNA analysis has become a more popular route for confirmation of identity of the deceased (Budimlija *et al.*, 2003). This relies on the ability to match the DNA of the victim to surviving family members or with DNA that can be retrieved from direct personal effects of the deceased, e.g. toothbrush, hairbrush, razor etc. This approach is also particularly useful for the matching and reassociation of body parts. In recent incidents, especially those that have involved explosives, forensic anthropology has proved particularly useful in the anatomical reassociation and identification of incomplete and fragmented remains as a preliminary and quick sorting approach that will be verified, or otherwise, by subsequent DNA information (Kahana *et al.*, 1997).

It is often thought that forensic anthropology and forensic osteology are one and the same thing, but this is not so. Whilst forensic osteology is solely restricted to the investigation of skeletal remains, forensic anthropology also considers soft tissue elements of identification, and is more appropriately a paramedical discipline dealing with the analysis of remains regardless of their degree of decomposition or fragmentation.

Therefore, forensic anthropology is a hybrid discipline that assimilates anatomical, osteological, ontogenetic, phylogenetic and anthropological information to assist the pathologist, odontologist and senior identification manager (SIM) in providing evidence to the court concerning the personal identity of the deceased.

Three types of cases are discussed below that involve anthropological expertise and serve to illustrate the practical role that the discipline can play in national and international incidents.

Case 1: UK homicide investigation

On 18 November 2001 a suitcase was found abandoned in a country lane near York (Figure 3.1). When opened, it contained the nearly naked remains of a woman arranged in a tight fetal position. There was some evidence of

Figure 3.1 The suitcase in which the body of J. H. J. was found. Courtesy of North Yorkshire Police and Forensic Science Service.

decomposition and her face was bound by sticky plastic tape. There was no circumstantial evidence to suggest her identity. The forensic pathologist confirmed the cause of death as asphyxiation but was unwilling to give a definitive age to the woman, or a racial origin. Forensic anthropology was therefore requested to assist in both these matters. Whilst the external appearance of the corpse did indicate Oriental persuasion, there was a moderate level of decomposition and skin slippage present that raised sufficient concerns to merit further investigation, and the binding effect around her face had caused some not-inconsiderable soft tissue disfigurement.

Her height was measured by the forensic pathologist to be 4 feet 11 inches and she was of gracile body composition with moderate volumes of body fat and musculature. Her hair was straight, of blue/black colouration and relatively coarse in texture. Her dentition indicated considerable care and attention by at least one dental practitioner, and the mode of reconstructive work was not consistent with normal UK practices. Her central maxillary incisors were moderately shovelled in appearance, with a significant degree of overbite. The zygomatic morphology and overall appearance of the facial skeletal elements (including an epicanthic fold) strongly suggested that she was indeed of at least partial Oriental genetic origin. This was confirmed in due course by DNA markers of ethnicity.

Her age at death was easier to assess as although adult, there were a number of indicators that placed her clearly within the early stages of her third decade. Looking at age prediction topographically and working in a craniocaudal direction, it was possible to consider the evidence in a progressive and confirmatory manner. The closure of the skull sutures could not be used to assist in this process as there was evidence of a developmental anomaly that resulted in parietal thinning. Therefore, there could be no guarantee that the sutures would conform to the expected pattern of closure although it should be noted that they were all open ectocranially with some indication of the commencement of endocranial synostosis. The laryngeal structures showed no evidence of ossification. The ossification of the ventral aspect of the ribs suggested an age of between 24 and 28 years. However it was the presence of epiphyseal slivers at the costal notches that was to prove most definitive with regards to age assessment (Figure 3.2). The sternal plate confers a tremendous amount of information concerning sex and age-related characteristics and it is frequently the 'Cinderella' bone when compiling a skeletal report. Small plaques of bone form in the costal notches between two adjacent sternebrae that assist with the unification of the sternal body. These flakes can appear in a person as young as 17 years of age and completion of growth is unlikely to occur prior to 25 years of age (Scheuer and Black, 2000). The sternum in this forensic case clearly showed that the distal sternal costal epiphyses were not united, placing the individual younger than 25 years of age but probably in excess of 20 years as the upper epiphyses had united and this process occurs in a craniocaudal progressive sequence. The remainder of the skeletal elements concurred with this prescriptive age estimation, including the pubic symphysis, iliac crest, sacrum, long bones, clavicle and scapula. Therefore, in summary, the Police were examining the unidentified remains of a young Oriental woman between the ages of 20 and 25 years.

Because of the disfigurement caused to her face through decomposition, it was decided to utilise the services of a forensic artist to produce an image that could be released to the public in an appeal for information. The forensic artist, quite correctly, drew what could be seen, and constructed an impression for dissemination by the Police (Figure 3.3). Through recourse back to the forensic anthropologist it was agreed that the image could not in fact be released as it did not take into account the soft tissue alterations caused by the tight plastic binding. This had resulted in a 'ballooning' of the lower lip and a crushing of the upper lip so that it developed a crenulated appearance where it interdigitated

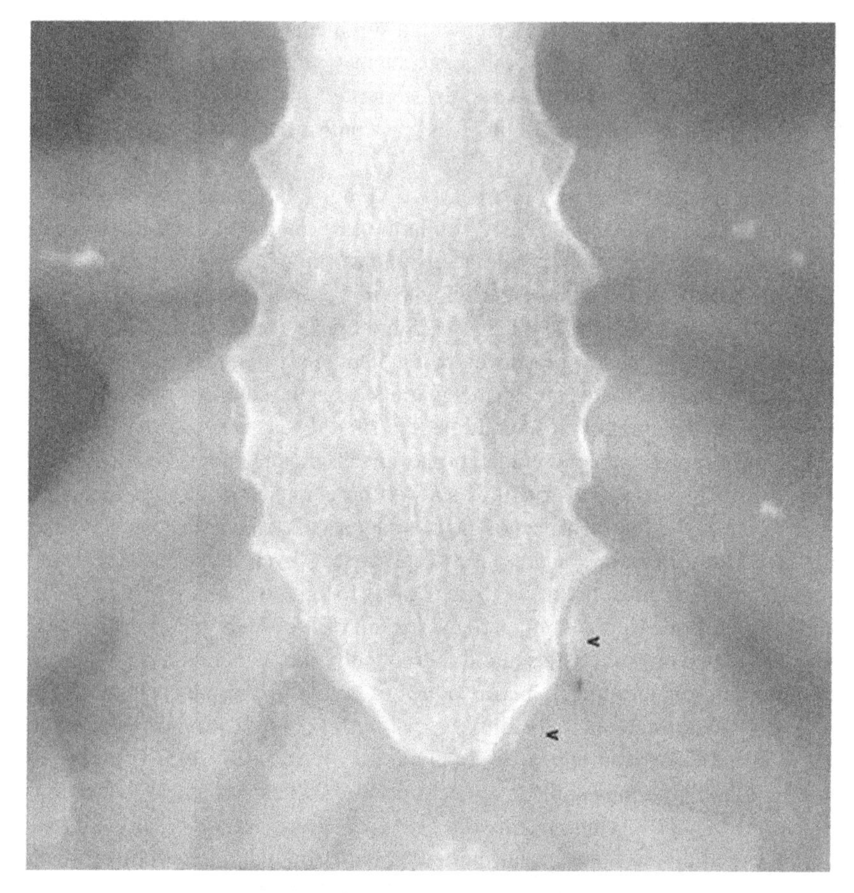

Figure 3.2 The small epiphyseal flakes can clearly be seen in the costal notches of the sternum (arrows). Courtesy of North Yorkshire Police and Forensic Science Service.

with the maxillary dentition. It was the opinion of the anthropologist, that if the image was released it would bear little resemblance to the missing person and so the image was withdrawn. This process highlights the importance of interdisciplinary communication to ensure that forensic specialists do not remain isolated within their individual disciplines.

The identity of J. H. J. (Figure 3.4) was finally confirmed through comparison of fingerprints and familial DNA comparison. She was a 21-year-old South Korean student who was studying French at Lyons University. She had travelled to London in October of 2001 on a sightseeing holiday and had stayed in a hostel in East London. Her landlord, also Korean (aged 31 years), was convicted of her murder in the central criminal court of the Old Bailey and given a life sentence.

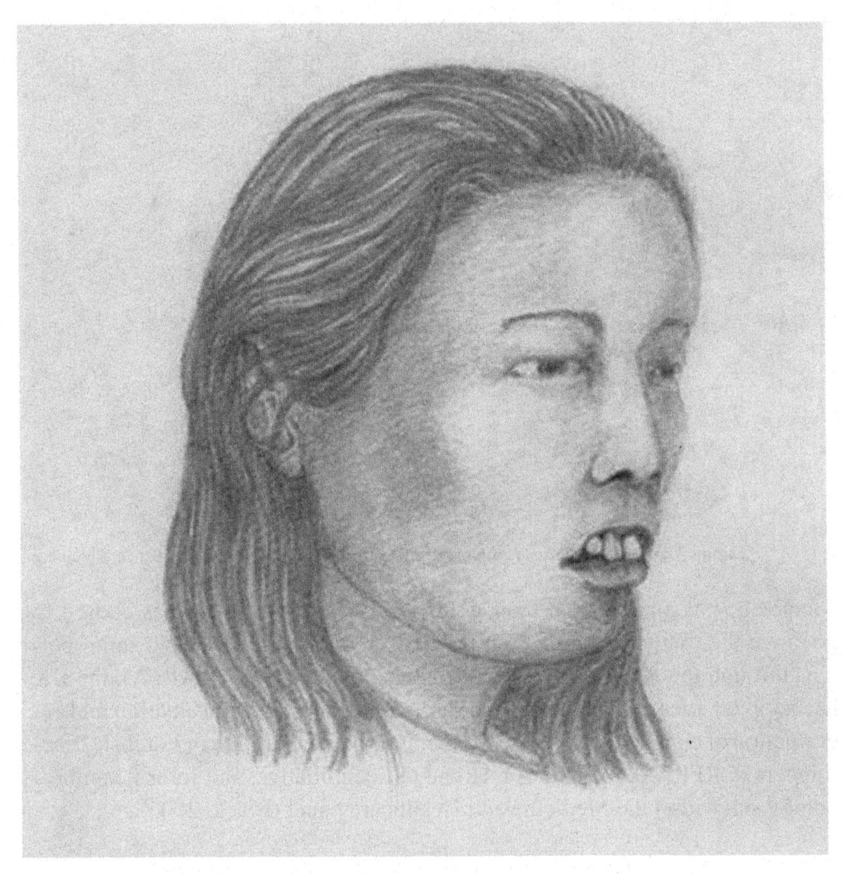

Figure 3.3 The image of J. H. J. produced by the facial artist. Courtesy of North Yorkshire Police and Forensic Science Service.

He was given a second life sentence for the murder of a second South Korean student whose body was found similarly bound in a cupboard of the hostel.

Homicide investigations utilise the service of forensic anthropologists with greater frequency than in the past and in particular where there is some doubt over the identity of the deceased, or indeed where the remains are severely decomposed, skeletonised, burned or fragmented. The process leading to identification requires that the anthropologist must be able to discern human from non-human remains and can identify every bone of the adult and juvenile human skeleton regardless of its degree of fragmentation, and be able to derive meaningful and realistic information from the evidence presented, irrespective of the state of presentation of the remains. For this reason it is clear that practitioners must be professionally accountable and be able to evidence their current

Figure 3.4 J. H. J. Courtesy of North Yorkshire Police and Forensic Science Service.

competency as an expert in their field. Whilst North America has chosen to go down the route of examination-based accreditation for forensic anthropology, through the American Board of Forensic Anthropology (ABFA), the UK has adopted a less formal approach through peer-reviewed evaluation and re-evaluation of competence, via the Council for the Registration of Forensic Practitioners (CRFP, 2005). This is a Home Office initiative, and forensic anthropology sits within the Medicine and Healthcare panel (Black, 2003).

Case 2: war crimes investigation

The 57-page document that comprised the Rambouillet Accords was signed by delegations from the UK, US and Albania on 18 March 1999, but not by the Serbians or the Russians (Glenny, 2000). The Accords called for NATO (North Atlantic Treaty Organisation) administration of Kosovo as an autonomous province within Yugoslavia. It called for Serbian withdrawal of military troops, and their replacement by a UN (United Nations) peacekeeping force. Slobodan Milosevic did not believe that NATO would carry out their bombing threats, but even if they did, he was of the opinion that it would be a relatively minor offensive and it would not lead to long-term alterations in the current status quo. International monitors from OSCE (Organization for Security and Co-operation in Europe) were ordered to withdraw from the country and NATO's bombing campaign began on 22 March and lasted until 10 June 1999. Within days of the ceasefire the Chief Prosecutor in The Hague (Louise Arbor) requested deployment of gratis forensic teams from NATO member states. By the beginning of the following week the first international team was in place (British Forensic Team – BFT) and the first massacre site within Kosovo was being forensically recorded by scientists and police officers from SO13 (Special Operations 13 – Anti-terrorist Branch).

It had been alleged that Milosevic and other top-ranking officials in the Serbian government, 'planned, instigated, ordered, committed or otherwise aided and abetted in a campaign of terror and violence directed at Kosovo Albanian civilians living in Kosovo in the FRY [Federal Republic of Yugoslavia]' (ICTY, 2001). In addition to the killings, by May 1999 it was reported that several tens of thousand Kosovo Albanians had been expelled from the country and thousands more were internally displaced (ICTY, 2001). Because the International Criminal Tribunal for the Former Yugoslavia (ICTY) is still ongoing, the site portrayed in this chapter has been anonymised, and its sole purpose for inclusion is to illustrate how forensic anthropology assisted in this type of investigation.

A witness survivor statement confirmed that earlier in that same year many villagers (men, women and children) had been rounded up at gunpoint and forced to walk the main highway to the southwest that led from Prizren to the Albanian border. At a certain point along this refugee route, over 40 men were separated from the main group and marched to an outhouse adjacent to a farm – the youngest was only 14 whilst the oldest was 82. They were herded through the front door of the building in Figure 3.5 and separated into two rooms. Two gunmen are alleged to have stood in the doorways and sprayed each room with Kalashnikov automatic fire. Accomplices reportedly stood outside the building and threw straw and combustibles through the open windows and torched the building.

This was a particularly important site for the tribunal as there was a survivor from this incident. He had been in the corner of one of the rooms and had been shielded from the bullets by the men who stood in front of him. When the room was torched, he was largely protected by the burning bodies that lay on top of him. He had to remain in this position, lying beneath the dead bodies of his neighbours and friends whilst they burned above him. He knew that if he alerted his captors to his presence then he too would be shot.

When the BFT arrived in June of that year they were confronted by two rooms with at least 40 deceased that were partly burned, partly buried under the asbestos tiles of the roof which collapsed during the fire, badly decomposed, commingled and partly dispersed due to the activity of scavenging animals. The pathologist present called for the assistance of forensic anthropology, and

Figure 3.5 The outhouse in Kosovo. The area of wall above the windows is blackened by the soot of the fire. The bodies were located in each of these two rooms to either side of the central corridor which leads from the front door. Courtesy of British Forensic Team, Kosovo.

from that time onwards, at least one anthropologist was present throughout the two-year rolling deployment undertaken by the BFT in Kosovo.

In this first investigation, the role of the forensic anthropologist was four-fold.

1. To identify human from non-human material

Although this sounds relatively straightforward, the intensity of the fire resulted in cremation of certain exposed areas of the corpses, rendering the remains often to little more than burned shards (Figure 3.6). These had to be identified and separated from potential confusing artefacts such as fragments of asbestos roof tiles, domestic utensils and non-human remnants (the outhouse had been used for animal storage for many years). Equally the effects of burning and melting on plastic refuse located within the outhouse resulted in some objects that were difficult to identify, and it was necessary for them to be eliminated from further forensic investigation. Radiography was not available in this early stage of the proceedings but it was introduced very shortly afterwards and manned by military operatives. This made for more rapid identification of both suspicious and unknown masses.

It should be appreciated that this work was not being undertaken with all the modern facilities of a clinical mortuary, and indeed much field forensic

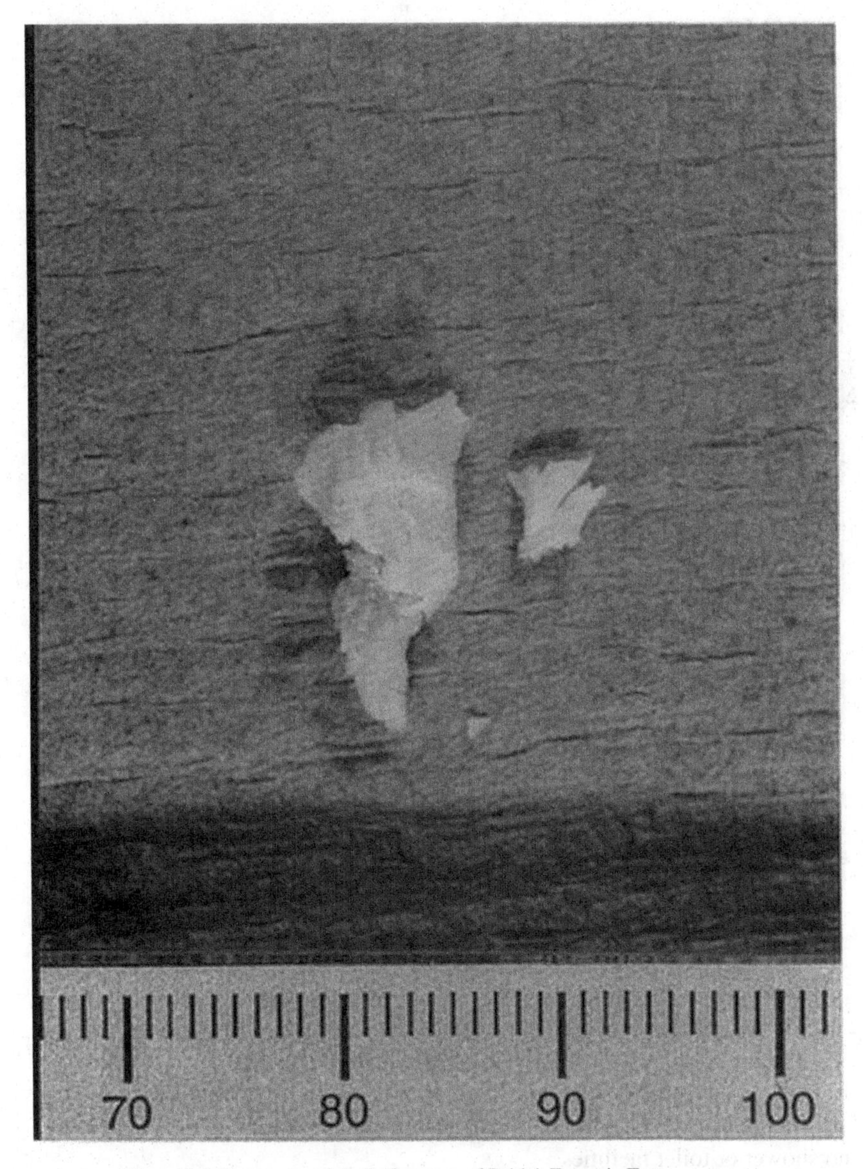

Figure 3.6 Fragments of skull. Courtesy of British Forensic Team.

work cannot adhere strictly to international protocols or standard operating procedures, as often the situation demands that the best be achieved with what is available within the security and time constraints imposed. In this case, the mortuary was a white 'scene of crime' tent erected adjacent to the outhouse, and much of the anthropology was conducted in the open air (Figure 3.7).

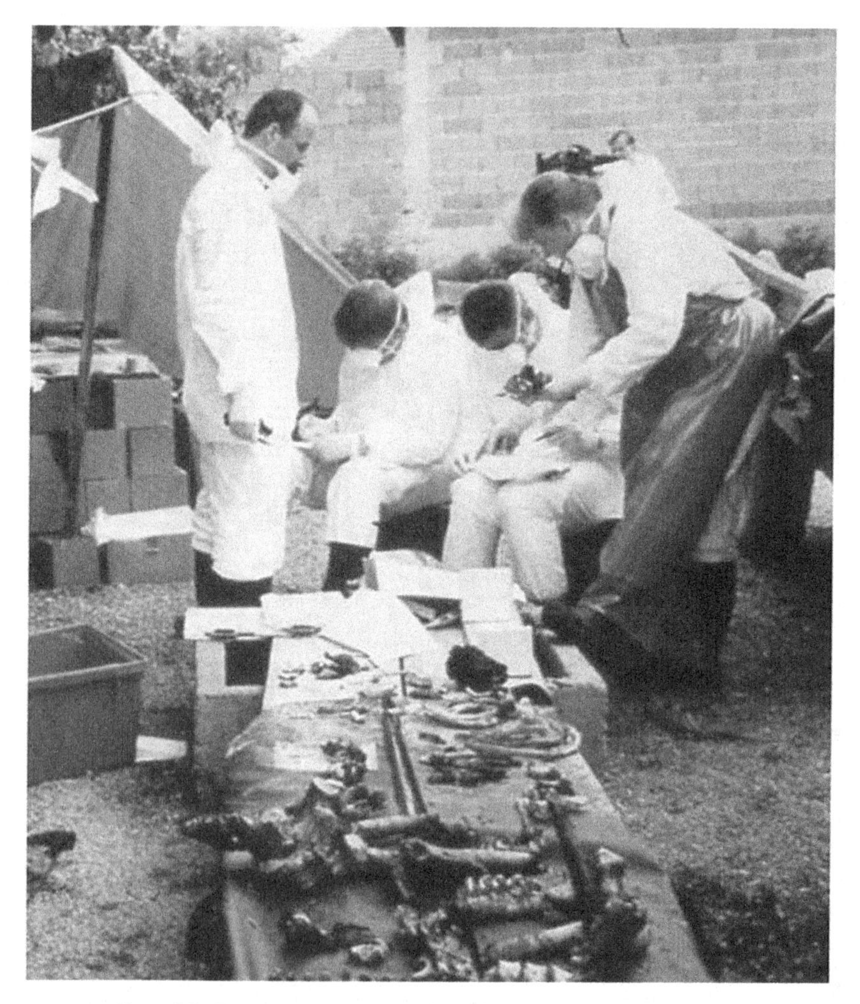

Figure 3.7 The external anthropological mortuary in Kosovo. Courtesy of British Forensic Team, Kosovo.

Electrical power came from a small generator; there was no running water, and air conditioning was achieved by keeping the flaps of the tent open! There were no shower or toilet facilities.

2. To be able to assign and reassign commingled remains to a single individual

The natural human survival instinct when confronted with a life-threatening force is to move as far away as possible from the source in a diametrically opposite direction. Thus, when the gunman stood at the door of the outbuilding, the men crowded together in the far opposite corner. When the closest individuals were shot, they would fall, leaving those behind exposed as the next line of massacre, therefore many bodies would be found in the same location where they fell. The forensic investigation occurred between three and four months after the incident and in that time the bodies had been burned, exposed to daily heat in excess of 30 °C and subjected to animal predation. Therefore the remains were badly decomposed, still encased in a writhing sea of maggot activity and commingled where they fell. Animal scavengers had tended to feed on extremities that could be removed (e.g. hands and feet), and these were often found quite some distance away from the site in adjacent fields. It was not possible to search for these parts as EOD (explosive ordnance disposal) had not been undertaken in the surrounding terrain. In fact, a device was found at this site with a tripwire across a path that led to a trigger embedded in a tree and attached to a ground-embedded explosive device (Figure 3.8).

The initial reassignation of remains occurred through obvious similarities of body parts and the exclusion of duplications. Much of the remains were still encased within clothing and therefore it was essential that delicate disentanglement of limbs was undertaken to ensure that body parts were not disassociated from the main trunk, and the integrity of the body was maintained (Skinner *et al.*, 2003). A body associated with three hands was obviously not correct and so common sense would prevail regarding the feasibility of matching. When a body, or a body part, was removed from the outhouse it was assigned a unique identification number. If the remains were relatively intact then they would pass to the pathologist who would undertake a postmortem examination to establish the cause of death. At this site, all deaths were by gun-shot injury. It was important to note the location, nature and number of the injuries, for the purposes of future prosecution.

3. To assist the pathologist in the identification of body parts for the purposes of evidential recording

The fragmentation of the human body, whether through blast injuries (bomb, gun shot etc.) or decomposition, renders some aspects of identification particularly difficult without in-depth anatomical knowledge. Anthropology assisted the pathologist in situations where body parts were retrieved and perhaps only a fragment of a bone or a mass of soft-tissue structures was present. Such detailed anatomical identification illustrates the necessity for a forensic anthropologist to have soft-tissue as well as hard-tissue experience. The positioning of a fascial plane in relation to a neurovascular bundle and muscle formation

Figure 3.8 Small device found at the scene. A – Device *in situ* in ground; B – Attachment to tree; C – Trip wire; D and E – Device. Courtesy of British Forensic Team, Kosovo.

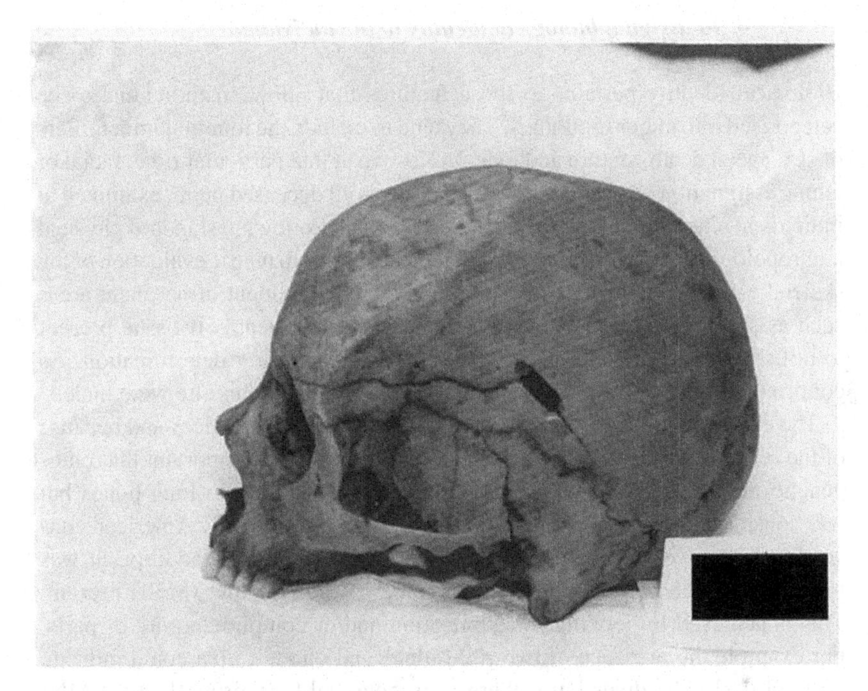

Figure 3.9 Gun-shot injury to left side of the skull showing radiating fracture lines. Courtesy of British Forensic Team.

can be sufficient for an anatomically trained anthropologist to identify both the side of the body from which the fragment of soft tissue arose and the specific location of that specimen. Needless to say, this task becomes increasingly more difficult as the remains decompose and tissue breakdown accelerates.

In skeletal terms, it is important to be able to identify the position of a particular rib or vertebra within the series to allow inferences to be made regarding the ballistic trajectory or the position of the victim at the time of their death. Equally it may be important to be able to identify single phalanges to accurately record defence injuries. Many of the deaths in this particular crime scene were attributed to gun-shot injury, and a considerable number of the fatalities arose through injuries to the skull and thorax. Therefore one of the primary roles of the anthropologist was to reconstruct the fragmented body parts and permit the pathologist to form opinions on the type of injury, cause of death etc. (Figure 3.9). The ethyl cyanoacrylate 'super glue' used by model makers is ideal for piecing together 'wet' sections of skull, especially when used with an accelerator.

4. To assign a biological identity to the individual

Biological identity pertains to those features that allow an individual to be categorised into major groupings. They tend to include the four major identifiers of sex, age at death, stature and race. In Kosovo at that particular time, racial or ethnic assignation was generally not an issue as all deceased being examined at indictment sites were likely to be ethnic Albanian. Following standard physical anthropological techniques, sex was determined by both metric evaluation of the skeletal elements present and by morphological assessment of pertinent areas such as skull and pelvis. There was generally insufficient soft tissue present to utilise evidence of internal or external genitalia for sex determination. As confirmed by the witness, all individuals recovered from this site were male.

The evaluation of age at death also followed standard practices and because of the relatively large number of young individuals it was important that ontogenetic markers were observed. Stature was calculated from long bones but was considered with a larger margin of error than usual as only American conversion tables were available for comparison at that time. In addition, it was important to determine a 'minimum number of individuals' (MNI) present. This is achieved by looking for duplication and/or complimentarity of parts. For example the presence of two right ulnae and one left ulna could indicate an MNI of 2 individuals but if there is no potential for pairing then the MNI would be 3 individuals. At this forensic crime scene the MNI was 35, and the assigned ages at death ranged from mid to late teenage years up to elderly males in excess of 60 years.

Where possible, each individual from the site would be assigned a sex, age at death and a stature. Confirmation of personal identity, i.e. leading to the name of the deceased, was difficult to achieve for a number of reasons. Although some individuals were associated with personal effects such as clothing or jewellery, these are at best circumstantial evidence of identity as it was not unknown for victims to be forced to swap clothing and documentation in an attempt to confuse forensic investigators. Under these circumstances, it is usually fairly evident that such a directive has occurred as the hurried and anxious nature of the operation leaves zips undone, buttons open or misaligned, laces untied etc. This was not the situation for this site and the witness confirmed that they were not requested to strip and exchange clothing. Therefore artefacts were given some weighting in the determination of identity.

The decomposition of the remains meant that fingerprints could not be taken and it would have been a relatively futile task as there were few opportunities for antemortem comparisons. The situation was similar for dental work. Many of the deceased simply had extractions rather than restorations, and record keeping by dentists was not reliable.

DNA samples from bone and teeth were retained on the chance that comparisons with living relatives would prove possible. Whilst DNA has proved to be particularly useful in Bosnia via the pioneering work of the ICMP (International Commission on Missing Persons), it often cannot be used to confirm personal identity but only familial connections. As many of the deceased originated from the same village and even from the same extended family, DNA analysis proved to be only one tool in the identification process (Schaefer and Black, 2005) that must be used in conjunction with forensic anthropology to achieve positive identification of the deceased.

In war crime situations, forensic anthropology can prove to be extremely helpful especially when remains are badly decomposed or the bodies are fragmented due to perimortem or even postmortem incidents. At this particular site, all the remains were removed by trailer to a communal burial site and laid to rest by the survivors who were principally the elderly, women and children (Figure 3.10). Those whose remains could be identified were released to family and buried in a named grave, however those who could not be identified retained their unique identifier tag, and the burial location was recorded to ensure that it could be retraced at a later date should this become appropriate or necessary if additional information (e.g. DNA matches) transpired.

Case 3: natural mass disasters

At 00:58:50 GMT on 26 December 2004 an earthquake measuring 9.0 on the Richter scale was recorded off the west coast of Sumatra. A fall in the ocean floor set up a tsunami that was to kill over 225 000 people. The wave hit the coasts of Sumatra, Thailand, Burma, India and Sri Lanka with no warning as it spread primarily around the Bay of Bengal and then across the Indian Ocean, where its effects were even recorded on the east coast of Africa in the Somali Republic, 2500 thousand miles from the epicentre. The international DVI (Disaster Victim Identification) effort that ensued is probably the largest, single incident, multinational forensic operation ever conducted, and it involved deceased from over 30 different countries. This work is still ongoing at the time of writing this text although full operational responsibility has now been handed over to the Thai authorities. However, the rolling international deployment was operational for a full 12 continuous months.

The enormous scale of the Asian Tsunami natural disaster, and the logistics for implementation of a global DVI response were almost unimaginable, but within only 8 months nearly 70 % identification of westerners had been achieved in Thailand and close to 100 % had been achieved in Sri Lanka. For comparison, after the World Trade Center disaster on 11 September 2001, identification

Figure 3.10 The Kosovo burial fields. Courtesy of British Forensic Team, Kosovo.

Figure 3.11 The deceased at a temple in Krabi. Courtesy of CIFA (Centre for International Forensic Assistance).

of 50 % of the 3025 deceased took 18 months to achieve (MMWR, 2002). Similarly, identification of the 202 deceased in the bombing of a nightclub in Bali, Indonesia, on 12 October 2002 took approximately 6 months to achieve (Lain *et al.*, 2003).

The success of the DVI initiative depended upon the provision of temporary mortuary facilities that could be staffed by gratis forensic teams. In the initial weeks following the disaster, over 30 international DVI teams (in excess of 600 practitioners) arrived in Asia to assist in the repatriation and identification effort. These teams included police officers, pathologists, odontologists, anthropologists, radiographers and mortuary technicians. Temples and other buildings were converted to provide temporary mortuaries in the Thai provinces of Phuket, Phang Na and Krabi. In the early stages, equipment and facilities were very basic, but within days electricity, water and refrigerated containers were operational. When the first teams arrived at the end of December, bodies brought to the temples were simply laid out in long lines with no cooling facilities or body bags (Figure 3.11). Decomposition is the fiercest enemy of the process of identification, and the rate of decomposition is proportional to both time and heat. The daily temperature of over 30 °C, in which bodies were openly exposed to the elements and insect infestation, ensured that by the time the international DVI teams were fully operational, decomposition was

Figure 3.12 Attempts to cool the bodies by laying dry ice. Courtesy of CIFA (Centre for International Forensic Assistance).

advanced. Local attempts to cool the remains and halt autolysis (tissue breakdown) were admirable but of limited success. Bodies were packed with dry ice but unfortunately this resulted not only in freezer damage to the corpses but also caused health and safety issues (Figure 3.12). A basic first-aid facility was provided at the mortuaries and a review of cases on a single day in the early phases of the operation logged 60 wound dressings, 45 persons experiencing headache, 28 eye washes, 2 tetanus vaccinations and 18 dry ice burns. In addition, the Thai authorities authorised early burial of bodies determined through visual inspection to be 'Oriental', in a laudable attempt to reduce the number of bodies at the temples and control potential environmental hazards. However, it was quickly realised that not only was visual identification of racial origin unreliable, but many of the foreign nationals were also 'Oriental'. All remains were subjected to a second postmortem examination.

On 12 January 2005 a Thailand Tsunami Victim Identification committee (TTVI) was formed to standardise procedures utilising the Interpol DVI guide. Until this time, the mortuaries had worked almost independently leading to a lack of information sharing and decentralisation of Personal Protective Equipment (PPE) and consumable provision. Postmortem data was recorded on the DVI Interpol forms, and comparisons with antemortem data were carried out at a separate building in Phuket designated as the IMC (Information Management Centre). Antemortem data was provided by relatives and friends of the deceased through consular or embassy interactions with the Royal Thai Police.

Figure 3.13 Dental evidence of identity. A mandible from a teenager with a full mandibular dental brace. Courtesy of CIFA (Centre for International Forensic Assistance).

Interpol utilised the comparative DVI software available from the Danish company *Plass Data*. This software enables investigators to input both antemortem and postmortem data, which is then analysed and compared to enhance the matching process and assist the identification of victims.

Between January and March, over 4000 postmortem examinations had been undertaken and over 2000 antemortem files had been listed. From these files over 1000 bodies were identified with over 80 % of these identifications arising from the comparison of dental information (Figure 3.13). Now that most of the dental identifications have been completed, the remainder of positive identifications will likely arise through DNA comparison, which is being undertaken by a number of laboratories throughout the world including the Tuzla laboratory of the International Commission on Missing Persons (ICMP) in Bosnia. When the steady flow of corpses showed no sign of abating, health and safety practical issues had to be addressed. Although there is some acceptance over the fact that an accumulation of a large number of corpses does not necessarily escalate into pandemic high-risk infections (PAHO, 2004), issues relating to personal protective equipment, disposal of waste and long-term storage of human remains had to be attended to. Late in March it was decided to consolidate mortuary

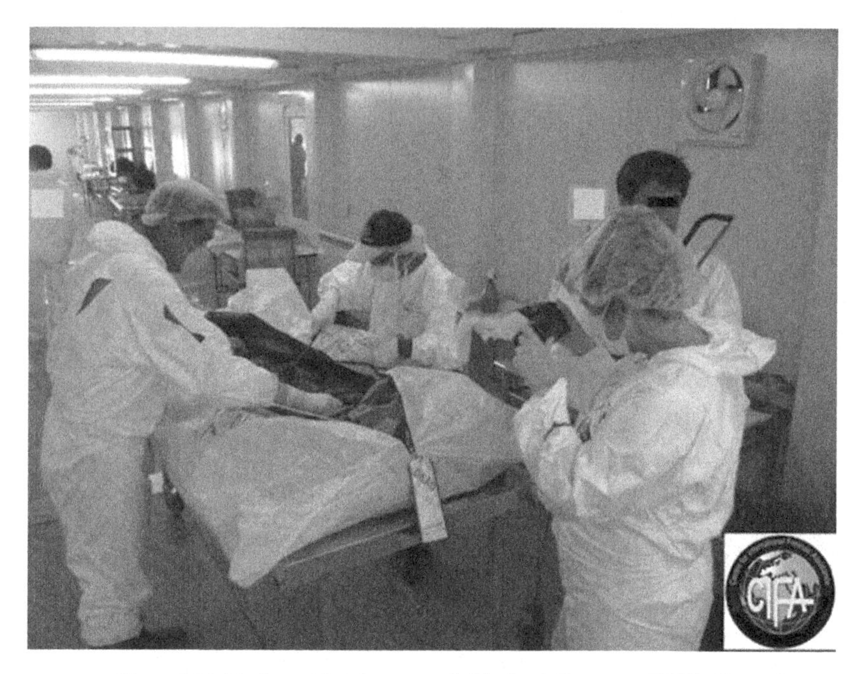

Figure 3.14 The international mortuary in Thailand. Courtesy of CIFA (Centre for International Forensic Assistance).

efforts by building new facilities at the site of the largest temporary mortuary in Wat Yan Yao in the northern Thai province of Phang Na (Figure 3.14). This was the largest of the temporary mortuary sites and had handled over 3000 corpses during the first three months of the operation whilst housing over 300 people in a working day.

The primary identifiers in this and many such disasters are fingerprints, DNA and dental information, although all basic body indicators are recorded. The Foreign and Commonwealth Office (2005) issued some directives about how this process would be conducted following the widely accepted Interpol DVI procedures. Postmortem data were collected in the mortuaries and recorded on the Interpol DVI postmortem forms which were then transferred to the IMC for transfer into the Plass Data system. Families were encouraged to contact their local police, embassy or consul to transfer any information pertaining to the deceased. This might include their dental records sourced from the family dentist, hair samples from a hairbrush left behind by the deceased, photographs, DNA buccal swabs from family members, hospital or medical records and any detailed information that could be verified. This antemortem data was then relayed to the IMC and also transferred to the Plass Data system. Antemortem DNA profiles were constructed, generally utilising local forensic resources, and these were also transferred into the Plass Data software for comparison with the postmortem profiles returned from ICMP and other international laboratories. It was necessary for all potential 'hits' to be verified by experts at the IMC before an identification was confirmed and the body released to relatives for repatriation.

The role of anthropology in this situation was in support of both the pathologists and the odontologists. However, the discipline is more likely to achieve higher prominence with the passage of time as it is inevitable that remains (ultimately all skeletal) will continue to be found around the rim of the Indian Ocean for many years to come. So, when the odontologists, pathologists and DNA experts have moved on to other disasters, it is likely that the anthropologists will continue to play a supportive role in the subsequent osteological analyses.

The similarities between the conditions experienced in Asia and those around New Orleans and the Gulf Coast in America following hurricane Katrina are undeniable. At the time of writing, the final death toll in America is unknown, but the issues surrounding confirmation of identity will undoubtedly be the same. The role of forensic anthropology through the DMORT teams (Disaster Mortuary Operational Response Team) will ensure that the discipline continues to provide assistance to justice and humanitarian issues.

In '*Memoria de Auschwitz*' (2003), Reyes Mate stated 'For a civilization to deserve that name, all of life must be valued, including the (absent) life of the dead'. This remains as true for the victims of the Holocaust as it does for the victims of the Asian Tsunami and hurricane Katrina. It has taken a long time for forensic anthropology to become acknowledged as a discipline of practical scientific value, but its pivotal involvement in major investigations in the last 10 years has ensured that it is no longer marginalised, and along with the other mortuary disciplines it is proud to play its role in serving justice.

The remains of the human may present at any stage of its ontogenetic development from the fetus through to advanced maturity. It may present as a recently deceased corpse, as a decomposing mass or as a skeleton. It may be intact, fragmented, crushed or burned. Therefore the forensic anthropologist must be able to assist in the identification of the deceased regardless of the age of the individual, state of decomposition or degree of fragmentation. It is a tall order for any practitioner to be able to achieve competence in each and every area and the admission and awareness of extent of experience is a vital realisation. The word 'forensic' originates from the word '*forensis*' which pertains to the court (forum) of Rome. The use of this adjective to describe the affiliation of any discipline implies that its practitioners are aware that their duty is to the court and therefore ultimately to justice. So whilst the forensic anthropologist may rarely grace the hallowed halls of our courts of justice, as often once identity is confirmed there is no further issue on this matter, they must never the less ensure that their evidence is always to courtroom standards. The temptation to let standards slip in instances where the forensic relevance is not immediately obvious is a dangerous precept. We can have no prior knowledge of where any case may ultimately lead, and therefore we must never lose sight of the fact that, like all forensic disciplines, our allegiance is to the court and to serve justice and truth.

The forensic anthropologist is now accepted as an integral part of a quite elaborate and highly specific team of experts whose prime purpose is to provide unbiased and objective information for the courts of law. Whilst ensuring this process is observed, they will along the way assist investigators and aid in the identification of the deceased, but this is not their prime directive, and this fact may be quite difficult for the inexperienced practitioner to retain in focus when embroiled in the processes of a deployment. The identification of the deceased is often a detailed and multidisciplinary task that cannot be hurried as the risk, and subsequent cost, of mistaken identity is far too high a price to pay for a fast result. The team is frequently subjected to pressures to release remains early to facilitate investigations and permit closure. However as James Poe (1921–80) reminded us, 'Speed is good only when wisdom leads' and that wisdom must come from the experienced practitioners who understand that their prime servitude is to the court.

Acknowledgements

Thanks are due to Mark Bates, (ex) Forensic Science Service, for requesting anthropological assistance in the investigation of the Yorkshire murder. Equally, the recognition by Professor P Vanezis of the role that forensic anthropology can play in war crimes investigation has played no small part in raising the profile of the discipline within the UK. Thanks are also due to CIFA (Centre for International Forensic Assistance) for permission to utilise photographic material obtained during their deployments on behalf of the Foreign and Commonwealth Office of the British Government to Sri Lanka and Thailand in the aftermath of the Asian Tsunami.

References

ABFA (2005). www.csuchico.edu/anth/ABFA/, accessed 17/06/2006.

Baraybar, J. P. and Gasior, M. (2004). Forensic anthropology and the most probable cause of death in cases of violations against international humanitarian law: an example from Bosnia and Herzegovina. *Journal of the Royal Society for the Promotion of Health*, **124**(6), 271–5.

Black, S. M. (2003). Forensic anthropology – regulation in the UK. *Science and Justice*, **43**(4), 187–92.

Budimlija, Z. M., Prinz, M. K., Zelson-Mundorff, A. *et al.* (2003). World Trade Centre human identification project: experiences with individual body identification cases. *Croatian Medical Journal*, **44**(3), 259–63.

Cabinet Office (2002). www.homeoffice.gov.uk/docs/id˙fraud-report.pdf, accessed 17/06/2006.

Collins, J. M. (2006). *Investigating Identity Theft: A Guide for Businesses, Law Enforcement, and Victims*. London: Wiley.

CRFP (2005). www.crfp.org.uk/, accessed 17/06/2006.

Edson, S. M., Ross, J. P., Coble, M. D., Parson, T. J. and Barritt, S. M. (2004). Naming the dead – confronting the realities of rapid identification of degraded skeletal remains. *Forensic Science Review*, **16**(1), 63–90.

Ferllini, R. (1999). The role of forensic anthropology in human rights issues. In *Forensic Osteology: Advances in the Identification of Human Remains*, 2nd edn., ed. K. Reichs. Springfield: Charles C. Thomas, pp. 287–301.

Foreign and Commonwealth Office (2005). www.fco.gov.uk/servlet/Front?pagename=OpenMarket/Xcelerate/ShowPage&c=Page&cid=1106653108591, accessed 17/06/2006.

Glenny, M. (2000). *The Balkans 1804–1999 Nationalism, War and the Great Powers*. London: Granta Books.

Haglund, W. D. and Sorg, M. H. (2001). *Advances in Forensic Taphonomy: Method, Theory and Archaeological Perspectives*. Florida: CRC Press.

ICRC (2004). www.icrc.org/Web/Eng/siteeng0.nsf/html/genevaconventions, accessed 17/06/2006.

ICTY (2001). www.un.org/icty/indictment/english/mil-ai010629e.htm, accessed 17/06/2006.

Interpol (2002). www.interpol.int/Public/DisasterVictim/guide/chapitre3.asp, accessed 17/06/2006.

Kahana, T., Ravioli, J. A., Urroz, C. L. and Hiss, J. (1997). Radiographic identification of fragmentary human remains from a mass disaster. *American Journal of Forensic Medicine and Pathology*, **18**(1), 40–4.

Kahana, T., Grande, A., Tancredi, D. M., Penalver, J. and Hiss, J. (2001). Fingerprinting the deceased: traditional and new techniques. *Journal of Forensic Sciences*, **46**(4), 908–12.

Kieser, J. A., Laing, W. and Herbison, P. (2006). Lessons learned from large-scale comparative dental analysis following the South Asian tsunami of 2004. *Journal of Forensic Sciences*, **51**(1), 109–13, doi: 10.1111/j.1556-4029.2005.00012.x.

Komar, D. (2003). Lessons from Srebrenica: the contributions and limitations of physical anthropology in identifying victims of war crimes. *Journal of Forensic Sciences* **48**(4), 713–16.

Lain, R., Griffiths, C. and Hilton, J. M. (2003). Forensic dental and medical response to the Bali bombing: a personal perspective. *Medical Journal of Australia*, **179**, 362–625.

Mate, R. (2003). *Memoria de Auschwitz: Actualidad Moral y Política*. Madrid: Trotta Editorial, p. 78.

MMWR (2002). Deaths in World Trade Center terrorist attacks – New York City, 2001. *Morbidity and Mortality Weekly Report*, **51**(Special Issue), 16–18.

Morgan, O., Tidball-Binz, M. and van Alphen, D. (2006). *Management of Dead Bodies After Disasters: a Field Manual for First Responders*. PAHO/WHO/ICRC. Available for download at www.paho.org/english/DD/PED/DeadBodiesFieldManual.htm, accessed 17/06/2006.

Nanavati, S., Thieme, M. and Nanavati, R. (2002). *Biometrics: Identity Verification in a Networked World*. London: Wiley.

PAHO. (2004). *Management of Dead Bodies in Disaster Situations*. Pan American Health Organization, Washington DC. Available at www.paho.org/english/dd/ped/manejocadaveres.htm, accessed 17/06/2006.

Schaefer, M. and Black, S. M. (2005). Comparison of ages of epiphyseal union in North American and Bosnian skeletal material. *Journal of Forensic Sciences*, **50**(4), 777–84.

Scheuer, J. L. and Black, S. M. (2000). *Developmental Juvenile Osteology*. London: Academic Press.

Shepherd, R. (2003). *Simpson's Forensic Medicine*. London: Hodder Arnold.

Sincock, P. (2006). The Rolex murder. In *An Introduction to Forensic Human Identification*, eds. T. J. U. Thompson and S. M. Black. Boca Raton: CRC Press, 459–72.

Skinner, M., Alempijevic, D. and Djuric-Srejic, M. (2003). Guidelines for international forensic bio-archaeology monitors of mass grave exhumations. *Forensic Science International*, **134**(2–3), 81–92.

Steadman, D. W. and Haglund, W. D. (2005). The scope of anthropological contributions to human rights investigations. *Journal of Forensic Sciences*, **50**(1), 23–30. doi: 10.1520/JFS2004214.

Thompson, T. J. U. and Puxley, A. (2006). Personal effects. In *An Introduction to Forensic Human Identification*, eds., T. J. U. Thompson and S. M. Black. Boca Raton: CRC Press, 365–77.

Walter, T. (2005). What is complicated grief? A social constructionist perspective. *Journal of Death and Dying*, **52**(1), 71–9.

Wilkinson, C. M. (2004). *Forensic Facial Reconstruction*. Cambridge: Cambridge University Press.

Zabeeh, F. (1968). *What is in a Name? An Inquiry into the Semantics and Pragmatics of Proper Names*. The Hague: Martinus Nijhoff.

4 *Biology and culture: assessing the quality of life*

RICHARD H. STECKEL

Social scientists generally agree on the considerable value of measuring and investigating long-term trends and socio-economic differences in the standard of living or quality of life. This essay compares and contrasts various approaches that scholars have developed and explains the unique and valuable contributions of biological measures, especially for historical studies. A concluding section ruminates on the continuing evolution of biological approaches, their potential for diffusion in the social sciences and whether it may some day be possible to formulate more comprehensive measures of well-being.

Although there is some overlap, the customary measures used by social scientists may be classified into three broad categories: material, health and psychological. The historical pioneers in these areas have been economists, demographers and psychologists, but in recent times other groups have joined and the boundaries are blurring as researchers increasingly recognize the interrelationships among these traditionally distinct ways of thinking about human well-being.

Material measures and their limitations

Economists are noted for defining and estimating a nation's income, an effort that began in the seventeenth century. In retrospect it seems simple enough, but generations pondered whether and how various types of physical output and the value of services, housing, capital (loans) and government activity should be included in the national accounts (Studenski, 1958). Table 4.1 displays the time path of significant developments in the field. The end result of two centuries of thought is Gross National Product (GNP), a measure of the market value of goods and services produced by a nation's economy during the year. Stimulated by planning and policy needs created by the Great Depression and the Second World War, this measure is now regularly tabulated for every country, and has been estimated by economic historians for some industrial nations from the mid nineteenth century onwards.

Between Biology and Culture, ed. Holger Schutkowski. Published by Cambridge University Press.

Table 4.1 *First attempts to estimate national income*

Country	Estimator	Date of preparation or publication	Approach
England	Petty	1665	Value of labour and proceeds of wealth; Expense of the people
France	Boisbuillebert	1697	Details unavailable
Russia	Hermann (Austrian)	1790	Based on per-capita consumption
US	Tucker	1843	Net value of material production
Austria	Czoernig	1861	Net value of the principal branches of production
Germany	Rumelin	1863	Income-tax statistics
Australia	Coghlan	1886	Net output
Norway	Unknown	1893	Unknown
Japan	Nakamura	1902	Net output
Switzerland	Geering and Hotz	1902	Income distributed
Netherlands	Bonger	1910	Income distributed
Italy	Santoro	1911	Net output
Bulgaria	Popoff	1915	Net output
Spain	Barthe	1917	Net output

Source: Compiled from Studenski (1958), Part One, 'History'.

Though very useful, these monetary measures recognize only market activity, and some, if not many, people in the household may not work outside the home. Therefore monetary measures provide little insight into the allocation of income to goods and services for various members within the household. Moreover we know that people devote considerable time and attach great value to goods and services produced at home. These activities are more important the further back one goes in time, especially to eras when households were substantially self-sufficient. How does one then reliably compare the income of a family earned in say, 1800, with that of a family of identical size and demographic composition but earned in 2005? A simple ratio of the two incomes would be very misleading. A long list of adjustments would have to be made, and obtaining the data for such adjustments can be difficult.

There are significant issues of comparability in monetary measures across time, which arise from measuring inflation or deflation. Not only do prices vary, but changing technology creates new goods and services, which complicates comparisons of material well-being based on income. It is important to know, but complicated to enumerate and value, precisely what goods and services a given income will purchase. Antibiotics, for example, were generally unavailable prior to the mid twentieth century, as were televisions, commercial jet airplanes, automobiles with radios, touch-tone telephones, TV dinners and so forth. How important were these and a host of other changes to peoples' material well-being?

The measurement problem is greater if the households live in different countries. One might plausibly use exchange rates to translate the value of output tabulated in one currency into that of another. But exchange rates are heavily a function of the products involved in trade. Textiles and coffee, for example, might be shipped between them, but houses, restaurant meals, and theatre performances are not. Nor are government services such as roads, fire protection and elementary education. Yet all these non-traded products are important for the material standard of living. Thus, exchange rates are often biased as a measure of what income will purchase in one country relative to that of another.

Biological measures and their advantages

Of course, people desire far more than material goods and in fact they are quite willing to trade or give up material things (i.e. buy fewer goods and services) in return for better physical or psychological health. Health is so important to the quality of life for most people that it is useful to refer to the 'biological standard of living.' Thus, life expectancy at birth, indicators of sickness or morbidity, and signs of psychological stress are regularly employed to help assess a country's social performance. In recent decades, various types of psychological surveys on happiness have entered the discussion.

Biological measures such as life expectancy, average stature and various skeletal lesions are usually more comparable indicators of social performance than are monetary measures or 'happiness' surveys; they mean about the same thing in one country or era as another. Research continues to expand on monetary measures but at a much slower pace than during the high point of the mid twentieth century; this approach encounters diminishing returns in adding new useful information. Nor have psychologists come forward with new approaches that have captured the attention of social scientists.

Academics have been fascinated, however, by new developments in biology. There are significant obstacles for non-specialists in locating and applying information in this exploding field, but at least some curious researchers should face the risks and make the effort to bridge the social and biological sciences. I believe this process should be a two-way street, with social scientists helping to define research questions and posing possible solutions.

Life expectancy at birth

The oldest and most widely used biological measure is based on the life table, which displays the probabilities of death between various exact ages, person-years lived at each age interval, and average remaining years of life at each age (life expectancy) for a synthetic cohort. Two types of raw data are needed to construct this table: deaths by age (also desirable are data by sex and perhaps other categories such as occupation) and a corresponding count of the population at risk. Vital registration ordinarily provides the first and censuses the second source of information. Scholars have used other sources to prepare historical life tables, including parish records of births and deaths, continuous population registers, and genealogies. The ten-volume set on the *History of Actuarial Science* (Haberman and Sibbett, 1995) charts the origins and evolution of the field.

Like early efforts to measure a nation's income, the life table had roots in intellectual discussions of the eighteenth century. At that time mathematicians and philosophers pondered generalizations or rules governing outcomes in games of chance, which could be applied to various life events and therefore to needs of the insurance industry to estimate contingent liabilities involving future benefit payments. In the late seventeenth century Edmund Halley (1693) prepared the first logically organized life table using data for the city of Breslau. He tacitly presumed that the age and sex composition of the population was stationary, and later investigators unwittingly imitated this error.

Using the mortality experience of two parishes in Carlisle, England for the years 1779–87, Joshua Milne (1815) prepared the first scientifically correct life table while he was chief actuary of the Sun Life Assurance Society. With this methodological breakthrough, a major obstacle to preparing life tables at the regional or national level was data on deaths by age. Some censuses existed to collect data by age and sex, but many nations had to create census agencies as well as procedures to administer death registration (Shryock and Siegel, 1975, p. 430). The rise of the public health movement in the late nineteenth century expedited the process. The United States was an early world leader in national census-taking (beginning in 1790), but registration of births and deaths lagged behind other countries that had industrialized by the late nineteenth century, in part because it was a state (or local) as opposed to a federal responsibility. Massachusetts was first among the states to pass a registration law (1842) but records were not reasonably complete until 1865. In 1900 the federal government created a Death Registration Area, which expanded to encompass all states by 1933. In 1900–2 the US Bureau of the Census began to publish life tables in the three-year period containing the decennial census, and more detailed tables (by state, year and so forth) evolved thereafter.

National income accounting diffused rapidly around the globe after the Second World War, but success with vital registration and census-taking was slower and less complete. Rapid population growth created a critical need for demographic estimates to assist economic and demographic planning. A small industry emerged in the 1950s and 1960s to formulate procedures to estimate various demographic measures from incomplete data (United Nations, 1967), such as model life tables and census-survival methods. Demographic historians applied some of these methods to reconstruct historical measures, the best-known example of which is *The Population History of England* (Wrigley and Schofield, 1981). In sum, a reasonably good picture of health from life tables is available from the late nineteenth century onwards for countries that industrialized by the turn of the century, and some useful information is available for virtually all countries beyond about 1970. Spotty but useful estimates are available for a few countries as far back as the eighteenth century and in exceptional cases, the sixteenth century.

Morbidity

Forming the methodology and creating administrative procedures to collect data for the life table were great leaps forward but researchers know the result measures only one dimension of health (Lilienfeld and Stolley, 1994, Chapter 6). Vigour and functional capacity while alive are also important, particularly if the population is aging or if people lived under demanding conditions that lead to illness or loss of functional capacity. New medical technology, health insurance, and intelligent use of these systems in combination with a healthy life-style may significantly prolong life, but one would also like to measure the health-quality of life. This is a challenge because there are numerous measures of morbidity and illness, and even if a standard is widely accepted, consistent collection of evidence over time and across space is usually difficult and expensive.

A couple of decades ago health economists devised the concept of quality-adjusted life years (QALY) to help estimate cost-benefit ratios from various health interventions (Drummond *et al.*, 1997). The measure takes into account both the quantity and the quality of life created by interventions such as cholesterol testing, pacemaker implants, kidney dialysis and so forth. With these methods one could estimate the cost per QALY to be used as a guide for resource allocation within the healthcare field. Household heads on a fixed budget are well aware of tradeoffs in purchases, and the issue here is to spend funds in a way that maximizes or at least significantly increases national health. The method places a weight from 0 to 1 on the time spent in different health states. A year in perfect health is worth 1 and death is assigned a 0 (some painful or agonizing states are considered worse than death and receive negative values). After considering the additional years of life created by various interventions, the result is a common currency that is useful for assessing benefits. The method has a number of practical and technical difficulties related to measuring the quality of life (assigning numerical values to morbidity), but physical examinations and surveys are ways to gain information. One popular survey (EQ-D5) asks the

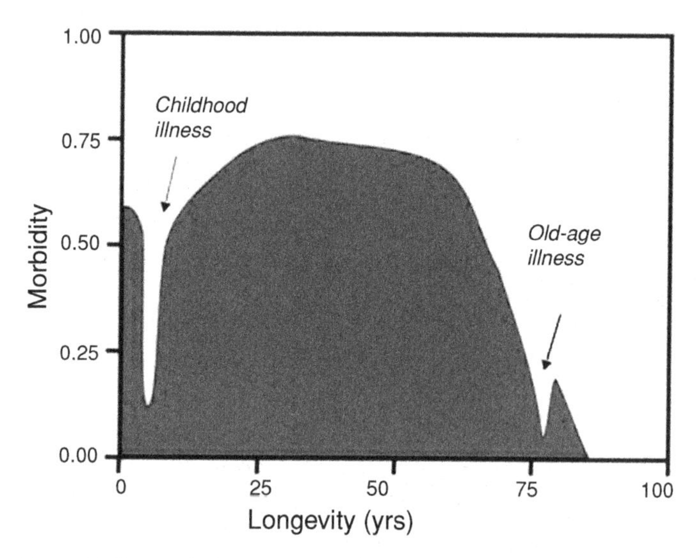

Figure 4.1 Hypothetical example of morbidity by age. Note: A higher number corresponds to less morbidity, and 1.00 refers to a complete absence.

extent to which individuals have functional problems in five areas: mobility, pain/discomfort, self-care, anxiety/depression, and pursuit of usual activities. If such data were available over the entire life-span of an individual one could construct a diagram such as Figure 4.1, in which quality-adjusted length of life is defined by the area under the curve.

Stature

J. M. Tanner's authoritative book on *A History of the Study of Human Growth* recounts the long history of studying body size and proportions (Tanner, 1981). Artists were among the first to quantitatively study human form for purposes of accurately rendering sculptures and paintings. What might be called scientific interest in heights began during the Enlightenment.

The history of national income accounting and that of auxology (the study of human growth) have two things in common: the first substantial efforts occurred in the seventeenth and eighteenth centuries and the early studies were sporadic, imprecise attempts made by individuals. Unlike national income, however, useful measurements of height and related attributes could be made on a small scale. Systematic data on both national income and life expectancy awaited large-scale government action while individuals made important progress in auxology before the end of the nineteenth century.

Table 4.2 *Milestones in auxology*

Place	Investigator	Year	Events or developments
Germany	Elsholtz	1654	Graduation Thesis on Anthropometria
Germany	Jampert	1754	Cross-section measurements of stature by age
Germany	Roederer	1754	Measured and weighed newborns
France	Montbeillard	1777	First longitudinal study from birth to adult
France	Villermé	1829	Studied environmental influences on growth
England	Chadwick	1833	First survey of factory children
Brussels	Quetelet	1842	First mathematical formulation of growth
England	Roberts	1876	Used frequency distributions to assess fitness; studied growth by social class
US	Bowditch	1877	School surveys; analysed velocity of growth
Italy	Pagliani	1879	Longitudinal studies; school surveys
England	Galton	1889	Studied inheritance of height; introduced regression coefficient
France	Budin	1892	First infant welfare clinic established
US	Boas	1892–1932	Tempo of growth; concept of developmental age; growth studies in anthropology; standards for height and weight
France	Godin	1903	Detailed growth surveillance
US	Baldwin	1921	Supervised the first large longitudinal study
England	Douglas	1946	First national survey of health and development
England	Tanner	1952	Models underlying clinical standards

Source: Compiled from Tanner (1981).

Table 4.2 charts milestones in anthropometry from the perspective of human biology. The table shows that initial steps were taken in the seventeenth and eighteenth centuries but progress was slow until the second quarter of the nineteenth century. The first person to use measurements for medical purposes may have been Sigismund Elsholtz, who tried to relate body proportions to health in the mid seventeenth century. In the next century early attempts at systematic anthropometry appeared in the form of Jampert's measurements of orphans of various ages, Roederer's study of newborns, and the growth table of Montbeillard's son from birth to maturity.

Substantial impetus to growth studies developed in the 1820s and 1830s when scholars realized that environmental conditions systematically influenced growth. The rise of auxological epidemiology can be traced to France, where Villermé studied the stature of soldiers; to Belgium, where Quetelet measured children and formulated mathematical representations of the human growth curve; and especially to England, where Edwin Chadwick inquired into the health of factory children. After examining the heights of soldiers in France and Holland and studying the economic conditions in their places of origin, Villermé concluded in 1829 that poverty was much more important than climate in influencing growth. The idea that human growth reflected health was put into action in reports on the stature of factory children that were submitted to Parliament in 1833. Legislation in that year incorporated stature as a criterion in evaluating minimum standards of health for child employment.

The greatest strides in the modern study of human growth occurred in the late 1800s and early 1900s with the work of Charles Roberts, Henry Bowditch, and especially, Franz Boas. Roberts's work in the 1870s increased the sophistication of judging fitness for factory employment with the use of frequency distributions of stature and other measurements, such as weight-for-height and chest circumference. Bowditch assembled longitudinal data on stature to establish the prominent gender differences in growth. In 1875 he supervised the collection and analysis of heights from Boston school children, a data set on which he later used Galton's method of percentiles to create growth standards. In a career that spanned several decades, Boas identified salient relationships between the tempo of growth and height distributions and in 1891 co-ordinated a national growth study, which he used to develop national standards for height and weight. His later work pioneered the used of statistical methods in analysing anthropometric measurements, and investigated the effects of environment and heredity on growth. The results of an explosion of growth studies in the twentieth century are contained in *Worldwide Variation in Human Growth* (Eveleth and Tanner, 1976).

Skeletons

It is unfortunate that Tanner's imposing book on the intellectual history of height research has not been duplicated in other fields such as physical anthropology, which examines the lesions and dimensions of bones in skeletal remains for insights into nutrition, disease, trauma, and activity patterns.[1] According to Moodie (1923, p. 21), whose book provides an excellent summary of the field up to the early 1920s, the term 'palaeopathology,' a branch of physical anthropology that studies skeletal lesions, was first defined in *A Standard Dictionary of the English Language*, published in New York in 1895. The science stems from John M. Clark's studies of pathological conditions among invertebrate fossils at the State Museum in Albany, New York. In 1913 Sir Marc Armand Ruffer introduced the term to the medical literature in his research on Egyptian mummies.

When Clark and Ruffer were studying fossils and mummies for signs of disease, numerous physical anthropologists and pseudo-scholars engaged in racial geometry and a subfield, craniology, which examined various features of the head such as the cephalic index (ratio of width to length), to define and classify human populations. Another interest was phrenology, or the 'science' of character divination whereby detailed measurements such as bumps or protrusions of the skull were thought to reveal psychological aptitudes and tendencies (Marks, 1995; Gould, 1996). Palaeopathology forged ahead with research by Hrdlička and Eaton on pre-Columbian Indians and reports by Garrison and Klebs on ancient Egyptians, and the work of many others, but in the mind of the public in the late twentieth century, the era is remembered for the debate over scientific racism and Nazi policies (Hrdlička, 1909; Eaton, 1916; Garrison, 1917; Klebs, 1917).

By the 1960s the field solidified around its original roots, inaugurated in textbooks by Wells, by Brothwell and Sandison and by others, but was notable for a focus on epidemiological or ecological approaches to disease and methods of life-style reconstruction (Wells, 1964; Brothwell and Sandison, 1967; Brothwell, 1981). These and later texts highlighted the advances made by individuals such as Lawrence Angel, and drew upon new techniques of carbon dating that were developed shortly after the Second World War and improved significantly using Accelerator Mass Spectrometry (AMS) in the late 1970s (Buikstra, 1990). Scholars also began to refine methods: analysis of chemical isotopes, which provided insights into diet; study of sexually dimorphic traits of the skeleton; macro and microscopic methods of aging individuals at death; radiologic assessment of nutrition; study of the micro or cellular structure of bone (histology) to identify specific diseases; activity-pattern analysis inferred from the macro and the cellular organization of bone; and more elaborate statistical methods. In these efforts there was a synergy between palaeopathology and forensic anthropology, or the study of bones in a legal context for clues about age at death, sex, state of health prior to death, and causes of injury or trauma. By the turn of the century a host of new textbooks and encyclopaedias were available to assist with research and classroom instruction (Buikstra and Ubelaker, 1994; Bass, 1995; Larsen, 1997; Aufderheide and Rodriguez-Martin, 1998; White, 2000).

Skeletons vastly extend the reach of anthropometric history by depicting important aspects of well-being over the millennia – aspects which embrace human activities from hunter-gathers onward to settled agriculture, the rise of cities, global exploration and colonization, and eventual industrialization. Skeletons are widely available for study in many parts of the globe, and unlike heights, they depict health over the life cycle. As a package the skeletal measures provide age and source-specific detail on nutrition and biological stress from early childhood through old age, and importantly, several indicators of health during childhood are typically visible or measurable on the skeletons of adults. In addition, remains also exist for women and for children, two groups often excluded from more familiar historical sources such as tax documents, muster rolls, and wage records. Skeletons provide a more extensive and complete picture of community health than available from stature alone, but their meaning is substantially enhanced when combined with contextual information from archaeology, historical documents, climate history, and Geographic Information Systems (GIS).

Skeletal measures of activity, health and biological stress

Unlike dental enamel, bones are living tissues that receive blood and adapt to mechanical and physiological stress. Habitual physical activity that requires exertion leads to a readily visible expansion of the related muscle attachments on the skeleton. The building process, articulated over a century ago by Julius Wolff, is rapid for growing children but also occurs among adults, albeit at a slower rate. If the action is repetitive in a particular direction, the bones adapt to the load by thickening in the direction of the plane of motion (Larsen, 1997). Hunter-gatherers who walked long distances, for example, had oval-shaped femurs, but these bones were nearly circular among settled agriculturalists that had diverse activity patterns. Similarly, professional athletes such as tennis players and baseball pitchers develop extensive muscles, tendons and bones in the shoulders and arms on the side they habitually use.[2]

Net nutrition has been a useful concept for understanding the environmental factors that influence human growth. The body is a biological machine that requires fuel for basal metabolism, to perform work and to combat infection, all of which claim dietary intake. If net nutrition is insufficient, child growth slows or ceases, and linear growth of the skeleton is stunted if the deprivation is chronic and severe. More generally the skeleton is an incomplete but very useful repository of an individual's history of health and biological stress, as explained below.

In principle, anthropologists could use any bone to estimate stature but the femur is often well preserved, is easily measured, and comprising about one-quarter of standing height, this bone has the greatest correlation with stature. A classic study estimated the relationship between the two variables using femur lengths of the deceased whose living height was known from muster rolls or other sources (Trotter and Gleser, 1952). The equations vary somewhat by sex (females are a few centimetres shorter than males for a given femur length), and accurate height estimates require anthropologists to draw upon sexually dimorphic characteristics of the pelvis and the skull that appear in adolescence. Growth plates obfuscate the bone lengths of juveniles, and height estimates are correspondingly problematic until the bony components of the femur fuse late in the teenage years (Tanner, 1978).

As a group, physical anthropologists collect hundreds of skeletal measures, many of which are very specialized in nature and some of which reflect rare or unusual forms of physiological stress. In designing a study of community health with a large number of collaborators, it is important to select general health indicators that are understood and reported by virtually all physical anthropologists, regardless of specialty. In a large collaborative study that investigated skeletal health over the past several thousand years, these included three indictors of health during childhood (stature, linear enamel defects, and skeletal signs of anaemia), two measures of decline among adults (dental decay and degenerative joint disease), and two that could affect any age group, but which are more prevalent among older children and adults (skeletal infections and trauma). Stature has already been discussed as a child health indicator, and the other indicators are briefly explained below.[3] Because bone tissue changes slowly over time, all of these indicators measure various aspects of chronic sickness, morbidity or loss of functional capacity.

Enamel hypoplasias (linear enamel defects)

Linear enamel hypoplasias (LEH) are readily visible lines or pits of enamel deficiency commonly found in the teeth (especially incisors and canines) of people whose early childhood was biologically stressful. They are caused by disruption to the cells (ameloblasts) that form the enamel. The disruption is usually environmental, commonly due to either poor nutrition or infectious disease or a combination of both. Although non-specific, hypoplasias are informative about physiological stress in childhood in archaeological settings.

Indicators of iron deficiency anaemia (porotic hyperostosis and cribra orbitalia)

Iron is essential for many body functions, such as oxygen transport to the body's tissues. In circumstances where iron is deficient – owing to nutritional deprivation, low body weight, chronic diarrhoea, parasite infection, and other factors – the body attempts to compensate by increasing red blood cell production. The manifestations are easily visible in the skeletons of young children in areas where red blood cell production occurs, such as in the flat bones of the cranium. The associated pathological conditions are sieve-like lesions called porotic hyperostosis and cribra orbitalia for the cranial vault and eye orbits, respectively. The lesions can also be caused by other factors, but iron deficiency is among the most common causes. In infancy and childhood, iron deficiency anaemia is associated with impaired growth and delays in behavioural and cognitive development. In adulthood, the condition is associated with limited work capacity.

Dental health

Dental health is an important indicator both of oral and general health, which is assessed in archaeological skeletons from dental caries, antemortem tooth loss, and abscesses. Dental caries is a disease process characterized by the focal demineralization of dental hard tissues by organic acids produced by bacterial fermentation of dietary carbohydrates, especially sugars. In the modern era, the introduction and general availability of refined sugar caused a huge increase in dental decay. In the more distant past, the adoption of agriculture led to a general increase in tooth decay, especially from the introduction of maize. The agricultural shift and the later use of increasingly refined foods have resulted in an increase in periodontal disease, caries, tooth loss, and abscesses.

Degenerative joint disease

Degenerative joint disease (DJD) is commonly caused by the mechanical wear and tear on the joints of the skeleton due to physical activity. Generally speaking, populations engaged in habitual activities that are physically demanding have more DJD (especially build-up of bone along joint margins and deterioration of bone on articular joint surfaces) than populations that are relatively sedentary. Studies of DJD have been valuable in documenting levels and patterns of activity in past populations.

Skeletal infections (osteoperiostitis)

Skeletal lesions of infectious origin, which commonly appear on the major long bones, have been documented worldwide. Most of these lesions are found as plaque-like deposits from inflammation of the periosteum, which is a vascular connective tissue with bone-forming capability that lines all bone surfaces (except joints). Substantial inflammation over a period of months or years creates swollen bone shafts and irregular elevations on bone surfaces. The lesions are often caused by *Staphylococcus* or *Streptococcus* bacteria that penetrate to the periosteum following injuries. Although these lesions can appear on any bone, the front of the tibia (shin) is especially vulnerable, as this section of bone often suffers minor injuries and has little muscle or other soft tissue to protect the periosteum. These lesions have proven very informative about patterns and levels of community morbidity because the infections fester and progress if the immune system is weakened by poor net nutrition.

Trauma

Fractures, weapon wounds and other severe injuries often create bone malformations that provide a record of accidents or violence. The visible record understates the actual level of trauma, however, because some forms do not involve the skeleton and even then healing might hide the injury. The latter occurs more often for children, who rapidly form new bone that can hide scars, and for breaks in which the bones happen to realign well. In pre-industrial societies that lacked X-rays, plaster casts and other devices such as pins or screws, however, the chances were slim for alignment that was both excellent and durable enough for healing without visible scars. The pattern of scars differs for accidents versus deliberate trauma. Violence often registers on the top or back of the head, the face, the forearm (raised to deflect a blow from above), or the back. Other forms include cut marks, bullet holes and embedded arrow points. Accidental injuries, common as ankle and wrist fractures, often reflect difficulty of terrain or the hazards of specific occupations. Deliberate skeletal trauma provides a barometer of domestic strife, social unrest, and warfare.

Comparative studies

Life expectancy

The twentieth century witnessed a vast expansion in population studies that were well grounded in evidence. By the middle of the century, scholars had formulated an influential generalization called the demographic transition (Kirk, 1996), which depicted progress from pre-modern regimes of high fertility and high mortality (in the neighbourhood of 3 to 3.5 %) to the postmodern situation in which both were low (about 1 to 1.5 %). Typically the mortality decline preceded the fall in fertility, and depending upon the country and time period, the difference may have been several decades or longer. The process of change tended to be more rapid in the twentieth as opposed to the nineteenth century, and those of the past half-century occurred even more quickly.

The health side of change is often called the mortality transition and the most recent, large compilations of evidence on the topic can be found in Riley, *Rising Life Expectancy: A Global History* and in Maddison, *The World Economy: A Millennial Perspective* (Riley, 2001; Maddison, 2001). Both books document and discuss possible explanations for change in the world of 1800, with one billion people and life expectancy of perhaps 25 years, to the present world of over six billion people and a life expectancy of about 66 years. By 1900 life expectancy across the world had risen slightly to more than 30 years, but important differences existed by region, with European countries and their colonial offshoots (plus Japan) having a 20-year advantage (46 versus 26 years) over the rest of the world, which had changed little if at all. Today there is even more variation across countries, where life expectancy differs by 2:1 (about 40 years to slightly over 80 years). Despite a situation that is dire relative to other modern countries, even those nations with the lowest life expectancy today are better off than the healthiest countries of two centuries ago.

There is little doubt that cost-effective public health measures played an important role by reducing exposure to pathogens via cleaner water, waste removal, sewage treatment, personal hygiene and chemical control of disease vectors. More controversial are explanations for improving health in Europe and its offshoots prior to 1900, before the public-health movement flourished and long before antibiotics and other advances in medical technology were available. One school of thought led by Thomas McKeown and by Robert Fogel emphasizes improving diets that stemmed from the agricultural revolution of the eighteenth and nineteenth centuries, which featured new crops and equipment as well as other changes such as enclosures, transportation improvements, and eventually the rise of free trade (Mckeown, 1976; Fogel, 2004). Others claim that rising incomes and/or decline in the virulence of pathogens were important.

Morbidity

Combining morbidity and length of life into quality-adjusted life years is an attractive idea (expressed in Figure 4.1), but it is difficult, time-consuming and expensive to conduct a national census of morbidity. Thus the resource costs of measuring morbidity are high relative to the life table because illnesses and disabilities are not only more common, but individual health is dynamic. To score functional capacities equivalent to the life table, medical experts would regularly have to evaluate all individuals. Instead, public health officials rely on physician reports of diseases, and survey information.

In the United States morbidity surveys began with Hagerstown, Maryland in 1921–4, but an ongoing programme did not begin until 1956. The National Center for Health Statistics interviews the non-institutionalized population for information on doctor visits, hospital stays, acute conditions, limits on physical activity and so forth, while other surveys gain data through physical examinations and various psychological and physiological tests (Lilienfeld and Stolley, 1994, Chapter 6). Numerous industrial countries such as Japan, the United Kingdom, and the Netherlands have similar surveillance systems (Alderson, 1988).

The most recent edition of *Historical Statistics of the United States* compiles dozens of morbidity statistics, including the incidence rates of many diseases. Figure 4.2 reports the example of measles from 1912 to 1998, and Figure 4.3 shows the average number of restricted activity days per person from 1967 to 1995. The National Health Interview Survey collected data on the latter, which

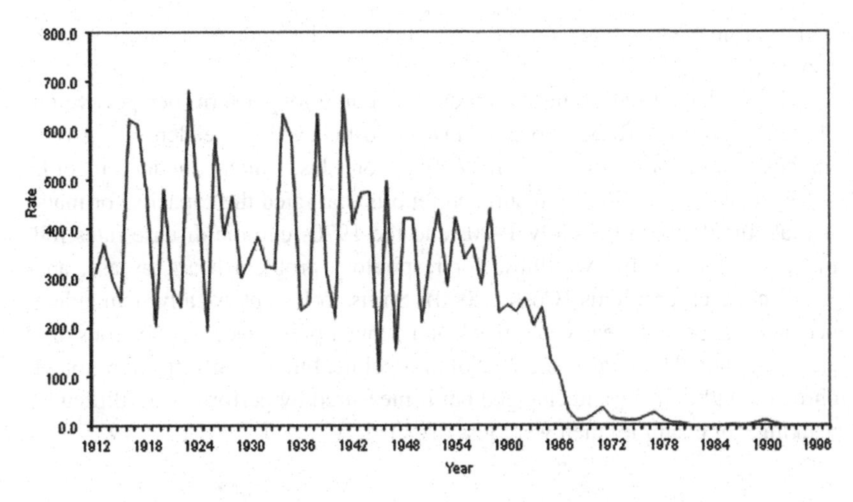

Figure 4.2 Incidence of measles in the United States per 100 000 population. Source: Steckel (2006), pp. 2–503.

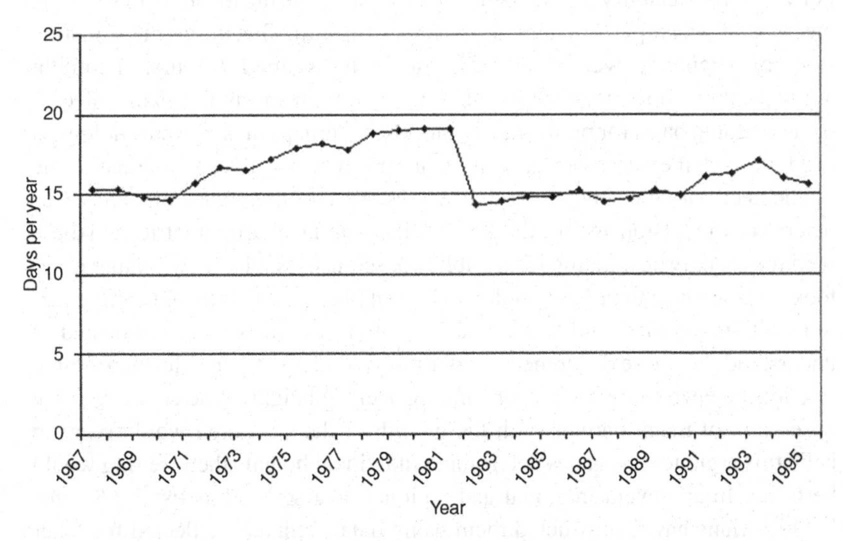

Figure 4.3 Restricted activity days per person in the United States. Source: Steckel (2006), pp. 2–588. Note: The definition changed between 1981 and 1982.

includes people unable or limited in carrying on the usual activity such as school, work and keeping house. Note the data are from interviews rather than physical examinations, and may be subject to cultural norms of what constitutes sickness or disability. Measles followed the pattern of several infectious diseases, for which immunizations led to abrupt declines in the middle of the twentieth century.

Scholars have used military records to obtain a longer-term perspective on chronic conditions. Robert Fogel and Dora Costa have been leaders in organizing data collection from the Civil War pension files, which contained records of physical examinations and surgeon reports that rated the capacity for manual labour. Between the early 1900s and the 1970s chronic disease rates fell markedly, notably by two-thirds for respiratory problem heart disease, and joint and back problems (Costa, 2000). Shifts to less physically demanding occupations explain nearly one-third, and a lower prevalence of infectious diseases accounted for nearly one-fifth of the decline. Interestingly, the duration of chronic conditions was unchanged but if measured by performance (difficulty walking, for example) men were less disabled.

Stature

For well over a century the life table has reigned as the most widely understood measure of social performance in the area of health. Given that the scientific study of height was well advanced by the early twentieth century, one might ask why stature does not occupy a similar position, with suitable data collected on an ongoing basis for public health purposes. Perhaps the large chronological head start of the life table gave it such an advantage that people saw little public need for new concepts and measures. Maybe acceptance of stature was deterred by revulsion against the Nazis and anything anthropometric, and these feelings are only now fading from public consciousness. Plausibly, stature never took hold because the larger public mistakenly believes that height differences reflect mainly genes, and while true for individuals, it is too complicated to understand that these differences tend to cancel in large groups, leaving a large role for the environment in understanding average heights. The concern, if not obsession, of many parents with the growth of their children would seem to belie this argument; if genes determined individual heights then fretting would be futile. In any event, regional and national data series do exist for heights but historians have constructed them using data originally collected for other purposes (mainly identification).

In the past 15 years scholars have completed several large historical studies or compilations of evidence on height with an interest in understanding the standard of living. Although there are many data sources such as slave manifests, muster rolls, convict records, passport applications and so forth, the most abundant source is military organizations, which routinely recorded heights for identification purposes, to assess fighting strength and to make uniforms, near the middle of the eighteenth century. Among the country studies are those on Austria-Hungary, England, and Japan (Floud *et al.*, 1990; Komlos, 1989; Mosk, 1996). Steckel and Floud organized a large effort for comparative study of England, France, the Netherlands, Sweden, Germany, the United States, Australia and Japan (Steckel and Floud, 1997). Komlos edited papers compiling evidence for numerous countries around the globe, and Steckel surveyed the state of the field as of the mid 1990s (Komlos, 1994; Komlos, 1995; Steckel, 1995). Thus, historical perspective is available for numerous countries. Moreover, the World Bank, the United Nations and other agencies now regularly collect height data as part of occasional surveillance programmes, to evaluate interventions, and to investigate socio-economic mechanisms that affect physical growth and child health.

Collectively these studies both confirm and contradict long-held beliefs about differences and changes in human well-being. Heights substantiate the poor health of cities relative to rural areas prior to 1900, a pattern long known from historical population studies. In nineteenth century Sweden, for example, average height was three to eight centimetres greater in rural areas compared with Stockholm, depending upon the time period and rural area (Sandberg and Steckel, 1988).

Comparing height patterns with traditional monetary measures of social performance across developing and developed countries in the second half of the twentieth century revealed a useful role for heights: assessing biological inequality. Steckel found that average height was not only a logarithmic function of average income at the national level, but that holding income constant, average height increased as the degree of income inequality declined (Steckel, 1983). That average heights are sensitive to inequality can be illustrated by the following thought experiment: take $100 from the richest family and give it to the poorest. Average height will increase because the children of the rich have more than the minimum resources for food, clothing, shelter and medical care and hence their heights will not decline from the loss in income. On the other hand the poor do not, and adding to their income will provide more basic necessities, increasing their heights and therefore the average in the country. From this insight researchers began to study occupational and regional differences in stature as a proxy for inequality. In late-eighteenth-century England, for example, the average heights at age 14 of poor boys admitted to the Marine Society were 20 centimetres below those of upper class boys who attended Sandhurst (Floud *et al.*, 1990).

Anthropometric history has garnered attention by uncovering surprising patterns of evidence that challenged traditional interpretations of the past and sometimes provided new insights for human biology. Some examples include the extraordinary growth depression in childhood and substantial recovery by

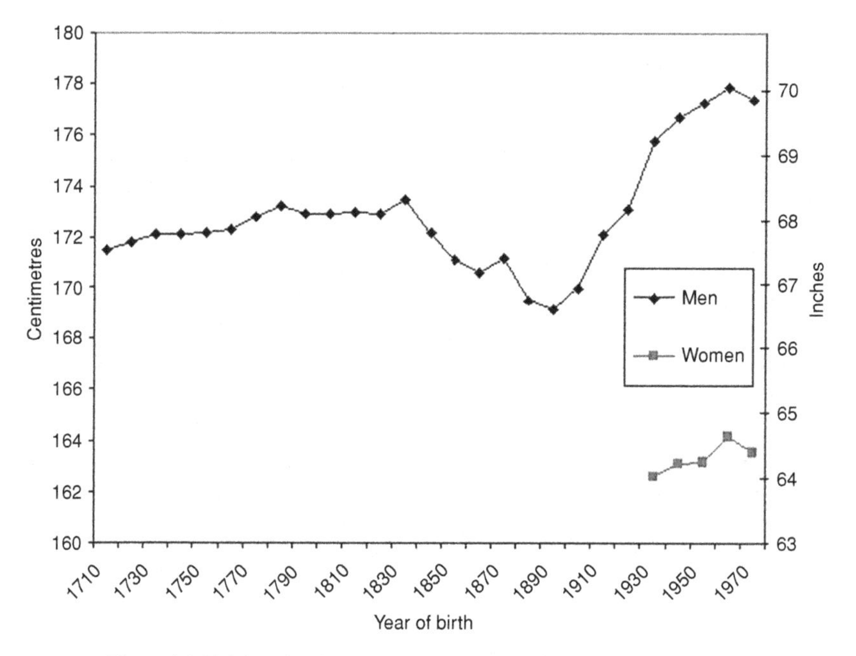

Figure 4.4 Heights of native born American men and women by year of birth. Source: Steckel (2006), pp. 2–503. Note: prior to the twentieth century, the sample is composed of whites and disproportionately of northerners. The heights in the middle of the nineteenth century are based on interpolations from the Ohio National Guard.

American slave teenagers. The pattern was associated with seasonal rhythms in diet, work and disease of pregnant slaves, which was followed by attenuated breastfeeding, and a low protein diet until slaves began working around age ten (Steckel, 1986a, b; 1987).

Economic historians were surprised to find that heights in America declined during the middle of the nineteenth century (Figure 4.4), which occurred during the midst of an industrial revolution and rapid economic growth. The United States and England were the only countries (for which evidence is available) to have experienced substantial and sustained height declines during industrialization prior to the late nineteenth century (Steckel and Floud, 1997).[4] Numerous explanations for the American case are now under investigation, including urbanization, the rise of public schools that spread diseases among children, higher food prices, growing inequality, and higher rates of inter-regional trade and migration that spread pathogens (Steckel, 1995; Komlos, 1998).

Data in Table 4.3 indicate that Americans were several centimetres taller than Europeans prior to the twentieth century. Farm land that was fertile and abundant (which contributed to a good diet), along with relative equality, low population

Table 4.3 *Long-term trends in the stature of adult men (cm)*

Approximate Date	Country						
	US	UK	Sweden	Norway	Netherlands	France	Austria/Hungary
1750	172	165	167	165	166		
1800	173	167	166	166	163	163	
1850	171	166	168	169	164	167	
1900	171	167	172	171	169	165	
1950	175	175	177	178	178	170	171

Sources: Steckel (2006); Floud *et al.* (1990); Sandberg and Steckel (1997); Kiil (1939); Drukker and Tassenaar (1997); Weir (1997); Komlos (1989).

density and long eras of peace, were contributing factors. On average within the United States, Americans of European descent, however, were not the tallest in the world, at least in the middle of the nineteenth century. This honour went to equestrian nomads who used horses to hunt and migrate across the Great Plains (Steckel and Prince, 2001). According to height data originally collected by Franz Boas, the men in eight of these tribes averaged 172.6 centimetres ($N = 1123$); the Cheyenne topped the list at 176.7 ($N = 29$), and the Arapaho were second at 174.3 ($N = 57$). It seems that the important explanatory factors were a high protein diet from bison, which was combined with diverse plant sources of food (including maize acquired through trade with farmers); a strong safety net that followed from an egalitarian society; very low population density; and regular migration that limited exposure to pathogens that accumulated in the waste of sedentary populations. The achievement was all the more remarkable because their numbers declined from epidemics of smallpox and measles and from wars or conflicts with European-Americans and with each other.

Skeletons

Scholars have completed few large-scale comparative studies of community health using skeletal data. Newness of the field (compared with life-table analysis and national income accounting), which limits the amount of material available for study, could be an obstacle but I suspect that the considerable effort required to collect data (relative to the resources available for the field) has been far more important. First the skeleton has to be unpacked and organized and then examined before various measurements can be taken. In the time that it takes to unpack, display, study and repack a skeleton, another researcher could have coded dozens of height observations from well-organized muster rolls.

Idiosyncratic coding schemes also limit study across sites. The variables collected by physical anthropologists and the details of measurement tend to vary across sites and schools of thought, and most researchers are interested in collecting and studying only a few of the variables that would be useful for assessing community health. While many if not most skeletons that have been excavated have been studied to some degree, meta analysis is generally not an option from the published literature. Instead researchers have had to make specific efforts for this purpose.

Palaeopathology at the Origins of Agriculture was the first significant publication of this type (Cohen and Armelagos, 1984). Contributors to the book assembled skeletal evidence of disease patterns that compared the health of hunter-gatherer societies with that of settled agriculturalists. This transition has been long celebrated by social scientists as a major advance in civilization, but to the surprise of those outside physical anthropology, the book reported that health deteriorated during the changeover. Critics and skeptics noted that the data were not truly comparable because contributors utilized varying coding schemes for individual records. Researchers included the frequency of common variables such as degenerative joint disease, skeletal infections and linear enamel hypoplasias, but the specific measures were not reported using the same scale, which complicated comparisons.

The Backbone of History: Health and Nutrition in the Western Hemisphere is a more recent project that built upon the work of Cohen and Armelagos by formulating broader ambitions to study not only the Neolithic revolution but health across a broad swath of time, space and ethnic groups (Steckel and Rose, 2002a). After agreeing on a coding scheme, collaborators pooled their evidence on seven skeletal features from 12 520 remains found at 65 localities that were collectively inhabited from 4000 BC to the early 1900s. Steckel and collaborators distilled the skeletal evidence into a health index, discussed in more detail below, that theoretically could range from 0 (most severe expression in all categories) to 100 (complete absence of lesions or signs of deficiency for every individual at the locality), but in practice that averaged 72.8 (s.d. = 8.0) and varied from 53.5 to 91.8 (Steckel *et al.*, 2002). Surprisingly, Native Americans were among the healthiest and the least healthy populations, with European Americans and African Americans (slaves excepted) falling near the middle of the distribution.

Summarizing community health using skeletons: a health index

For both life tables and national income accounts, which are so familiar and widely used, there is an accepted procedure for distilling complex data into a single number: life expectancy at birth and GNP per capita, respectively. At this stage of research on skeletons, however, numerous simplifying assumptions and approximations are required to distil diverse skeletal data into a single number for comparative ranking and study of populations.[5] Ideally both life expectancy and morbidity would be available, so that one might roughly approximate a measure such as quality-adjusted life years. Unfortunately, many sites in *The Backbone of History*, the largest comparative skeletal study undertaken to date, lack reliable estimates of life expectancy. Therefore the health index discussed here includes only morbidity as expressed in the frequency and severity of skeletal lesions, but the index could be modified to incorporate length of life. A positive correlation between morbidity and mortality is likely, however, which mitigates the lack of data on life expectancy in ranking health across sites.

The index was estimated from the 12 520 skeletons of individuals who lived at 65 localities in the Western Hemisphere over the past several thousand years (Steckel and Rose, 2002a). For each individual, the seven skeletal measures discussed above were graded on a scale of 0 (most severe expression) to 100 (no lesion or deficiency). Age-specific rates of morbidity pertaining to the health indicators during childhood (stature, LEH and anaemia) were calculated by assuming that conditions persisted from birth to death, an assumption justified by knowledge that childhood deprivation is correlated with adverse health as an adult.[6] The duration of morbidity prior to death is in fact unknown for the remaining four components (infections, trauma, DJD, and dental decay) and will be the subject of future research, but was approximated by an assumption of 10 years. Results were grouped into age categories of 0–4, 5–9, 10–14, 15–24, 25–34, 35–44 and 45+.

Next, the age-specific rates for each skeletal measure were weighted by the relative number of person-years lived in a reference population that is believed to roughly agree with pre-Columbian mortality conditions in the Western Hemisphere (Model West, level 4), and the results were multiplied by life expectancy in the reference population (26.4 years) and expressed as a percentage of the maximum attainable (26.4, which corresponds to a complete lack of skeletal defects or lesions). The seven components of the index were then weighted equally to obtain the overall index.

Numerous assumptions underlying the index can be challenged, modified and refined, which cannot be pursued in a short paper. It would be appropriate to weight the elements of the index, such as dental decay and trauma, by their functional consequences, but this is complicated by the nature of the social safety net, medical technology and other factors that vary in unknown ways across societies. Thus, equal weighting is questionable but it is also difficult to justify an alternative scheme given the present state of knowledge. In addition, the index is an additive measure that ignores interactions, but having both a skeletal infection and trauma could have been worse than the sum of their independent effects on health.

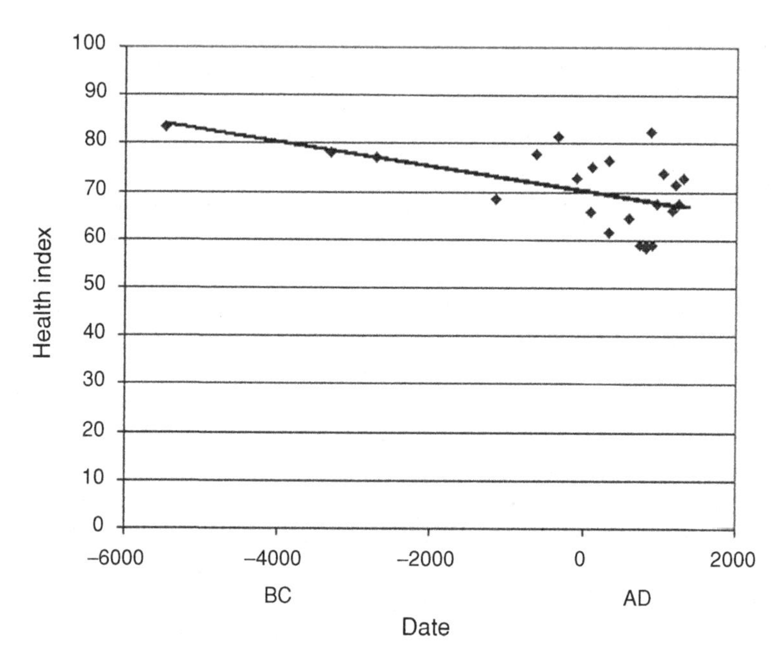

Figure 4.5 Pre-Columbian time trend in the health index. Source: Steckel and Rose (2002b), p. 565.

The most intriguing finding from this project was a long-term decline in the health index in pre-Columbian America. The downward trend over time (Figure 4.5) is statistically significant according to the following estimated linear regression equation:

$$\mathrm{HI} = 65.41 + 0.0025\ \mathrm{YBP},\ N = 23,\ R^2 = 0.53,\ t = 2.89.$$

where HI stands for the health index, and YBP represents years before present (before 1950). The coefficient of the time variable indicates that the index fell, on average, by 0.0025 points per year from roughly 7500 years BP to about 450 years BP, which amounts to 17.5 points over roughly seven millennia. A decline of this magnitude represents a significant deterioration in health and is larger than the difference between the most and least healthy groups who lived in the Western Hemisphere.

Unfortunately, the observations are concentrated in the two millennia before the arrival of Columbus, an era when there was clearly a great deal of diversity in health across sites. The highest value for the index did occur at the oldest site, but two sites in the later era also scored above 80. The least healthy sites (scores under 65), however, were all concentrated within 2000 years of the present.

The key question that this discussion raises is why did health decline? In order to tease apart the potential contributors to decline in health, Steckel and Rose performed a sequence of regression analyses that examined the statistical connection between various ecological categories (e.g. terrain, climate, vegetation, settlement) and health. This analysis revealed that climate – as readily measurable in categories of tropical, subtropical, and temperate – bore no relevance to the health index. The climate result was unanticipated and will be studied again using more refined measures of climate.

On the other hand, most other ecological variables appear to have some relevance. One of the most important to emerge is size of the resident community. Groups living in paramount towns or urban settings had a health index nearly 15 points (two standard deviations) below that expressed for mobile hunter-gatherers and others not living in large, permanent communities. Clearly, there was something about living in a large community that was deleterious to health. And, of course, there is abundant information regarding the health hazards posed by the unsanitary conditions people faced in large pre-modern communities. Such environments were conducive to the spread of infectious disease and other maladies. Moreover, these larger communities were fuelled by an agricultural economy.

Diet was also closely related to the change in the health index, with performance being nearly 12 points lower under the triad of corn, beans and squash compared with the more diverse diet of hunter-gatherer groups. Because the transition to settled agriculture usually occurred with the rise of large communities, it is difficult to obtain precise measures of their separate effects on health.

Among the other environmental conditions that systematically affected health (as measured by the health index) was elevation. People who lived above 300 metres scored about 15 points lower in the index. The exact mechanism for this relationship is unknown, but it is likely that a richer array of foods was available (with less work effort) at lower elevations.

Vegetation surrounding the site may have affected health via the type and availability of resources for food and shelter. Forests, for example, provided materials for the diet, fuel and housing, and also sheltered animals that could have been used for food. Semi-deserts posed challenges for the food supply relative to more lush forests or grasslands, but the dry climate might have inhibited the transmission of some diseases. The net effect of these forces favoured forests and semi-deserts as opposed to open forests and grasslands, where the health index was about 9 points lower.

Flood plain or coastal living provided easy access to aquatic sources of food, and enabled trade compared with more remote, interior areas, but trade may have promoted the spread of disease. Uneven terrain found in hilly or mountainous areas may have provided advantages for defence, but could have led to more accidents and fractures. Apparently the net benefit to health favoured coastal areas, where the health index was about 8 points higher compared with non-coastal regions.

Breaking the sample into two chronological periods, pre and post 1500 years before the present, investigators found that people increasingly lived in less healthy ecological environments. It is unclear why this may have happened, but one possibility is that population growth may have directed settlement into less desirable areas, where greater work effort was required to provide food. Another possibility is that over time, more complex, hierarchical societies emerged, leading to greater biological inequality. Investigators will consider possible explanations in future research.

Attempting to measure happiness

Methodology

Up to this point we have assumed, along with many social scientists, that 'objective' measures of well-being such as income, life expectancy and average height correlate well with a group's quality of life, or more vaguely, 'happiness.' But is this true? More fundamentally, how could one investigate this question?

When post Second World War economic success was well underway in the 1950s and 1960s some people began to ponder whether, or at least the extent to which, social welfare improved with economic growth (Galbraith, 1958; Mishan, 1967). Soon thereafter Richard Easterlin assembled additional empirical evidence to investigate whether economic growth improved the human lot (Easterlin, 1974). At the time such questions were daring if not impertinent, at least among many economists who had tacitly assumed all along that higher income was the primary vehicle to improve human well-being. Since that time a small interdisciplinary industry has emerged to measure and investigate determinants (for a good survey of this literature see Offer (2003)). Some people have created alternatives or compliments to the national income accounts, such as the Measure of Economic Welfare and the Human Development Index that adjust for the ecological, psychological and social costs of producing goods and services (Nordhaus and Tobin 1972; UNDP, 1993). A second approach defines social norms and assesses whether they are met by various indicators such as income, inequality, life expectancy, suicide rates and so forth. Finally, some researchers try to assess mental states directly using surveys of subjective well-being. Collectively these approaches indicate that income is an important, but certainly not the only, substantial determinant of well-being or happiness. Moreover, measures of human welfare or satisfaction are often historically contingent; in *The Wealth of Nations* Adam Smith identified people who would have been ashamed to appear in public without a linen shirt, and while this is irrelevant in industrialized countries today, one can imagine equivalent forms of dress that play a similar role. Thus happiness or life satisfaction is an elusive concept, but in a given setting one can try to measure how social indicators influence it. Lastly, if research agendas underway are successful, one could assess mental states directly by measuring brain activity and levels of various biochemicals.

Some findings

Richard Easterlin has been a leader in evaluating the link between various indicators of well-being and happiness as learned from surveys. His most recent work examines happiness over the life cycle, including the stage of life at which people are happiest and the factors influencing the life cycle pattern (Easterlin, 2006). In the process he tests the set-point model of psychologists, which argues that each person has an inherent, biologically determined level of happiness that they maintain in the face of changing objective circumstances. A weaker version recognizes that happiness can change but most people have useful coping mechanisms to help balance their lives and maintain their level of life satisfaction. Economists emphasize that objective circumstances are powerful in shaping if not determining happiness. Easterlin uses the US General Social Survey from 1972 to 1994 to piece together the full life-course (ages 18–81) using 21-year segments of the life cycle. He then measures the effects of age, financial situation, health, employment and family situation to explain reported levels of happiness. Actual and predicted happiness follows an inverted-U over the life cycle (peaking in the 50s), but this is the net effect of several forces that operate differently with age. Specifically, the pure effect of age is positive on happiness over the life cycle. Happiness increases with perceived financial security, which is U-shaped with age, whereas health has a positive effect on happiness but health declines constantly as people get older. The proportion of people married increases from the late teens through the 30s but thereafter declines as the chance of losing a spouse increases through death or divorce, and so the combined pattern has an inverted U shape with age. Job satisfaction follows a similar inverted-U pattern over the life cycle. In sum, happiness tends to increase with age if people remain married and healthy.

Cross-national happiness surveys

Since the 1990s the World Values Survey has sampled hundreds of thousands of individuals in some 80 countries across the globe, asking among other

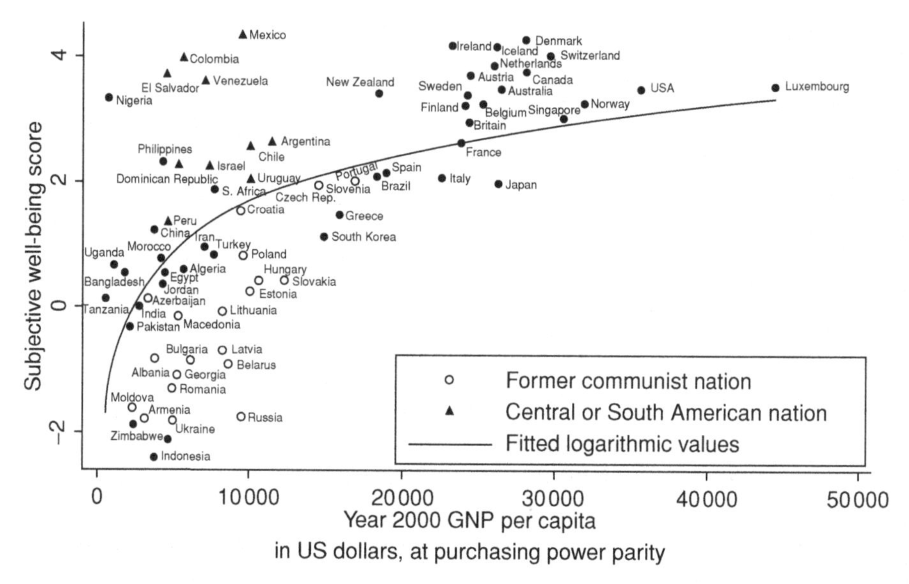

Figure 4.6 Subjective well-being and GNP per capita, 1999–2002.
Sources: Calculated from data in www.worldvaluessurvey.org (World Values Survey) and in http://pwt.econ.upenn.edu/php_site/pwt_index.php (Penn World Tables).

things their level of happiness, classified into categories of: very happy; quite happy; not very happy; not at all happy; and don't know. Here I used results for the 1999–2002 interview wave (available at www.worldvaluessurvey.org/) and then linked the results with data on per capita GNP in each country (http://pwt.econ.upenn.edu/). Figure 4.6 displays the scatter diagram, in which happiness is scaled from -2 to $+4$ and GNP is reported in US dollars. Three interesting patterns emerge. First, happiness increases with per capita GNP but at a decreasing rate. A log function fits the data reasonably well. Second, two groups of countries stand out in the comparisons. Latin Americans report levels of happiness that are much higher than expected from their GNP, whereas former communist countries are much less happy given their GNP. These results are consistent with Easterlin's finding that objective circumstances affect happiness.

Frontiers uniting biology and social sciences

Neuroeconomics

After many decades of separation, Economics and Psychology have regained common ground in the emerging field of behavioural economics (Camerer *et al.*,

2005). Behavioural decision research is a branch of psychology informed by neuroscience, which uses images of brain activity and other methods to infer how the brain operates, which is relevant to understanding optimizing behaviour and other aspects of resource allocation. Utility maximization has been part of economic theory since the nineteenth century, but generations of economists have despaired of learning details inside the brain's 'black box.' New studies of the brain and the nervous system, however, are allowing direct measurement of connections between the mind and action, which has cast doubt on some theories of decision-making and led to formulation of new ones.

One puzzle confronted by this growing field is the influence of emotions on decisions (Cohen, 2005). Do emotions explain forms of behaviour thought to be irrational? How and why do people sometimes deviate from optimality? To what extent can people use higher cognitive processes to override emotional responses that do not serve their best interest? Neurological studies suggest the brain operates as a confederation of systems and while these parts usually work cooperatively, sometimes they compete to guide behaviour. Human evolution may help explain why emotion and cognition vie for control, which can lead to suboptimal or 'irrational' outcomes.

Crudely, the brain is divisible into the neocortex or outer surface and the evolutionarily older and deeper subcortical structures such as the striatum and the brainstem, which are thought to be critical to emotional processing of information (Dalgleish, 2004). In contrast, higher reasoning, such as deliberate thought, language and planning, occupies the prefrontal cortex. Researchers have combined this knowledge of brain anatomy with methods such as positron emission tomography (PET scanning) and functional magnetic resonance imaging (fMRI), which allow them to infer brain activity during the course of decision-making.

Behavioural economists have used these tools to investigate numerous examples of apparently inconsistent behaviour, or at least decisions that contradict rational principles. Consider the case of the ultimatum game, in which partners are given a single, confidential chance to divide an endowment. One partner proposes a split and the other may accept or reject, and if the offer is accepted each player shares the proposed amount, but if the offer is rejected, neither receives anything. Numerous experiments show that offers of less than 20 % are routinely rejected even though this means getting nothing, a result that holds across a wide range of cultures. Behavioural economists have hypothesized that low offers generate an emotional response while high offers create a reasoned reaction, which is substantiated by activity in emotion-centred or reason-centred portions of the brain of partners in the course of rejecting or accepting offers. Rejecting an offer may seem irrational but it could be consistent with evolutionarily older circumstances when interaction was limited to the same few people. Through biological adaptation over thousands of generations, self-interest could have led people to react instinctively to very low offers. Because interaction was repetitive within the tribe or group, reputations were at stake, and discouraging exploitative behaviour was consistent with self preservation. If correct, decision-making that affects the quality of life must be understood in the context of human evolution.

Biosensors to monitor and manage health

Physicians have long assessed illness or morbidity by discussing symptoms, conducting physical tests, and analysing bodily fluids. In the last-mentioned procedure, doctors frequently collect samples of blood or urine, which they send to a central laboratory that returns results anywhere from hours to days later, often at considerable cost. A single test might cost several thousand dollars (or more) if the analysis requires a large volume of expensive reagents and many hours of skilled labour. Miniaturization substantially reduces the amount of costly reagents. New technologies are creating labs-on-a-chip to analyse miniscule amounts of fluids that deliver diagnoses cheaply and quickly at the point of care, and, in some cases, conducted by the patient at home. A companion technology, such as patch-type insulin pumps, now automates the delivery of some medications. Many applications of micro (one millionth of a metre, or one micron) technology are now available, and others built with nano-scale (billionth of a metre) operative parts are being developed, and if forecasts are fulfilled, devices of this scale will be implanted in the body to revolutionize health monitoring and treatment within the next one or two decades.[7] The rapid pace of technological change presents many legal, social, ethical, political and economic issues, and many influential groups in society should be aware of these developments lest they be caught flat-footed or otherwise unprepared to interpret, adapt and direct this technology in society's best interests.

Developed over a decade ago, blood-glucose monitors are perhaps the most familiar of the micro-scale devices (Wang, 2002). The Glucose Watch Biographer is a wrist-wearable monitor that extracts glucose electro-osmotically and can provide readings at 10-minute intervals for up to 13 hours. The current version of the device can signal very low or very high glucose levels and create an electronic diary of more than 8500 readings (www.glucowatch.com/us/consumer/frame_set.asp). Disposable test kits are now available to detect tuberculosis, HIV, pregnancy, sexually transmitted infections, and hepatitis B. A pocket-sized apparatus called the i-STAT Portable Clinical Analyzer needs only three drops of blood to deliver 'accurate electrolyte, chemistry, blood gas and haematology results in only two minutes' (www.heska.com/istat/pr_info.asp). More generally, microfluidic instruments are now in use for four common laboratory techniques, including blood chemistry, immunoassays, nucleic-acid amplification and flow cytometry (Whitesides, 2006). The last-mentioned is the preferred method to count cells with specific physical or chemical characteristics, which are needed, for example, to determine white cell counts.

Such devices hold great promise for medicine, not only in rich countries. Their impact on health may be greatest in the developing world where the centralized laboratory of the developed world is unavailable or does not apply. In low-resource settings, technicians can be taught to use them with little training, and researchers believe it will be possible to manufacture disposable instruments at low cost. Many experts predict that it will become routine for microfluidic diagnostic systems to detect pathogens causing maladies such as influenza, acute fevers or diarrhoea, and provide guidance on treatment (Yager *et al.*, 2006).

Miniature total analysis systems, commonly called lab-on-a-chip devices, evolved from the generation of sensors and computer chips that emerged in the late 1960s (Focus, 2006a; Whitesides, 2006). Such systems contain all the necessary elements for treating a fluid sample, including intake, transport, mixing, separation and detection (measuring results). A significant breakthrough occurred in the late 1980s after research groups developed micropumps, flowsensors and concepts for integrated fluid treatments. These ideas and devices demonstrated that all the steps traditionally performed in a laboratory could be miniaturized to the micro scale and combined onto a single device. An apparatus of this size can manipulate individually objects approximating the size of red blood cells. By the mid 1990s the military sponsored research to develop portable bio/chemical detection systems, and thereafter commercial interest emerged in chemical analysis, environmental monitoring, and medical diagnostics. Research continues on downscaling to sub-micrometer and nano structures, the latter having the potential of individual control over large molecules such as DNA.

In the 1980s lithography was a popular method of manufacturing the substrate housing, a process borrowed from the semiconductor industry, but researchers have learned that etching, laser ablation or injection moulding on glass, polymers, ceramics and other materials provides desirable optical features and chemical compatibility within the capillary structures of the unit (Jakeway *et al.*, 2000; Bilitewski *et al.*, 2003). Within these channels and compartments, centrifugal force, capillary action or electrical fields are commonly used to move the liquids into desired locations for mixing, washing and analysis. Over the years, researchers have developed a variety of detection systems to analyse samples treated with various reagents, but the most common approach is based on lasers that excite fluorescent dyes to emit light that is focused by lenses on photomultiplier tubes (Bilitewski *et al.*, 2003).

Numerous in vitro (outside the body) applications of microfluidics are now available and much research is underway to expand the list of appliances, and to significantly reduce their costs. Researchers debate the types and potential dates of availability for in vivo (inside the body) devices. Built at nano-scale dimensions, such mechanisms could circulate in the blood or be implanted into body tissues. The pharmaceutical industry anticipates that within five years there will be several systems that will automatically administer drugs and sense drug levels, as well as various medical diagnostic tools, such as cancer tagging mechanisms (www.nano.gov/html/res/faqs.html#money). Christopher Wiley offers a futuristic vision of the possibilities for these tools to manage health and control aging (Wiley, 2005). He imagines a world of medical sensors and therapeutic devices that will 'analyze, understand and precisely control the molecular machinery of the human body. This will allow for the detection and correction of any undesired structural changes (disease or aging) at the finest level of detail and the earliest possible time.' Of course imagination is essential to guide research, but most people who actually build devices needed to reach such goals are more tempered in their assessment, pointing out the many obstacles that would have to be overcome (Focus, 2006a). Among these are a source of power for the device, a means to upload data, regenerating the equipment with reagents or medications, disposal of waste, and so forth.

On the other hand, over a billion dollars has been spent on research in the area and far more is anticipated in the future. There is intense competition for success and the rewards it would bring. In the United States the National Science Foundation has taken the lead in developing the technology, but nearly two-dozen agencies now participate (www.nsf.gov/crssprgm/nano/reports/sp_report_nset_final.pdf). US Federal investment in nanotechnology research and development increased from $116 million in fiscal year 1997 to a request of $849 million in fiscal year 2004 (National Nanotechnology Initiative). Globally, spending increased 700 % between 1997 and 2003, from $423 million to $3.024 billion (www.nano.gov/html/res/IntlFundingRoco.htm). Nano-bio efforts are strong in Asia, where China spent $230 million from 2001 to 2005, and Korea launched a 10-year plan in 2001 to spend $1.485 billion (Editor, 2005).

Much of the nanotechnology research has been conducted by engineers, physicists and chemists, who struggle with the practical side of design, manufacture, and operation of such devices. If the most optimistic visions for the technology in monitoring and managing health are to be realized, it seems clear that greater participation by biologists and physicians, and ultimately by social scientists, will be required. Some people have asked 'where are the biologists?' (Focus, 2006b). Input from this community and from physicians will be important for learning what chemical components of bodily fluids are significant for monitoring health. For example, what concentrations of various proteins or other chemicals in the blood are reliable signals of high stress levels? Of impending heart attack, various cancers, epileptic seizure, or inflammation in specific organs? Considerable longitudinal data will be needed to answer such questions, and micro or nanodevices would be ideal to collect it. Traditional laboratory procedures are simply too slow and expensive. Researchers will need extensive longitudinal records on hundreds and possibly thousands of chemical components of blood from thousands of people to learn which ones are good markers for health status. Moreover, such data on chemical precursors of changes in health states will be needed to design interventions such as drug therapy. Future health care will rest increasingly on early detection or anticipating disease, when treatment options are larger and more effective.

It is clear that adoption of nanotechnologies hinges on the outcomes of various economic and efficacy tests. Whether or when the most optimistic visions for the technology are realized, social scientists must be aware of technological developments if they are to help guide research on applications to monitor and improve a society's social performance. Enough has already been achieved in nanotechnology, and far more is likely with the scale of investment now underway, to make it worthwhile for social scientists to ponder the implications of various realistic technology scenarios. Economists, for example, have invented and overseen operation of numerous measures of this sort, such as gross domestic product, the consumer price index and unemployment rates. Similarly, demographers and sociologists have created life tables and crude measures of morbidity such as days lost from work or school. If visions for nanosensory systems are realized, improved measures of morbidity will be within society's grasp. One could imagine, for example, monthly or even daily reports on a country's state of health much like we receive on per capita income or jobs, but based on information gathered by, and uploaded from, nano-scale devices imbedded in the bodies of a national sample of individuals.

Like most profound new technologies, these miniature devices raise a host of issues that society should ponder (Whitesides, 2006). The competition and pace of change is such that if intellectual leaders do not consider the implications, an unmanageable cascade of questions and problems will be forced upon them. Out of ignorance, commentators, newsmakers and organizations may react unwisely to control or limit research and the production of devices that would perform in society's best interests if the technology was properly understood. Foremost among these questions are those of privacy, equity, economic dislocation, intellectual property, pollution, and ethics. Nanoresearchers and their sponsors certainly want to avoid the upheavals that have dogged genetically modified foods, an area of research about which social scientists and the general public remained largely uninformed until the new products appeared on the shelf. Nanotechnology presents legitimate risks and concerns, and the public must be educated to judge the benefits and costs, and if necessary, be prepared to intelligently regulate the development of these remarkable devices. One must recognize, however, that control (or even prohibition) within a single or a small group of countries may be ineffective in the longer term if the ideas or the technology can migrate to other places. The challenge is to anticipate and to address the issues while still reaping the benefits.

One fear is that biosensors could be created and implanted, either willingly or surreptitiously, to probe the body so as to detect diseases or conditions, and then when the data are broadcast to a data storage device, the information could be intercepted by parties not privileged to that information (Rodrigues, 2006). Potential employers or health insurance companies could use such information to screen job applicants or deny policies. Should society prevent this technology because it might be misused? What kind of legislation and enforcement are best to prevent abuses?

Several major enabling technologies of the past half-century, such as computer hardware and the Internet, went forward with little or no patent protection, but the US Bayh-Dole Act of 1980 encouraged universities to file patents on inventions created with federal research funds (Lemley, 2005). They have been vigorous in doing so with nanotechnology since the 1990s. Some lawyers and economists wonder whether this legislation has created a thicket of patents that will stifle innovation, a phenomenon called the 'tragedy of the anti-commons' (Vaidhyanathan, 2006).

Economic historians have documented that important new technologies tend to increase inequality by region, occupation or social class. Even if the technology is eventually widely beneficial, the better educated and the wealthy are the early adopters who benefit more in the short run. Rapid change also threatens economic interests, and can even lead to violence, which is well known from the example of the Luddites who smashed textile machinery of the Industrial Revolution. Physicians seem like unlikely candidates for this activity if nano-devices replace some of their work, but it is clear that various powerful groups will lobby their legislators to limit, control or otherwise shape the technology in their interests.

Contamination fears also concern, if not alarm, some groups. It has been suggested that nanotechnology will create novel types of matter that threatens the environment (Mnyusiwalla *et al.*, 2003). In this vein, the ETC group just announced winning designs for a nano-hazard symbol at the World Social Forum of some 30 000 people in Nairobi, January 2007 (www.etcgroup.org/en/materials/publications.html?id=604). Other people anticipate a different type of pollution, in the form of health risks to workers engaged in manufacturing, or the escape of toxic byproducts (Colvin, 2003). Although there is no evidence at this point to justify such fears, people who lead nanotechnology research must be prepared to contend with public concern over the issue.

Perhaps the most fundamental question provoked by using nanotechnology to improve the human condition is 'what does it mean to be human?' (Grunwald, 2005). The public readily accepts artificial joints and medications that cure disease, but at some point people will wonder whether tiny devices that operate at the molecular level to instantly repair cell damage then change the meaning of 'human.' Should these devices be allowed in athletic competition? What about ordinary jobs? Should we build devices that could extend life almost indefinitely? How should the legal system treat people who commit crimes but undergo 'molecular therapy' that reorganizes their brain cells to alter memory or change personality? My reading of the literature suggests that it will be many years, if ever, before we face some of these questions. Nonetheless, the state of public debate has been infected with bipolar characterizations of the technology, with some who foresee utopia built on extremely tiny machines while others lament the prospect of 'grey goo' consuming the world (Macnaghten *et al.*, 2005). It would be timely for well-informed heads to join the conversation.

Conclusions

For over three centuries scholars have struggled to measure and analyse personal and national well-being. It is a complicated subject and despite great leaps forward much remains to be understood. Over the past century, researchers have made considerable progress in defining and implementing objective measures such as GNP, life expectancy, and morbidity. Happiness remains elusive, particularly at the individual level, but on average subjective feelings are systematically related to various indicators of well-being. In my view, the next great research frontier will use nano-size biosensors to measure brain activity and biochemicals in a search for patterns and determinants of well-being and happiness.

Notes

1. See, however, a collection of papers recently published on the history of bioarchaeology (Buikstra and Beck, 2006).
2. For discussion and references to the literature, see White (2000); Goodman and Martin (2002). Julius Wolff was a nineteenth century surgeon who formulated what skeletal biologists call Wolff's Law, which is the principle that mechanical stress determines the architecture of bone. Specifically he claimed that 'Remodeling of bone . . . occurs in response to physical stresses – or the lack of them – in that bone is deposited in

sites subjected to stress and is resorbed from sites where there is little stress.' Quoted from Salter (1999).

3. For a discussion of variables see Buikstra and Ubelaker (1994). For additional information on the meaning of variables see text and references in Larsen (1997) and White (2000).
4. Heights in several European countries declined during the late 1830s and the 1840s in connection with harvest failures and/or rising food prices. By the late nineteenth century the public health movement noticeably diminished the consequences for health of events associated with industrialization.
5. A short paper necessarily conveys only a flavour of the methodology; for additional details and justification see Steckel *et al.* (2002). Presumably future research will lead to more appropriate assumptions and an improved health index.
6. The effect of fetal and early childhood health on adult health is sometimes called the Barker hypothesis (Barker, 1998). For a general discussion see Fogel and Costa (1997), pp. 56–57.
7. The scales typically correspond to dimensions of operative parts conveniently expressed in whole numbers of one to three digits. By way of comparison, a human red blood cell is about seven microns and contains roughly 270 million haemoglobin molecules, whereas a DNA double helix molecule has a diameter of around two nanometres.

References

Alderson, M. R. (1988). *Mortality, Morbidity, and Health Statistics*. New York: Stockton Press.

Aufderheide, A. C. and Rodriguez-Martin, C. (1998). *The Cambridge Encyclopedia of Human Paleopathology*. Cambridge: Cambridge University Press.

Barker, D. J. P. (1998). *Mothers, Babies, and Health in Later Life*. Edinburgh: Churchill Livingstone.

Bass, W. M. (1995). *Human Osteology: A Laboratory and Field Manual*. Columbia: Missouri Archaeological Society.

Bilitewski, U., Genrich, M., Kadow, S. and Mersal, G. (2003). Biochemical analysis with microfluidic systems. *Analytical and Bioanalytical Chemistry*, **377**, 556–69.

Brothwell, D. R. (1981). *Digging up Bones: The Excavation, Treatment and Study of Human Skeletal Remains*. Oxford: Oxford University Press.

Brothwell, D. R. and Sandison, A. T. (eds.) (1967). *Diseases in Antiquity: A Survey of the Diseases, Injuries and Surgery of Early Populations*. Springfield: Charles C. Thomas.

Buikstra, J. E. (1990). *A Life in Science: Papers in Honor of J. Lawrence Angel*. Kampsville: Center for American Archeology.

Buikstra, J. E. and Beck, L. A. (eds.) (2006). *Bioarchaeology: The Contextual Analysis of Human Remains*. Amsterdam: Elsevier.

Buikstra, J. E. and Ubelaker, D. H. (1994). *Standards for Data Collection from Human Skeletal Remains*. Fayetteville: Arkansas Archeological Survey.

Camerer, C., Loewenstein, G. and Prelec, D. (2005). Neuroeconomics: how neuroscience can inform economics. *Journal of Economic Literature*, **43**, 9–64.

Cohen, J. D. (2005). The vulcanization of the human brain: a neural perspective on interactions between cognition and emotion. *The Journal of Economic Perspectives*, **19**, 3–24.

Cohen, M. N. and Armelagos, G. J. (1984). *Paleopathology at the Origins of Agriculture*. New York: Academic Press.

Colvin, V. L. (2003). The potential environmental impact of engineered nanomaterials. *Nature Biotechnology*, **21**, 1166–70.

Costa, D. L. (2000). Understanding the twentieth-century decline in chronic conditions among older men. *Demography*, **37**, 53–72.

Dalgleish, T. (2004). The emotional brain. *Nature Reviews Neuroscience*, **5**, 583–9.

Drukker, J. W. and Tassenaar, V. (1997). Paradoxes of modernization and material well-being in the Netherlands during the nineteenth century. In *Health and Welfare During Industrialization*, eds. R. H. Steckel and R. Floud. Chicago: University of Chicago Press, pp. 331–78.

Drummond, M. F., Stoddart, G. L. and Torrance, G. W. (1997). *Methods for the Economic Evaluation of Health Care Programmes*. Oxford: Oxford University Press.

Easterlin, R. A. (1974). Does economic growth improve the human lot? some empirical evidence. In *Nations and Households in Economic Growth*, eds. P. A. David and M. W. Reder. New York: Academic Press, pp. 89–125.

Easterlin, R. A. (2006). Life cycle happiness and its sources: intersections of psychology, economics and demography. *Journal of Economic Psychology*, **27**(4), 463–82.

Eaton, G. F. (1916). *The Collection of Osteological Material from Machu Picchu*. New Haven: Connecticut Academy of Arts and Sciences.

Editor (2005). Nano-bio efforts in Asia. *Asia Pacific Nanotech Weekly*, **3**, 1–3.

Eveleth, P. B. and Tanner, J. M. (1976). *Worldwide Variation in Human Growth*. Cambridge: Cambridge University Press.

Floud, R., Wachter, K. W. and Gregory, A. (1990). *Height, Health and History: Nutritional Status in the United Kingdom, 1750–1980*. Cambridge: Cambridge University Press.

Focus (2006a). Labs-on-a-Chip: origin, highlights and future perspectives on the occasion of the 10th μTAS conference. *Lab on a Chip*, **6**, 1266–73.

Focus (2006b). Where are the biologists? *Lab on a Chip*, **6**, 467–70.

Fogel, R. W. (2004). *The Escape from Hunger and Premature Death, 1700–2100: Europe, America, and the Third World*. Cambridge: Cambridge University Press.

Fogel, R. W. and Costa, D. L. (1997). A theory of technophysio evolution, with some implications for forecasting population, health care costs, and pension costs. *Demography*, **34**, 49–66.

Galbraith, J. K. (1958). *The Affluent Society*. Boston: Houghton Mifflin.

Garrison, F. H. (1917). *An Introduction to the History of Medicine*. Philadelphia: W. B. Saunders.

Goodman, A. S. and Martin, D. L. (2002). Reconstructing health profiles from skeletal remains. In *The Backbone of History: Health and Nutrition in the Western Hemisphere*, eds. R. H. Steckel and J. C. Rose. New York: Cambridge University Press, pp. 11–60.

Gould, S. J. (1996). *The Mismeasure of Man*. New York: Norton.

Grunwald, A. (2005). Nanotechnology – a new field of ethical inquiry? *Science & Engineering Ethics*, **11**, 187.

Haberman, S. and Sibbett, T. A. (1995). *History of Actuarial Science*. United Kingdom: Pickering and Chatto.

Halley, E. (1693). An estimate of the degrees of the mortality of mankind, drawn from curious tables of the births and funerals at the city of Breslaw; with an attempt to ascertain the price of annuities upon lives. *Philosophical Transactions of the Royal Society of London*, **17**, 596–610 and 654–6.

Hrdlička, A. (1909). Tuberculosis among certain Indian tribes of the United States. *Bureau of American Ethnology Bulletin*, **42**.

Jakeway, S. C., de Mello, A. J. and Russell, E. L. (2000). Miniaturized total analysis systems for biological analysis. *Journal of Analytical Chemistry*, **366**, 525–39.

Kiil, V. (1939). *Stature and Growth of Norwegian Men During the Past Two Hundred Years*. Oslo: I Kommision hos Jacob Dybwad.

Kirk, D. (1996). Demographic transition theory. *Population Studies*, **50**, 361–87.

Klebs, A. C. (1917). Paleopathology. *Johns Hopkins Hospital Bulletin*, **28**, 261 6.

Komlos, J. (1989). *Nutrition and Economic Development in the Eighteenth-Century Habsburg Monarchy: An Anthropometric History*. Princeton: Princeton University Press.

Komlos, J. (1994). *Stature, Living Standards, and Economic Development: Essays in Anthropometric History*. Chicago: University of Chicago Press.

Komlos, J. (1995). *The Biological Standard of Living on Three Continents: Further Explorations in Anthropometric History*. Boulder: Westview Press.

Komlos, J. (1998). Shrinking in a growing economy? The mystery of physical stature during the industrial revolution. *Journal of Economic History*, **58**, 779–802.

Larsen, C. S. (1997). *Bioarchaeology: Interpereting Behavior from the Human Skeleton*. New York: Cambridge University Press.

Lemley, M. A. (2005). Patenting nanotechnology. *Stanford Law Review*, **20**, 101–30.

Lilienfeld, D. E. and Stolley, P. D. (1994). *Foundations of Epidemiology*. New York: Oxford University Press.

Macnaghten, P., Kearnes, M. B. and Wynne, B. (2005). Nanotechnology, governance, and public deliberation: what role for the social sciences? *Science Communication*, **27**, 268–91.

Maddison, A. (2001). *The World Economy: A Millennial Perspective*. Paris: Development Centre of the Organisation for Economic Co-operation and Development.

Marks, J. (1995). *Human Biodiversity: Genes, Race, and History*. New York: Aldine de Gruyter.

McKeown, T. (1976). *The Modern Rise of Population*. New York: Academic Press.

Milne, J. (1815). *A Treatise on the Valuation of Annuities and Assurances on Lives and Survivorships*. London: Longman, Hurst, Rees, Orme, and Brown.

Mishan, E. J. (1967). *The Costs of Economic Growth*. New York: F. A. Praeger.

Mnyusiwalla, A., Daar, A. S. and Singer, P. A. (2003). Mind the gap: science and ethics in nanotechnology. *Nanotechnology*, **14**, R9–R13.

Moodie, R. L. (1923). *Paleopathology; An Introduction to the Study of Ancient Evidences of Diseases*. Urbana: University of Illinois Press.

Mosk, C. (1996). *Making Health Work: Human Growth in Modern Japan*. Berkeley: University of California Press.

Nordhaus, W. D. and Tobin, J. J. (1972). Is economic growth obsolete? In *Conference on the Measurement of Economic and Social Performance*, ed. M. Moss. New York: NBER.

Offer, A. (2003). Economic welfare measurements and human well-being. In *The Economic Future in Historical Perspective*, eds. P. A. David and M. Thomas. Oxford: Oxford University Press, pp. 371–99.

Riley, J. C. (2001). *Rising Life Expectancy: A Global History*. Cambridge: Cambridge University Press.

Rodrigues, R. (2006). The implications of high-rate nanomanufacturing on society and personal privacy. *Bulletin of Science, Technology and Society*, **26**, 38–45.

Salter, R. B. (1999). *Textbook of Disorders and Injuries of the Musculoskeletal System; An Introduction to Orthopaedics, Rheumatology, Metabolic Bone Disease, Rehabilitation and Fractures*. Baltimore: Williams & Wilkins.

Sandberg, L. G. and Steckel, R. H. (1988). Overpopulation and malnutrition rediscovered: hard times in 19th-century Sweden. *Explorations in Economic History*, **25**, 1–19.

Sandberg, L. G. and Steckel, R. H. (1997). Was industrialization hazardous to your health? Not in Sweden! In *Health and Welfare During Industrialization*, eds. R. H. Steckel and R. Floud. Chicago: University of Chicago Press, pp. 127–59.

Shryock, H. S. and Siegel, J. S. (1975). *The Methods and Materials of Demography*. Washington: USGPO.

Steckel, R. H. (1983). Height and per capita income. *Historical Methods*, **16**, 1–7.

Steckel, R. H. (1986a). A peculiar population: the nutrition, health, and mortality of American slaves from childhood to maturity. *Journal of Economic History*, **46**, 721–41.

Steckel, R. H. (1986b). A dreadful childhood: the excess mortality of American slaves. *Social Science History*, **10**, 427–65.

Steckel, R. H. (1987). Growth depression and recovery: the remarkable case of American slaves. *Annals of Human Biology*, **14**, 111–32.

Steckel, R. H. (1995). Stature and the standard of living. *Journal of Economic Literature*. December, **33**, 1903–40.

Steckel, R. H. (2006). Health, nutrition and physical well-being. In *Historical Statistics of the United States: Millennial Edition*, eds. S. B. Carter, S. Gartner, M. R. Haines *et al.* New York: Cambridge University Press, pp. 499–508.

Steckel, R. H. and Floud, R. (1997). *Health and Welfare During Industrialization*. Chicago: University of Chicago Press.

Steckel, R. H. and Prince, J. (2001). Tallest in the world: Native Americans of the Great Plains in the nineteenth century. *American Economic Review*, **91**, 287–94.

Steckel, R. H. and Rose, J. C. (eds.) (2002a). *The Backbone of History: Health and Nutrition in the Western Hemisphere*. New York: Cambridge University Press.

Steckel, R. H. and Rose, J. C. (2002b). Patterns of health in the Western Hemisphere. In *The Backbone of History: Health and Nutrition in the Western Hemisphere*, eds. R. H. Steckel and J. C. Rose. New York: Cambridge University Press, pp. 563–79.

Steckel, R. H., Sciulli, P. W. and Rose, J. C. (2002). A health index from skeletal remains. In *The Backbone of History: Health and Nutrition in the Western Hemisphere*, eds. R. H. Steckel and J. C. Rose. New York: Cambridge University Press, pp. 61–93.

Studenski, P. (1958). *The Income of Nations; Theory, Measurement, and Analysis: Past and Present; A Study in Applied Economics and Statistics*. New York: New York University Press.

Tanner, J. M. (1978). *Fetus Into Man: Physical Growth from Conception to Maturity*. Cambridge: Harvard University Press.

Tanner, J. M. (1981). *A History of the Study of Human Growth*. Cambridge: Cambridge University Press.

Trotter, M. and Gleser, G. C. (1952). Estimation of stature from long bones of American whites and Negroes. *American Journal of Physical Anthropology*. December, **10**, 463–514.

UNDP (1993). *Human Development Report*. New York: United Nations Development Programme.

United Nations (1967). *Methods of Estimating Basic Demographic Measures from Incomplete Data*. New York: United Nations.

Vaidhyanathan, S. (2006). Nanotechnologies and the law of patents: a collision course. In *Nanotechnology: Risk, Ethics and Law*, eds. G. Hunt and M. Mehta. London: Earthscan, pp. 225–6.

Wang, J. (2002). Portable electrochemical systems. *Trends in Analytical Chemistry*, **21**, 226–32.

Weir, D. R. (1997). Economic welfare and physical well-being in France, 1750–1990. In *Health and Welfare During Industrialization*, eds. R. H. Steckel and R. Floud. Chicago: University of Chicago Press, pp. 161–200.

Wells, C. (1964). *Bones, Bodies, and Disease: Evidence of Disease and Abnormality in Early Man*. New York: Frederick A. Praeger.

White, T. D. (2000). *Human Osteology*. San Diego: Academic Press.

Whitesides, G. M. (2006). The origins and the future of microfluidics. *Nature Reviews Neuroscience*, **442**, 368–73.

Wiley, C. (2005). Nanotechnology and molecular homeostasis. *Journal of the American Geriatrics Society*, **53**, S295–S298.

Wrigley, E. A. and Schofield, R. S. (1981). *The Population History of England, 1541–1871: A Reconstruction*. Cambridge: Harvard University Press.

Yager, P., Edwards, T., Fu, E. *et al.* (2006). Microfluidic diagnostic technologies for global public health. *Nature*, **442**, 412–18.

5 *Ecology, culture and disease in past human populations*

DONALD J. ORTNER AND HOLGER SCHUTKOWSKI

Introduction

The last 10 000 years of human prehistory and history brought changes in the human biocultural environment on a scale never before experienced by humans. The shift from hunting/gathering to an agriculture-based economy led to dramatic concomitant changes in patterns of social organisation and the appropriation of nature. These changes posed and continue to pose some of the most difficult challenges that human groups have ever encountered (Ortner, 2001). During the Holocene, humans domesticated both animals and plants, invented irrigation, developed a sedentary lifestyle, and the establishment and growth of towns began a population explosion that threatens the stability of society today (e.g. Harris, 1996; Sutton and Anderson, 2004). These innovations had a lasting impact on the cultural and natural environments in which human groups operate, although a few forager societies continue a hunting/gathering economy today. But even these societies are affected by the problems created by their agriculturist cousins (Johnson and Earle, 2000; Moran, 2000; Panter-Brick *et al.*, 2001).

Before proceeding further, clarification of the terms *cultural* and *natural environment* is necessary for the purpose of this paper. Particularly in humans, the distinction between these two components of the environment in which human societies evolve is inexact at best. Much more so than any other species, humans make use of their capacity to modify and control the environment, both cultural and natural, to an unprecedented extent.

In this chapter the term *cultural environment* tends to include those aspects of the environment that relate to the relationship between people and the accumulated cultural traits and innovations that a society uses to survive and, ideally, thrive in a natural environment, e.g. socio-political organisation, modes of subsistence or belief systems. By contrast, the *natural environment* tends to consist of the non-human animal life, plants, air, water and soil constituents on which people depend for their survival. We emphasise the modifier 'tends' because

Between Biology and Culture, ed. Holger Schutkowski. Published by Cambridge University Press.

of the inevitable overlap between these natural constituents and the cultural environment. For example, water is a crucial part of the natural environment. However, humans contaminate water supplies and, to some extent at least, control the supply and flow of water through dams, storage in cisterns and the development and use of irrigation systems. Furthermore, the term natural environment is used with the understanding that there is no longer such a thing as pristine nature, at least not in those areas humans have set foot on, as all levels of human activity deliberately modify existing natural conditions. Thus, *natural* takes the meaning of quasi-natural.

The new social innovations developed during the Holocene brought with them a host of new problems to which our ancestors had to adjust. Increased exposure to existing and new pathogens, both biological and chemical, greatly increased the risk of disease. Domestication of animals made the transition of zoonotic infectious agents to human groups inevitable (e.g. Lancaster, 1990, pp. 51 ff.; Groube, 1996). Irrigation created an ideal environment for several human pathogens to flourish, e.g. *Plasmodium* or *Schistosoma* (e.g. Boelee *et al.*, 2002). Increased use of well water enhanced the risk of potentially toxic contaminants such as fluorine. Expanding use of marginal environments led to dietary deficiencies of essential elements such as iodine and the resultant hypothyroidism that affects some individuals living in these conditions (Ortner, 2003). Culinary practices, e.g. the widespread consumption of unleavened bread in the Middle East, can create health impairments, in this case as a result of zinc deficiency (Sandström, 1989).

The transition to a more sedentary agricultural economy took place at different times in different places, although the earliest evidence of this transition is found in the Near East (Smith, 1995, p. 50). But even in this geographical region the transition was not universal. The picture is also complicated by the development of nomadic pastoralism as an important adjunct to agriculture. Sheep and goats were domesticated very early in the development of agriculture, and provided a crucial food resource that could thrive on agriculturally marginal land and on the by-products of agriculture that were unsuited for human consumption.

In the following paragraphs we will present examples of harmful and beneficial human interactions with environmental resources, explore some of the factors in human disease ecology, and speculate on the effect these had on human groups during the past 10 000 years. There is an emerging corpus of data on palaeopathology in archaeological human skeletal samples that has enhanced, and will continue to enhance, our knowledge of disease in past human groups (e.g. Aufderheide and Rodríguez-Martín 1998; Steckel and Rose, 2002; Ortner, 2003). However, this corpus remains incomplete at this time and there are as yet unanswered questions about what can be inferred from evidence of disease in archaeological human skeletal samples (e.g. Wood *et al.*, 1992). Nevertheless the knowledge we now have does at least allow us to explore some aspects of the complex relationship between human groups and their environment, and begin to define some of the questions and issues that need to be clarified.

Factors in human disease ecology

In any attempt to understand the role of disease in human groups, an important factor that tends to be overlooked is that a pathogen is only one of several necessary components of pathogen-induced disease. Whether or not the disease occurs often depends on other factors, including the general health and diet of the individual and the genetically influenced potential of the body to destroy, isolate and/or remove pathogens before they can cause serious disease. The emergence and prevalence of diseases in humans is thus heavily influenced by biocultural factors. Like any other organism, humans are but one of many constituent parts of ecosystems, and hence are tied into fundamental ecological relationships, such as flows of matter and energy, within the habitats they occupy. Yet, although they are affected by the natural environment, human groups, especially during the Holocene, have demonstrated a remarkable capacity to modify the environment through cultural innovation – a characteristic of their ecological niche (Schutkowski, 2006, p. 22).

These flexible and versatile innovations, along with associated longer-term biological adaptations, have created the biocultural environment in which humans live today. Human presence in virtually every land habitat on earth testifies to the success of this evolutionary development. Nevertheless this expanding exploitation of natural resources did not occur without costs or without an increasing emphasis on innovation to minimise these costs through cultural adjustments made in response to environmental challenges. Less apparent, but undoubtedly important as well, have been the adjustments made in the genetic aspects of human biology through natural selection in various environments and different cultural contexts. There are even striking examples where both have co-evolved, for example in the emergence of the selective advantage of lactose tolerance in pastoralist societies (Durham, 1991, p. 226 ff.; Simoons, 2001). Relevant factors in human disease ecology thus comprise elements of culture and biology alike: the biology of the human host and various pathogens set against the cultural and natural environments in which they develop and adjust. None of these factors occur in isolation from the others and typically there is considerable interaction between them in influencing the presence and prevalence of disease.

Cultural environment

Increasing population density was one by-product of the success of the agricultural economy. Greatly increased food resources raised the carrying capacity and stimulated reproduction, much as occurs elsewhere in the animal kingdom when available food is unusually abundant for some reason. The important difference is, however, that this was the result of learning how to cope with delayed returns and to operate pre-emptive and retrospective control over resources (Ellen, 1994). Agriculture, with the associated animal domestication, undoubtedly resulted in a more abundant source of calories most of the time for most of the affected people. However, dependence on a few crops increased the risk of malnutrition resulting from famine or from an unbalanced diet (Cohen, 1989). Tilling the soil and close contact with domestic animals increased exposure to infectious agents. Ergot, for example, is a potent crop pest, especially in wet and cool environments, and its alkaloid has fertility-reducing capacity. Large-scale consumption of ergot-infested grain, believed to be common before the spread of the potato as a new and widely accepted cultivar, may thus have had an impact on population growth in medieval and early modern times. Irrigation provided the potential of exploiting marginal land or enhancing yields, but increased exposure to water-borne pathogens such as *Schistosoma*. The clearing of forests to provide agricultural land is known to create breeding grounds for the *Anopheles* mosquito, and has thus facilitated the spread of malaria in those areas where climate helps it remain endemic. Domesticated animals offered a more stable source of protein, but were also carriers for diseases such as tuberculosis, brucellosis and echinococcosis, even though latest research suggests that the transmission of *Mycobacterium tuberculosis* can occur from human to cattle as well (Ocepek *et al.*, 2005). Many of the diseases transmitted by direct contact and prevalent in human populations today originate from interactions with the animals that were the original hosts to the pathogen. The ability of pathogens to cross species boundaries leaves the new host with little or no capacity for competent immune reaction in the first place, and pandemic proportions of morbidity and mortality can easily be reached, as the spread of the HIV has demonstrated, and as avian flu is likely to show (May *et al.*, 2001).

With a sedentary residence pattern becoming common during the Holocene, property, and protection of property, including land, domestic animals and crops, became an important dimension of human society. Crop surpluses provided the resources stimulating trade relationships with distant human societies that helped to stabilise access to food resources. However, this created enhanced pathways for infectious pathogens to be transmitted across the natural barriers that would have existed in a hunting/gathering economy. The advent of urbanisation provided effective population sizes large enough to host infectious diseases not sustainable in smaller groups. Interpopulation contact on a larger scale led to transitions in human disease ecology which first saw an equilibration of infectious disease pathogens across Eurasia in early historic times, followed by a wholesale trans-oceanic export of diseases in the wake of exploration and conflict (Crosby, 2003). This finally culminated in a global exchange of disease agents such as we face today, as a result of almost unlimited mobility (McMichael, 2004). Such epidemiological transitions would also have been associated with the prevalence of typical or likely ensembles of pathogens (Barrett *et al.*, 1998; Cohen and Crane-Kramer, 2003). During the Palaeolithic, mostly macroparasites and isolated cases of chickenpox and herpes would have survived in small mobile bands, while with increasing sedentism, food production and storage, zoonotic infections and vector-borne diseases would have become more frequent. High prevalence of infectious disease, e.g. smallpox, typhus, measles or bubonic plague, characteristic of many countries until the Industrial Revolution, would have begun to decline, at least in Europe, largely from the eighteenth century onwards, accompanied by a steady reduction in mortality. Today, in contrast, we are seeing a return of infectious diseases thought to have been under control, and a significant number of new diseases which seem to adapt and evolve quicker than remedies can be found (McMichael, 2001).

Natural environment

The diversity of biomes and their latitudinal organisation has a profound effect on biogeographical gradients in species richness, and the prevalence of human pathogens follows this distribution by being correlated with latitude. A number of agents causing parasitic and infectious diseases, such as viruses, helminths, protozoans, arthropods, micro- and macroparasites, are more prevalent in tropical regions than they are in temperate latitudes. This pattern is most likely a function of climatically based energy availability, which generates and maintains gradients in species richness (Guernier *et al.*, 2004). Thus, the exposure of human hosts to varying pathogen densities is driven by large-scale ecological processes. Besides the natural response mechanisms that constitute host-pathogen interaction and the co-evolution of virulence and immunity, there appear to be sophisticated socio-cultural mechanisms, too, that affect the biological propensity for adequate immune response. More frequent occurrence of polygyny in latitudes with high pathogen stress, i.e. an environment with potential detrimental health effects, reflects regional congruence of pathogen richness and marriage patterns. This suggests that human societies are capable of culturally responding to the diversity in potential health hazards by adjusting the genetic variability of their offspring (Low, 1990; Cashdan, 2001; Schutkowski, 2006, pp. 200ff.).

Humans are able to respond to changes in climate and weather with physiological adaptation and acclimatisation. There are limits to this capacity, though, under conditions of more dramatic climate and weather change, e.g. in terms of short-term fluctuations or weather extremes. The effects of these extremes are visible in casualties from heat waves or weather-related natural disasters. The unusually hot summer of 2003 caused at least 27 000 more deaths than in comparative years in Europe; weather-related natural disasters claimed some 600 000 deaths in the 1990s, most of them in poor countries. Longer-term changes affect the incidence of climate-sensitive diseases, such as diarrhoea, malaria or protein-energy malnutrition, causing more than 3.3 million casualties in 2002 alone (WHO, 2006). A spread of vector-borne diseases into the subtropics and higher latitudes is being forecast as a result of anthropogenic global warming (Epstein *et al.*, 1998), facilitated by the disruption of sensitive ecological equilibria due to habitat loss (Chivian, 2001). The predictable outcome is an increase in mass fatalities. These differences in disease responses are reflected in general systems properties. Whilst climate, ecosystem and human society are capable of recovering from occasional severe stress, prolonged changes or recurrent extreme fluctuations can overwhelm the resilience of an ecosystem and thus severely affect human health (Epstein, 2000).

But it is the ambient environmental conditions in a given ecological setting that may have an immediate effect on well-being. Water is a major resource crucial to life everywhere. In many environmental contexts both the purity of the water and its mineral content or lack thereof are critical dimensions of the natural environment. Fluorine is an element that is often found in water supplies. In concentrations less than about 1.5 parts per million, fluorine in the diet has beneficial effects on both bone mineral and teeth. When the concentration exceeds that amount, toxic effects can occur and these can result in severe morbidity and disability (Resnick, 2002a, p. 3434).

The modern island country of Bahrain, in the Arabian Gulf off the coast of Saudi Arabia, has little rainfall and in antiquity depended on well water for much of its water supply. One of the dangers of well water is that toxic materials from the surrounding soil can easily leach into the water and reach hazardous levels. Excessive amounts of fluorine in the water supply was a chronic problem in Bahrain, but this region of the world is just one of many (Figure 5.1) where soluble fluorine-containing soil constituents leach into the water supply and cause morbidity.

Excessive amounts of fluorine can cause dental defects in children and skeletal problems accompanied by neurologic disorders in some adults. Pain and disability can be severe, and premature death can occur in some cases (Resnick 2002a, p. 3434). A poorly understood dimension of this disease is that not all

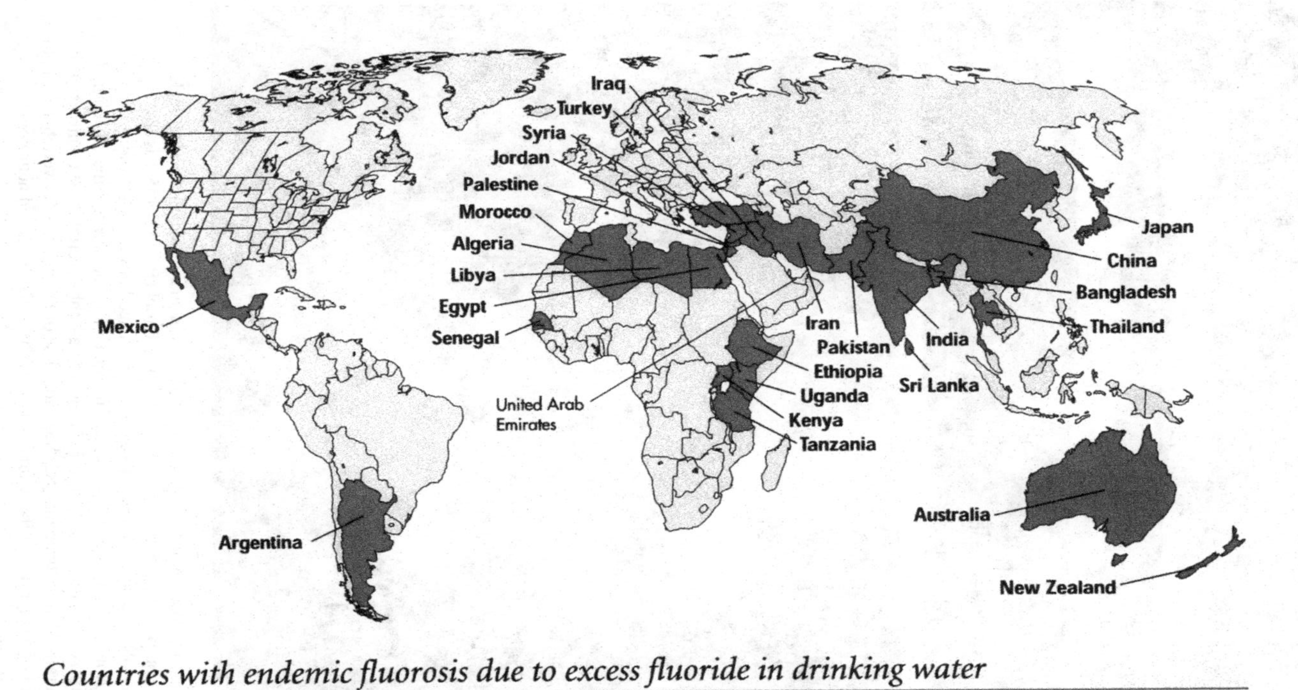

Figure 5.1 World distribution of fluorosis. (UNICEF, www.unicef.org/wes/fluoride.pdf)

(A)

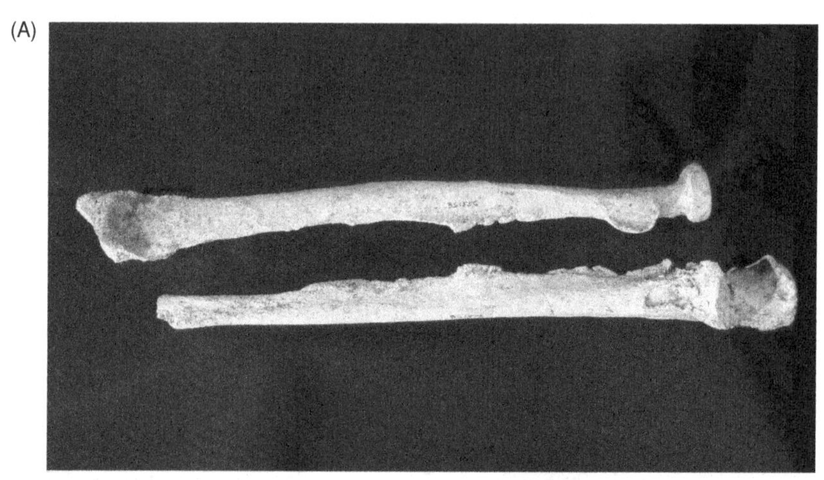

(B)

Figure 5.2 Skeletal manifestations of fluorosis in an archaeological adult male skeleton between 45 and 50 years of age. The burial (B South 1556) is from an archaeological site at Hamad Town, Bahrain, dated between 2000 and 1700 BC. (A) Anterior view of the right radius and ulna. Note the partial mineralisation of the interosseus ligament. (B) Posterior view of the lumbar vertebrae showing extensive mineralisation of ligaments. (C) Superior view of the first lumbar vertebra. Note the greatly reduced size of the neural canal. This very likely caused neurologic impairment.

(C)

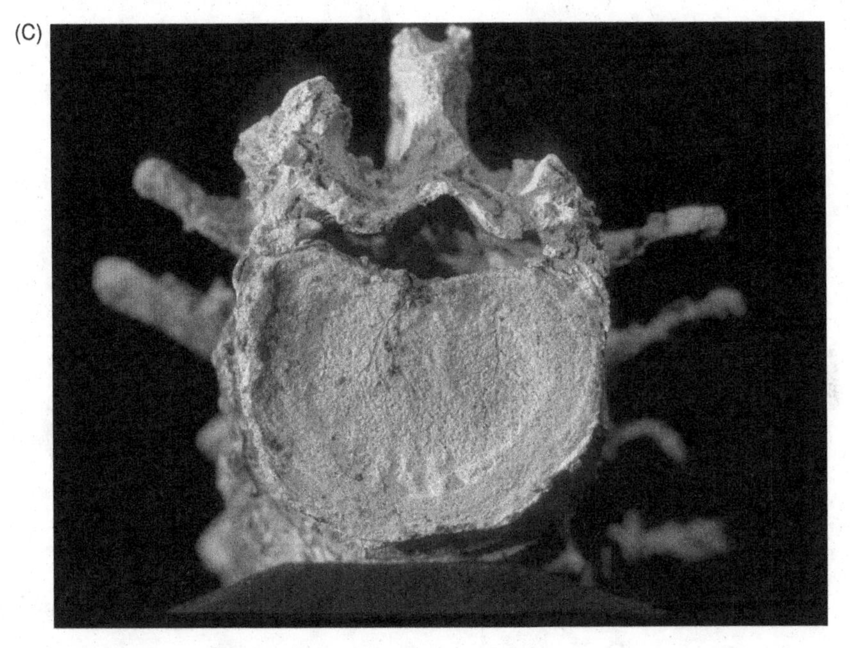

Figure 5.2 (*cont.*)

people living in an environment with toxic levels of fluorine in the water supply, and presumably using the same water resource, become ill.

The effect of fluorosis is apparent in several skeletons excavated from archaeological sites in Bahrain. One of these is from a Dilmun site (Hamad Town) dated between 2000 and 1700 BC. One skeleton is from a male about 45–50 years of age at the time of death (Burial B South 1556). There is widespread pathological mineralisation of tendons and ligaments where they attach to bone. Some of this mineralisation occurs in the soft connective tissue of the vertebral canal, where it encroached on the spinal cord with likely neurologic dysfunction (Figures 5.2A–5.2C). Severe physical disability associated with neurological disorders is a common symptom in modern cases of this disease (Frohlich *et al.*, 1989; Littleton, 1999).

Although several cases of fluorosis have been identified in archaeological skeletal remains from Bahrain, most of the skeletons recovered during excavations do not show bony manifestations. Of course, people without skeletal abnormalities resulting from fluorosis may still have been adversely affected by toxic levels of fluorine in the drinking water. Undoubtedly many individuals with morbidity resulting from toxic levels of fluorine did not have visible skeletal involvement.

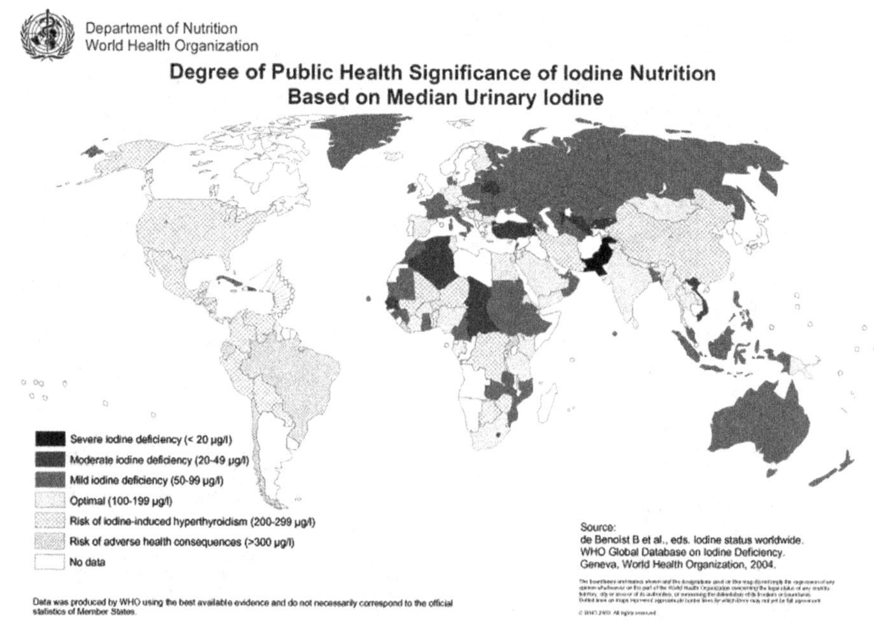

Figure 5.3 World distribution of iodine deficiency in 2004 (de Benoist *et al.*, 2004).

The element iodine is another important micronutrient. Like fluorine it can be toxic if consumed in excessive amounts (Andersson *et al.*, 2005). However, by far the most common problem is a dietary deficiency of iodine (Figure 5.3, WHO data and map). Iodine is a necessary element in the synthesis of thyroid hormone (thyroxine). Although iodine deficiency occurs worldwide, it is particularly prevalent in many of the high mountain regions where water is usually glacial runoff which tends to be devoid of iodine.

For some individuals living in this environment, thyroid hormone cannot be produced in adequate amounts and this results in endemic hypothyroidism, the most severe manifestation of which is cretinism. This disorder usually begins in utero. This early onset often results in diminished mental capacity, even in subacute cases (Andersson *et al.*, 2005) and retarded growth with the characteristic anatomy of the cretin dwarf. The human skeleton is affected in these individuals, with the greatest impact on the postcranial skeleton (Ortner and Hotz, 2005). The vertebrae are significantly affected (Schinz *et al.*, 1951–1952), more so than in achondroplasia, but otherwise cretin and achondroplastic dwarfs have similar postcranial skeletal abnormalities. The rib cage is malformed and the long bones are short but relatively normal in diameter. Abnormal joint development leads to premature, and often severe, osteoarthritis (Figures 5.4A and 5.4B). Given the global distribution of iodine deficiency today (Table 5.1), it is surprising

(A)

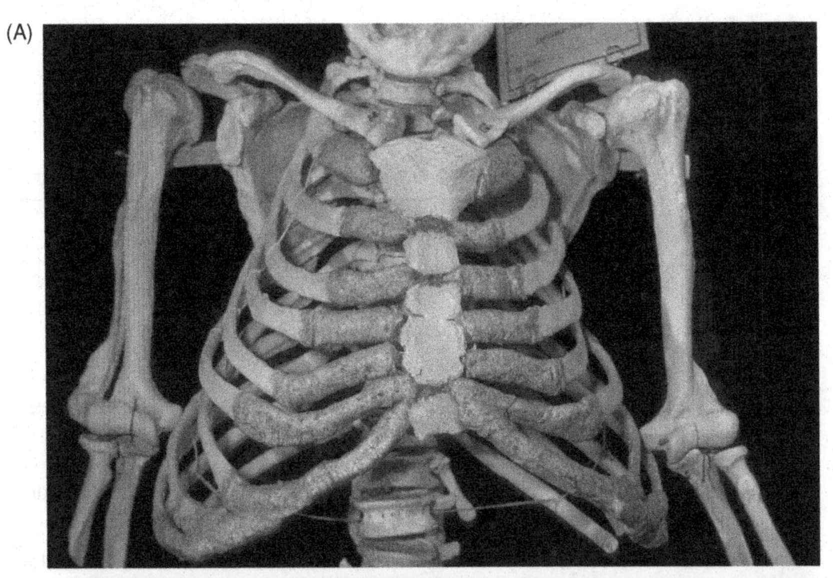

(B)

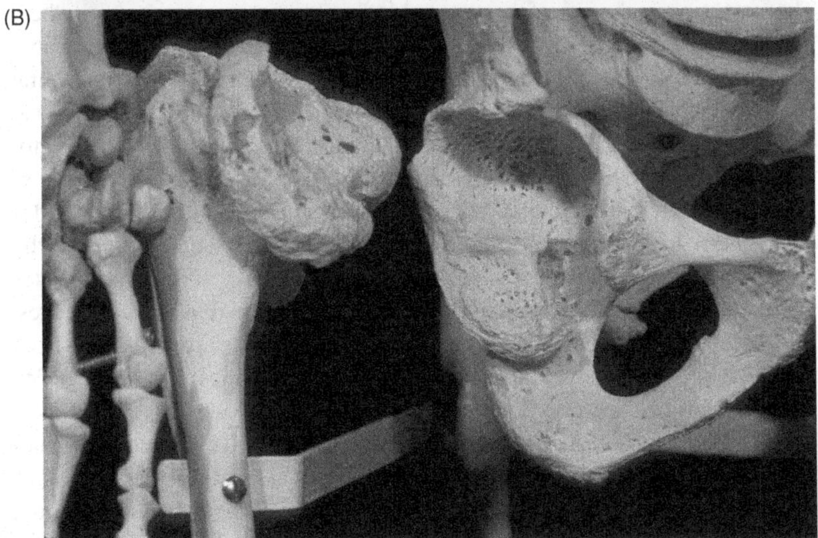

Figure 5.4 Skeletal manifestations of sporodic hypothyroidism. Modern case (Galler Collection 85 (4059), currently curated at the Natural History Museum, Bern, Switzerland). Male 80 years of age. (A) Rib cage and partial upper extremity. Note the abnormal shape and size of the rib cage and the abnormally short humeri. (B) Detailed view of the anterior right hip showing severe osteoarthritis secondary to abnormal development of the femoral head.

Table 5.1 *Prevalence of hypothyroidism (goitre) by WHO regions in 1999 (WHO, 2001).*

WHO region	Population in millions	Population affected by goitre in millions	Population affected by goitre % of the region
Africa	612	124	20 %
The Americas	788	39	5 %
Southeast Asia	1477	172	12 %
Europe	869	130	15 %
Eastern Mediterranean	473	152	32 %
Western Pacific	1639	124	8 %
Total	5858	741	13 %

that very little evidence of endemic hypothyroidism has been reported in archaeological human remains.

Some ecosystems also provide an ideal environment for infectious pathogens. The marshy coastal zones of the eastern Mediterranean region are an area where malaria, a disease caused by a protozoan, *Plasmodium falciparum*, has probably been present for many millennia (Angel, 1966). However, not only such natural environments, but also habitats altered by human activity, influence the prevalence of this disease. Forest clearings can attract the *Anopheles* mosquito vector in higher numbers than closed landscapes, either by providing patches of standing water as suitable breeding grounds or by allowing higher biting rates (Vittor *et al.*, 2006). In addition, mosquitoes appear to feed more often on humans who have been infected by the transmissible gametocyte stage of malaria parasites than on uninfected or those infected by the asexual, non-transmissible stages. This means that the gametocyte, which is the only transmission stage from human to mosquito, manipulates the biting behaviour of the mosquito vector to increase transmission (Lacroix *et al.*, 2005).

The anaemias, caused by a specific gene variant (sickle-cell anaemia (S gene) and thalassaemia (Thal gene)), are associated with areas where the *Anopheles* mosquito vector for malaria thrives. In these genetic anaemias, the abnormal haemoglobin caused by either of the abnormal genes provides some immunity to malaria. This is best expressed in the heterozygous manifestation of these gene loci which provides considerable immunity to malaria but not severe anaemia. The homozygous expression of the genes produces severe anaemia and early death in most but not all cases. Some individuals with sickle-cell anaemia may be asymptomatic for extended periods of time (Resnick, 2002b, p. 2149). Both sickle-cell anaemia and thalassaemia produce identifiable bone changes in some children, as haematopoietic marrow enlarges to provide space for increased demand for red blood cell formation. In the skull this results in enlargement of

Figure 5.5 Porous hypertrophic remodelling of the cranial vault in a child who had anaemia from the American Southwest (NMNH 327074). The child was three years of age and from the site of Pueblo Bonito, New Mexico, USA, that is dated between 950 and 1250 AD.

the diploë at the expense particularly of the outer table, although the inner table may also be affected in severe cases (Ortner, 2003). This creates the porous, hypertrophic morphology caused by marrow hyperplasia in the cranium.

Iron deficiency anaemia is caused by inadequate availability of dietary iron for red blood cell formation. It is a very common public health problem in the world today. There has been considerable debate about the extent to which this deficiency can produce skeletal lesions (Ortner, 2003). Certainly there is evidence of skeletal changes associated with anaemia in areas like the prehistoric American Southwest (Figure 5.5), where there is no evidence of genetic anaemias being present in pre-Columbian Native American groups (Ortner, 2003, pp. 374–5). One hypothesis proposed for iron deficiency anaemia is a high dependence on maize (American corn) among Native American groups living in the American Southwest (El-Najjar *et al.*, 1976). Corn contains phytate which binds dietary iron and prevents its utilisation. The crucial point to be made regarding the skeletal evidence of anaemia in human remains from archaeological sites in the American Southwest is that it reflects increased demand for haematopoietic marrow and red blood cell formation. Whether this is exclusively caused by a deficiency in dietary iron or some combination of iron deficiency and another disorder causing excessive bleeding remains to be resolved. However, this does represent a possible example of an adverse result of too much dependence on a single crop.

Pathogen biology

Two basic evolutionary patterns can be identified in host-pathogen interaction, relating to the emergence and the prevalence of infectious diseases. Conventional models of epidemic outbreak or emergence of infectious diseases tend to claim that the pathogen's basic reproductive number, which is a measure of the average number of secondary infections arising from a single individual in a fully susceptible population, would have to be greater than one for a disease to become manifest in a (human) population (Anderson and May, 1991). Changes in ecological conditions would provide a pathogen with the opportunity to enter into a new (human) host population from the animal reservoir and to subsequently trigger human-to-human transmission. Recent modelling (Antla *et al.*, 2003) suggests, though, that while this may still be the case, the transmission of a pathogen from the animal reservoir into a human population may take place even if the basic reproductive number is smaller than one. The emergence of an infectious disease results from various stages of stochastic genetic change in the pathogen. This increases the chain of disease transmission and allows the pathogen to adapt to the new host, i.e. co-evolve, and thus eventually cause an outbreak of a new disease.

This process of host–parasite co-evolution involves complex relationships between pathogenic organisms and human groups. Infectious pathogens that kill the host die as well, so that high virulence and rapid death of a host is generally not a good co-evolutionary strategy for the pathogen. However, if infectious organisms can rapidly infect new hosts, high virulence is possible, since new generations of pathogens have already invaded new hosts as ancestral organisms die with their host. In this situation selective pressure for less virulent strains to become dominant is minimal or absent. Plague, cholera and dysentery are examples of infectious diseases that are very virulent and cause rapid death, but move quickly between hosts.

However, reduced virulence is a more common pathogen condition. In this host-pathogen relationship, a slow progression of the disease along with delayed transmission between hosts tends to result in reduced virulence and chronic disease, since longer host survival enhances the probability of infecting a new host. This latter expression of infectious disease is the most common co-evolutionary strategy and the one most frequently encountered in archaeological human skeletal remains. Most archaeological evidence of infectious disease reflects a long co-evolution of host and pathogen in which the host population becomes increasingly able to eliminate the pathogen or, at least, survive with minimal morbidity and relatively normal mortality. Acute infectious diseases with severe morbidity accompanied by rapid death very rarely affect the skeleton.

Human biology

Scurvy, caused by vitamin C deficiency, illustrates a somewhat different dynamic interaction between culture and human biology. Unlike most mammals, humans lost the ability to synthesise vitamin C during the hominoid evolutionary process, almost certainly before the development of the major hominid species. They share this trait with other higher primates, whose diet is largely based on a variety of plant resources and therefore minimises the need for synthesising vitamin C. Similarly, the broad-spectrum economy of hunting/gathering societies would have provided a varied diet that significantly reduced the potential harm arising from this biological trait.

With the advent of agriculture, dietary diversity declined and scurvy probably became a more serious problem. Today scurvy tends to be rare in developed countries. In less developed areas of the world the prevalence of scurvy is associated with malnutrition in societies that are disrupted by warfare and genocide, i.e. when socio-political conditions are unlikely to allow for the provision of fresh fruit and vegetables in sufficient quantities. Among refugees in some African countries the prevalence of scurvy ranges from 1 to 44 % (Prinzo, 1999). Among the early agriculturists the prevalence is likely to have been significant, particularly in those societies living in marginal habitats.

Scurvy can result in severe morbidity and death, but the onset tends to be gradual, and months of dietary deficiency can occur before death. In children there are skull manifestations that are relatively pathognomonic in many cases (Ortner and Ericksen, 1997). Skull features have also been described for the adult (Maat, 1982).

In children, skeletal features are the result of a vascular response to chronic bleeding associated with defective collagen in blood vessels that develop during the period when vitamin C is deficient in the diet. Chronic bleeding stimulates an increase in vascularity needed to eliminate blood clots that form. Unfortunately, the new blood vessels are also abnormal which compounds the problem.

When the bleeding occurs adjacent to bone, this increased vascularity stimulates abnormal porosity in the cortical bone underlying the site of the chronic bleeding. If bleeding is more severe, reactive bone formation may develop in addition to cortical bone porosity. In the cranium, porosity most commonly occurs bilaterally and symmetrically in bone of the orbital roof and bone underlying muscles associated with mastication, such as the greater wing of the sphenoid (Figure 5.6A and 5.6B). However, the cranial vault can be affected as well (Figure 5.7). In archaeological skeletal samples, prevalence rates for children with cranial lesions associated with scurvy (Table 5.2) range from zero to 38 % (Ortner *et al.*, 1999; 2001). One of the remarkable cultural adjustments to the need for dietary vitamin C occurred among the Inuit people in the Arctic ecosystem. The traditional diet of the Inuit contains very few plant sources for vitamin C, and they obtained most of what they needed from eating raw or near-raw meat. Especially high concentrations of vitamin C occur in the liver of animal food sources and the liver of both seals and caribou were eaten by the Inuit (Prinzo, 1999). However, they have learned that bear liver is toxic and to not eat this otherwise rich source of protein because of the morbidity they came to associate with the consumption of this meat. There are high concentrations of vitamin A in bear liver. Unlike excess amounts of vitamin C, overdoses of vitamin A are toxic, with serious morbidity and even death resulting shortly after eating even fairly small amounts of bear liver.

Discussion and conclusions

The extraordinary explosion of human population size during the Holocene (Butzer, 1971; Hassan, 1983) provides compelling evidence that the transition to an agricultural economy represented a major adaptive change in human biocultural evolution. It is equally evident, however, that this transition was, and continues to be, associated with significant challenges to human biological adaptability and to the ability to develop cultural innovations needed to address the cultural dimensions of these challenges.

Treatment of vitamin C deficiency (scurvy) by some Native American groups illustrates an effective response to one disease (Vogel, 1970, pp. 78, 250). In this case the effect of potions containing vitamin C on a scorbutic child is rapid, usually within a few hours. This made the link between cause and effect in the treatment of disease much easier to identify. Nevertheless this link was not made by Europeans until the eighteenth century and even then the response was slow and characterised by controversy.

In this context, one thinks of the cultural aversion among the Inuit people to eating bear liver. The very high concentration of vitamin A in this protein source leads to hypervitaminosis A, a potentially lethal metabolic disease. Morbidity is dose dependent but can develop within 24 hours, with even small amounts of bear liver ingestion. In this situation the link between a potentially toxic food source and morbidity is more obvious than that which occurs in scurvy.

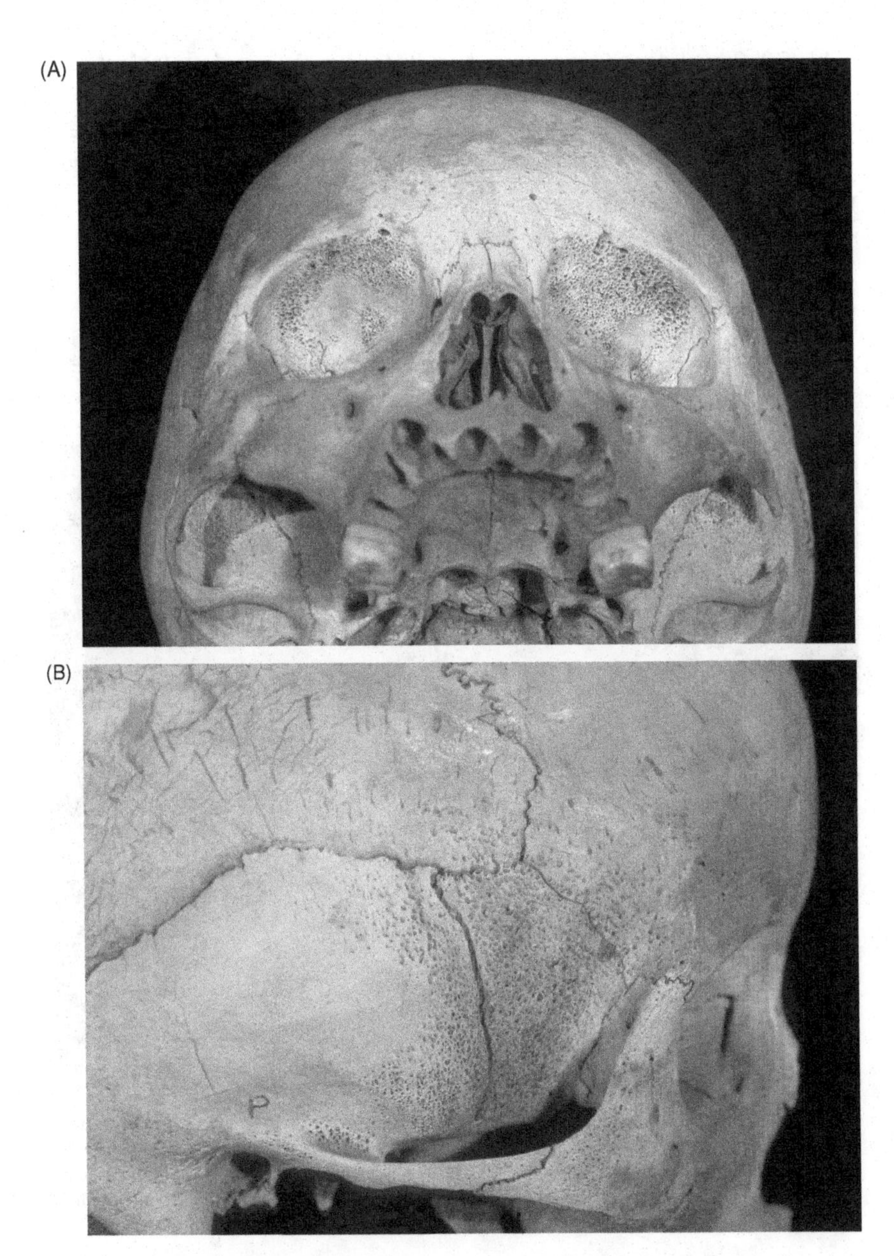

Figure 5.6 Cranial manifestations of scurvy in a child about 11 years of age from the site of Pachacamac, Peru, South America. The site is dated between 900 and 1450 AD. (A) Porous and hypertrophic lesions of the orbital roof. (B) Porous and slightly hypertrophic lesion of the right greater wing of the sphenoid and adjacent bones. An almost identical lesion is apparent on the left side.

Table 5.2 *Prevalence of scurvy in subadults from various archaeological skeletal samples in Peru and the USA.*

Region	Archaeological dates	Prevalence of scurvy	Total subadults
Peru	500–1530 AD	38 (10 %)	363
Midatlantic, USA	1400–Early Historic	10 (6 %)	170
Plains, USA	1650–1830 AD	0 (0 %)	54
Southeast, USA	Early Historic	6 (38 %)	16
Southwest, USA	950–1800 AD	7 (2 %)	317
Total USA	22 (4 %)		567

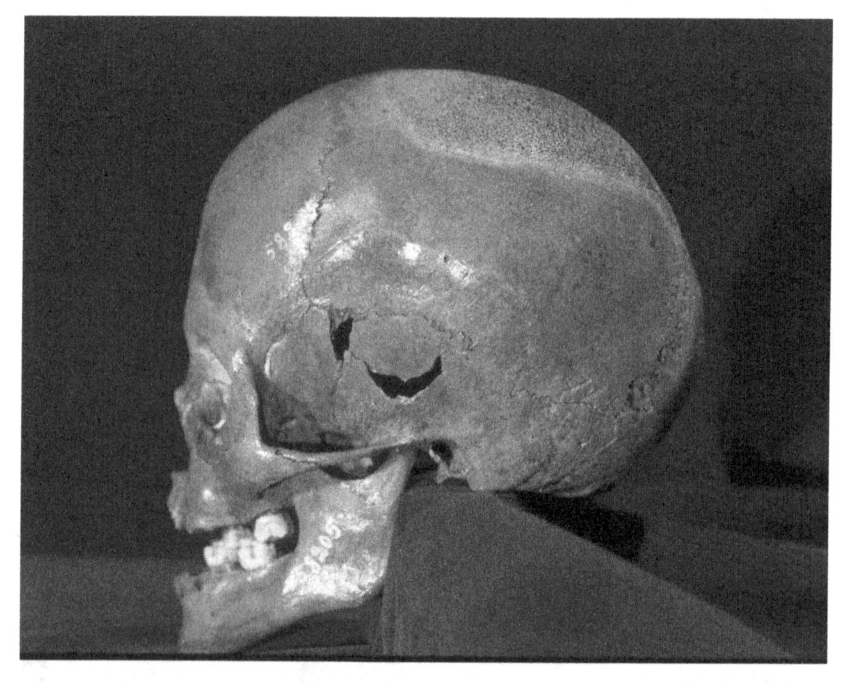

Figure 5.7 Skull vault lesions in a probable case of scurvy from the site of Chichen Itza, Yucatan. The cranium is from a child between six and seven years of age. The site is dated between 800 and 1200 AD. The burial is curated at the Peabody Museum, Harvard University (Peabody No. 58205), and permission to publish the photograph was granted by the Museum.

Nevertheless, there clearly was a fairly sophisticated understanding of the relationship between cause, eating bear liver, and effect, illness and death.

In the case of fluorosis, the process of establishing the linkage between cause and effect would be more difficult because morbidity does not occur quickly and some individuals ingesting the same water may not have significant disease. This is certainly the case for skeletal fluorosis, which does not affect most of the burials in the Bahrain human skeletal sample. It would even hold for dental fluorosis which, even though perhaps more readily visible, would not necessarily have been connected to toxic levels of fluorine in drinking water.

We also need to emphasise that many treatments for disease were, and continue to be, ineffective if not harmful, despite widespread belief that they are useful therapies. This simply emphasises just how difficult it is to make links between cause and effect. Even in an age when science and the scientific method have become major cognitive tools and frames of reference, inadequate and incorrect linkages between cause and effect are common.

It is equally important to recognise that potentially beneficial as well as potentially harmful cultural innovations relative to human use of environmental resources need not be the result of a cognitive process in which the relationship between cause and effect are clearly understood. On the beneficial side of innovation, all that needs to happen is for the trait to aid in human adaptation to a given environmental context; why it does so need not be understood by the people benefiting from the innovation. A case in point is the widespread use of limewater for cooking maize in Meso- and South America. Adding calcium compounds during the preparation increases the relative bioavailability of amino acids from the glutelin portion of the maize grain, thus enhancing the nutritional value of a staple food with an unbalanced composition of essential nutrients (Katz *et al.*, 1974). The cultural justification for this culinary practice is far from being scientific, but rather that the additive makes better dough for tortillas and has therefore been adopted as part of the cuisine.

However, cognitive processing of environmental, cultural and biological information is a mental process that has a very long evolutionary history (Potts, 2004); the potential for this type of processing is deeply implanted in our genetic heritage. Through cultural innovation, human societies have evolved often complex methods to provide the food and other resources necessary to sustain life. The cognitive processes involved in these innovations certainly involve, in some cases at least, a clear understanding of cause and effect. If, for example, drinking a herbal tea results in rapid recovery of a seriously ill child, the link between cause and effect may be fairly easy to establish. The development of a treatment for scurvy by some Native American groups probably is a good example of this type of innovative process when the link between cause and effect is more easily made.

In other situations the process may be much more complex. For example, the complex sequence needed to convert poisonous bitter manioc into a staple food resource is a step-by-step process in which each stage must be carefully completed before the roots can be eaten. Here the innovation might have been developed through a series of random experiments, but at least the possibility, if not the probability, exists that experimentation involving complex cognitive processes with fairly high-end understandings of cause and effect were part of the development. The psychobiology of this, from the detection of its toxicity to the eventual incorporation of manioc into the cuisine, has been outlined as a multi-feedback process between innovative individuals and the cultural fabric of a society (Rozin, 1982). This raises the possibility that the mental potential for innovation based on complex experimentation and recognition of the relationship between cause and effect was an important evolutionary adaptation that occurred relatively early in the development of the hominins. We argue this because the development of complex processes for extracting resources in various environments is not only widespread but also a necessity in recent human groups. Subsisting on an omnivorous diet creates the dilemma of having to explore and exploit a large variety of foodstuffs at the risk of picking unpalatable or even toxic ones. To cope with this requires a unique combination of social and instrumental intelligence which is characteristic of higher primates (Casimir, 1994).

A similar situation exists for iodine deficiency. In most groups living in areas where soil and/or water are deficient in iodine, relatively few individuals will have serious and debilitating manifestations of the disease. This, of course, makes an understanding of what is causing morbidity in some individuals and not in others much more difficult (see above).

One of the intriguing questions for both fluorosis and hypothyroidism is why all the individuals living in the deficient environment are not equally affected. As with most diseases that affect humans, morbidity depends on the individual's biological response to the pathogen challenge, and a major factor in determining this response is the individual's genetic heritage. This, of course, provides the basis for evolutionary adjustments within human populations, and is likely to be a major factor in the variation apparent in the prevalence of morbidity caused by environmental pathogens. But while this appears to be the biological blueprint against which we can observe the constant co-evolution of host and pathogen, it is the inherent cultural capacity of humans to modify and shape natural into managed environments that had, and continues to have, profound consequences for the emergence and prevalence of diseases. Only quite recently (e.g. McDade, 2005; Smith *et al.*, 2005) has the incorporation of ecological concepts into disease control and public health begun to address the fundamental relationships that govern our ability to persist in the widest possible range of habitats, yet at the increasing price of induced pathology. To further explore the ecology of our biocultural existence will thus greatly add to a better understanding of the underlying processes, present and past.

Acknowledgements

The authors wish to acknowledge with appreciation the contributions of many people who assisted in some way in the preparation of this report. This paper was first presented at a symposium organised at the University of Bradford, Bradford UK on 'Biological Anthropology at the Interface of Science and Humanities' held 23–4 May 2003. The symposium was part of the celebration associated with the establishment of the Biological Anthropology Research Centre in the Department of Archaeological Sciences at the University. The principal organiser was Dr. Holger Schutkowski of that Department, and the first author of this chapter and the other participants are indebted to him and his colleagues for conceiving and organising the conference. To our colleagues who also gave presentations during the conference we express our thanks for the stimulating ideas which shaped in important ways the ideas presented in this paper.

References

Anderson, R. and May, R. (1991). *Infectious Diseases of Humans: Dynamics and Growth*. Oxford: Oxford University Press.

Andersson, M., Takkouche, B., Egli, I., Allen, H. E. and de Benoist, B. (2005). Current global iodine status and progress over the last decade towards the elimination of iodine deficiency. *Bulletin of the World Health Organization*, **83**; 518–25.

Angel, J. L. (1966). Porotic hyperostosis, anemias, malarias, and marshes in the prehistoric Eastern Mediterranean. *Science*, **153**, 760–3.

Antla, R., Regoes, R. R., Koella, J. C. and Bergstrom, C. T. (2003). The role of evolution in the emergence of infectious disease. *Nature*, **426**, 658–61.

Aufderheide, A. and Rodríguez-Martín, C. (1998). *The Cambridge Encyclopedia of Human Paleopathology*. Cambridge: Cambridge University Press.

Barrett, R., Kuzawa, C. W., McDade, T. and Armelagos, G. J. (1998). Emerging and re-emerging infectious diseases: The third epidemiologic transition. *Annual Review of Anthropology*, **27**, 247–71.

de Benoist, B., Andersson, M., Egli, I., Takkouche, B. and Allen, H. (eds.) (2004). *Iodine Status Worldwide. WHO Global Database on Iodine Deficiency*. Geneva: World Health Organization.

Boelee, E., Konradsen, F., and van der Hoek, W. (2002). *Malaria in Irrigated Agriculture*. Silverton: International Water Management Institute.

Butzer, K. (1971). *Environment and Archeology*. 2nd edn. Chicago: Aldine Atherton, Inc.

Cashdan, E. (2001). Ethnic diversity and its environmental determinants: effects of climate, pathogens and habitat diversity. *American Anthropologist*, **103**, 968–91.

Casimir, M. J. (1994). Die Evolution der Kulturfähigkeit. In *Vom Affen zum Halbgott*, eds. W. Schiefenhövel, C. Vogel, G. Vollmer and U. Opolka. Stuttgart: Trias, pp. 43–58.

Chivian, E. (2001). Environment and health: 7. Species loss and ecosystem disruption – the implications for human health. *Canadian Medical Association Journal*, **164**, 66–9.

Cohen, M. N. (1989). *Health and the Rise of Civilization*. New Haven: Yale University Press.

Cohen, M. N. and Crane-Kramer, G. (2003). The state and future of palaeoepidemiology. In *Emerging Pathogens. The Archaeology, Ecology, and Evolution of Infectious Disease*, eds. C. L. Greenblatt and M. Spigelman. Oxford: Oxford University Press, pp. 79–91.

Crosby, A. W. (2003). *The Columbian Exchange. Biological and Cultural Consequences of 1492*. London: Praeger.

Durham, W. H. (1991). *Co-evolution. Genes, Culture and Diversity*. Stanford: Stanford University Press.

Ellen, R. (1994). Modes of subsistence: hunting and gathering to agriculture and pastoralism. In *Companion Encyclopaedia of Anthropology*, eds. T. Ingold. London: Routledge, pp. 197–225.

El-Najjar, M., Ryan, D., Turner II, C. and Lozoff, B. (1976). The etiology of porotic hyperostosis among the prehistoric and historic Anasazi Indians of Southwestern United States. *American Journal of Physical Anthropology*, **44**, 477–88.

Epstein, P. R. (2000). Is global warming harmful to health? *Scientific American*, August 2000, 50–7.

Epstein, P. R., Diaz, H. F., Elias, S. *et al.* (1998). Biological and physical signs of climate change: Focus on mosquito-borne disease. *Bulletin of the American Meteorological Society*, **79**, 409–17.

Frohlich, B., Ortner, D. and Al Khalifa, H. (1989). Human disease in the ancient Middle East. *Dilmun*, **14**, 61–73.

Groube, L. (1996). The impact of diseases upon the emergence of agriculture. In *The Origins and Spread of Agriculture and Pastoralism in Eurasia*, ed. D. R. Harris. London: University College, pp. 101–29.

Guernier, V., Hochberg, M. E. and Guégan, J.-F. (2004). Ecology drives the worldwide distribution of human diseases. *PLoS Biology* **2**, 740–6.

Harris, D. R. (ed.) (1996). *The Origins and Spread of Agriculture and Pastoralism in Eurasia*. London: UCL Press.

Hassan, F. A. (1983). Earth resources and population: An archeological perspective. In *How Humans Adapt: A Biocultural Odyssey*, ed. D. J. Ortner. Washington: Smithsonian Institution Press, pp. 191–216.

Johnson, A. W. and Earle, T. (2000). *The Evolution of Human Societies*. 2nd edn. Stanford: Stanford University Press.

Katz, S. H., Hediger, L. and Valleroy, L. A. (1974). Traditional maize processing techniques in the New World. *Science*, **184**, 765–73.

Lacroix, R., Mukabana, W. R., Gouagna, L. C. and Koella, J. C. (2005). Malaria infection increases attractiveness of humans to mosquitoes. *PLoS Biology*, **3**, 1590–3.

Lancaster, H. O. (1990). *Expectations of Life*. New York: Springer.

Littleton, J. (1999). Paleopathology of skeletal fluorosis. *American Journal of Physical Anthropology* **109**, 465–83.

Low, B. S. (1990). Marriage systems and pathogen stress in human societies. *American Zoologist*, **30**, 325–39.

Maat, G. (1982). Scurvy in Dutch whalers buried at Spitsbergen. In *Proceedings of the Paleopathology Association's Fourth European Meeting, Middleberg, Antwerpen*, eds. G. Haneveld and W. Perizonius, pp. 82–93.

May, R. M., Gupta, S. and McLean, A. R. (2001). Infectious disease dynamics: what characterizes a successful invader? *Philosophical Transactions of the Royal Society London, Series B*, **356**, 901–10.

McDade, T. W. (2005). The ecologies of human immune function. *Annual Review of Anthropology*, **34**, 495–521.

McMichael, A. J. (2001). *Human Frontiers, Environments and Disease*. Cambridge: Cambridge University Press.

McMichael, A. J. (2004). Environmental and social influences on emerging infectious diseases: past, present and future. *Philosophical Transactions of the Royal Society London, Series B*, **259**, 1049–58.

Moran, E. F. (2000). *Human Adaptability. An Introduction to Ecological Anthropology*. 2nd edn. Boulder: Westview Press.

Ocepek, M., Pate, M., Žolnir-Dovč, M. and Poljak, M. (2005). Transmission of Mycobacterium tuberculosis from human to cattle. *Journal of Clinical Microbiology*, **43**, 3555–7.

Ortner, D. J. (2001). Human palaeobiology: disease ecology. In *Handbook of Archaeological Sciences*, eds. D. R. Brothwell and M. A. Pollard. London: John Wiley & Sons, Ltd, pp. 225–35.

Ortner, D. J. (2003). *Identification of Pathological Conditions in Human Skeletal Remains*, 2nd edn. Amsterdam: Academic Press.

Ortner, D. J. and Ericksen, M. F. (1997). Bone changes in the human skull probably resulting from scurvy in infancy and childhood. *International Journal of Osteoarchaeology*, **7**, 212–20.

Ortner, D. J. and Hotz, G. (2005). Skeletal manifestations of hypothyroidism from Switzerland. *American Journal of Physical Anthropology*, **127**, 1–6.

Ortner, D. J., Kimmerle, E. and Diez, M. (1999). Skeletal evidence of scurvy in archeological skeletal samples from Peru. *American Journal of Physical Anthropology*, **108**, 321–31.

Ortner, D. J., Butler, W., Cafarella, J. and Milligan, L. (2001). Evidence of probable scurvy in subadults from archeological sites in North America. *American Journal of Physical Anthropology*, **114**: 343–51.

Panter-Brick, C., Layton, R. H. and Rowley-Conwy, P. (eds.) (2001). *Hunter-Gatherers. An Interdisciplinary Perspective*. Cambridge: Cambridge University Press.

Potts, R. (2004). Paleoenvironmental basis of cognitive evolution in Great Apes. *American Journal of Primatology*, **62**, 209–28.

Prinzo, Z. W. (1999). *Scurvy and its Prevention and Control in Major Emergencies*. Geneva: World Health Organization (WHO/NDH/99.11).

Resnick, D. (2002a). Disorders due to medications and other chemical agents. In *Diagnosis of Bone and Joint Disorders*. 4th edn, ed. D. Resnick. Philadelphia: W. B. Saunders Company, pp. 3423–55.

Resnick, D. (2002b). Hemoglobinopathies and other anemias. In *Diagnosis of Bone and Joint Disorders*, 4th edn, ed. D. Resnick. Philadelphia: W. B. Saunders Company, pp. 2147–87.

Rozin, P. (1982). Human food selection: The interaction of biology, culture and individual experience. In *The Psychobiology of Human Food Selection*, ed. L. M. Barker. Westport: AVI Publishing Company, pp. 225–54.

Sandström, B. (1989). Dietary pattern and zinc supply. In *Zinc in Human Biology*, ed. C. F. Mills. Berlin: Springer, pp. 351–63.

Schinz, H., Baensch, W., Friedl, E. and Uehlinger, E. (1951–2). *Roentgen Diagnostics: Skeleton*, vols. 1 and 2 (J. Case, Trans.). New York: Grune and Stratton.

Schutkowski, H. (2006). *Human Ecology. Biocultural Adaptations in Human Communities*. Berlin: Springer-Verlag.

Simoons, F. J. (2001). Persistence of lactase activity among northern Europeans: a weighing of evidence for the calcium absorption hypothesis. *Ecology of Food and Nutrition*, **40**, 397–469.

Smith, B. D. (1995). *The Emergence of Agriculture*. New York: Scientific American Library.

Smith, K. F., Dobson, A. P., McKenzie, F. E., *et al.* (2005). Ecological theory to enhance infectious disease control and public health policy. *Frontiers in Ecology and the Environment*, **3**, 29–37.

Steckel, R. H. and Rose, J. C. (eds). (2002). *The Backbone of History*. Cambridge: Cambridge University Press.

Sutton, M. Q. and Anderson, E. N. (2004). *Introduction to Cultural Ecology*. Oxford: Berg.

Vittor, A. Y., Gilman, R. H., Tielsch, J. *et al.* (2006). The effect of de-forestation on the human-biting rate of Anopheles darlingi, the primary vector of falciparum malaria in the Peruvian Amazon. *American Journal of Tropical Medicine and Hygiene*, **74**, 3–11.

Vogel, V. (1970). *American Indian Medicine*. Norman: University of Oklahoma Press.

WHO (2001). *Assessment of Iodine Deficiency Disorders and Monitoring their Elimination*, 2nd edn. Geneva: World Health Organization (WHO/NHD/01.1, (www3.who.int/whosis/micronutrient).

WHO (2006). *Climate and Health*. Fact sheet 266. Geneva: World Health Organization.

Wood, J., Milner, G. Harpending, H. and Weiss, K. (1992). The osteological paradox. *Current Anthropology*, **33**, 343–70.

6 *The fossil evidence of seasonality and environmental change*

GABRIELE A. MACHO

Introduction

The overall effects of temperature on habitat fragmentation, vicariance, species turnover, dispersal and directional morphological change have long been recognised, but the specific effects of abiotic factors on the course of hominin evolution remain unclear (Vrba *et al.*, 1995; Bromage and Schrenk, 1999). To explain some of the unresolved evolutionary patterns observed in hominin evolution, Potts (1998a, b) posed the 'Variability Selection Hypothesis', according to which periods of increased long-term variability in climatic conditions, rather than global change per se, may have triggered key evolutionary stages, such as bipedality, encaphalisation, sociality and cultural innovations. This suggestion has received widespread attention (e.g. Trauth *et al.*, 2005). However, it was not until recently that the validity of the Variability Selection Hypothesis came under scrutiny and was found to be too generalised (Bonnefille *et al.*, 2004). In their innovative study, to which Potts is a co-author, Bonnefille and collaborators (2004) analysed pollen from Hadar, Ethiopia. Their results suggest that during its lifetime *Australopithecus afarensis* must have experienced considerable environmental variability, including a large biome shift, up to 5 °C directional cooling and an increase in annual rainfall of about 200–300 mm. Such changes are considerable and should have affected the species' ecological niche, its biology and life history. Apparently it did not. This led the authors to conclude that the ability of *A. afarensis* to withstand such high variability '. . . did not depend on an enlarged brain or stone toolmaking . . .' (p. 12 128). This is contrary to expectations raised by the Variability Selection Hypothesis (Potts, 1998a, b) and seems to imply that other, perhaps more important, factors may have been overlooked when formulating this hypothesis.

All species have habitat preferences and the effects of various habitats on life-history strategies are well documented (e.g. Partridge and Harvey, 1988; Stearns, 1992; Charnov, 1993). Local conditions and predictability of the environment, in particular, will affect a species' demography (e.g. Ross, 1992;

Between Biology and Culture, ed. Holger Schutkowski. Published by Cambridge University Press.

Bronikowski *et al.*, 2002), susceptibility to stress (Macho and Williamson, 2002) and mortality rate (King *et al.*, 2005). In other words, although there are overall relationships between life-history strategies and habitat, stochastic environmental fluctuations are likely to have a dramatic effect on survival and the reproductive success of a species. From an evolutionary perspective, such stochastic events constitute a powerful force which may lead to demographic bottlenecks and/or extinction of local populations and even species (e.g. Ramakrishnan *et al.*, 2005). Conversely, neontological studies on primates (Reader and MacDonald, 2003), birds (Sol *et al.*, 2005) and cetaceans (Rendell and Whitehead, 2001) have shown that it is such short-term unpredictability that favours species with relatively larger brains, greater potential for innovation and greater mobility. It logically follows that brain size may be selected for during periods of uncertainty. If correct, the evolutionary consequences of smaller-scale climatic fluctuations may be more important than previously assumed. Rather than/or in addition to focusing on global climatic changes (Vrba *et al.*, 1995; Potts, 1998a, b; deMenocal, 2004), reconstructing the local conditions and smaller-scale fluctuations in environmental conditions may have the potential to throw further light on our evolutionary past. This has already been mooted and, over the last few years, researchers have begun to reconstruct spatially and temporally restricted climates through analyses of pollen (Bonnefille *et al.*, 2004), diatoms (Trauth *et al.*, 2005) or paleosols (Wynn, 2004). However, the resolution of these studies is still small, and does not allow inferences about the environmental conditions encountered by individuals/populations of a species. In order to overcome these limitations and to contribute to the debate about the influence of climate on hominin evolution, we have, over the last decade, explored new research tools for the determination of small-scale environmental changes, such as the pattern of rainfall seasonality (Macho *et al.*, 1996, 2003). Here these data are compiled and contrasted with climatic changes with the aim to explore whether the combined environmental factors may shed light on key stages in hominin evolution.

Climate and hard tissue

Smaller-scale fluctuations in abiotic conditions have long been known to be reflected in the growth patterns of various plants and animals, and can be used to make inferences about past environments (e.g. Berry and Barker, 1968). Mammalian teeth grow incrementally and retain a permanent record of their development within the hard tissue, thus carrying the potential to be used as tools in palaeoecological reconstructions. However, other than is the case in shells, trees etc., where the amount of material deposited will give an indication

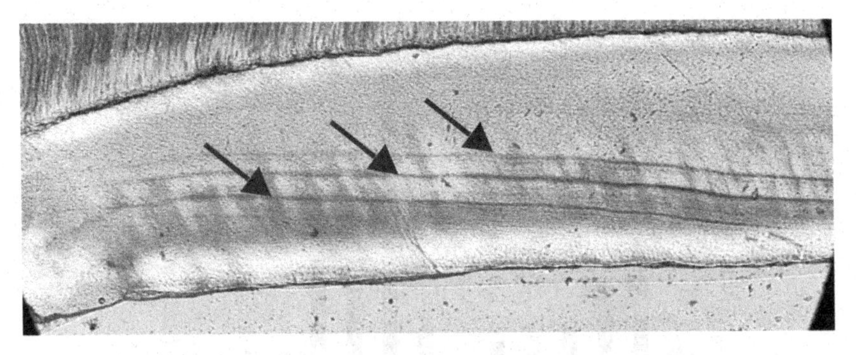

Figure 6.1 Histological section of a mammalian tooth is shown to indicate the appearance of stress lines utilised in the present study.

of the resource availability, teeth are disrupted in their normal development by periods of stress when resources become scant (Figure 6.1). These metabolic disturbances, while unspecific, are commonly associated with fever, infections and nutritional deficiencies (e.g. Rose, 1977; Rudney, 1983); all of these causes are, however, directly or indirectly related to food availability, which is dependent largely on precipitation. More severe disturbances might become manifest as linear enamel hypoplasias (LEH). Despite a plethora of studies on LEH, no study has yet explored the link between these disturbances and climate in modern humans, although this is not the case for apes and monkeys (e.g. Skinner, 1986; Guatelli-Steinberg and Skinner, 2000; Hannibal and Guatelli-Steinberg, 2005). Nonetheless, it is evident that even modern humans respond to seasonal variation in rainfall. For example, the pattern of childhood mortality found in rural Kenya today mirrors the pattern of rainfall with great fidelity (Arudo *et al.*, 2003), although the two curves are offset by about two months (Figure 6.2). For palaeoecological reconstruction it is irrelevant whether or not the curves are synchronised, as long as the shapes of the curves are the same. As regards controlled experiments, there is currently only one study that explores the influence of rainfall on the development of stress lines in teeth of hamadryas baboons (Dirks *et al.*, 2002), while the evidence presented in our own work is more indirect. It revealed however that, although the effects of environmental fluctuations would be felt by animals across all trophic levels, their manifestations as stress lines differed depending on the habitat exploited by the animals (Macho *et al.*, 1996, 2003; Macho and Williamson, 2002). For example, browsers/mixed feeders in Kenya typically exhibit three periods of stress during the annual cycle. The spacing between lines corresponds with the periods between onset of successive rainy seasons, while the additional line can be related to the middle of the dry season, when food sources become scarce and the animals have to switch to fall-back foods and/or draw on their fat reserves. The probabilities of

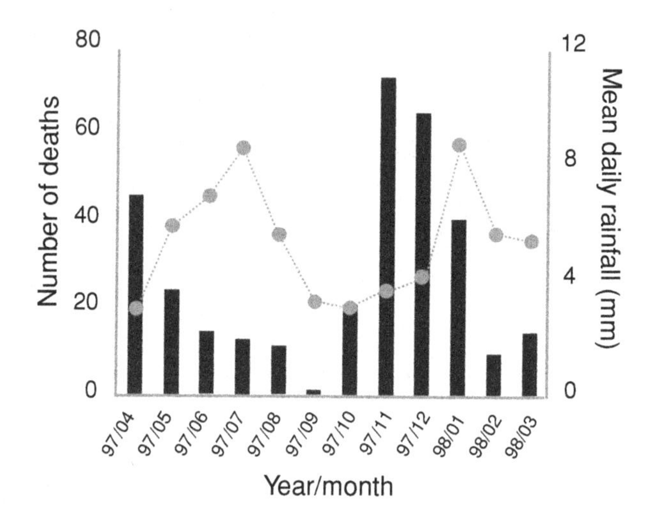

Figure 6.2 In modern humans, like other mammals, there is a strong correspondence between mortality rate and seasonal rainfall. The graph is modified from Arudo *et al.* (2003) for mortality in children under five years of age in rural Kenya.

finding comparable patterns of recurrent stress lines between years and between specimens are extremely low, and unlikely to occur due to chance, even when only a few teeth are available for study. This makes browsers/mixed feeders a particularly useful group for the present purpose, given that histological sections cannot be prepared on a large number of fossils. Specifically, by compiling all data available on browsers/grazers over four distinct time periods spanning from 3.9 Ma to the present, we enquire (a) whether temporal trends in seasonality can be observed, (b) whether these trends correlate with other aspects of climatic fluctuations and (c) whether such interactions between large-scale climatic fluctuations and small-scale biotic factors could have contributed to the morphological and cognitive evolution of modern humans.

Material and methods

Data were collated from previous research, where details about sample composition, preparation and histological methods can be found (Macho *et al.*, 1996, 2003). For the present analyses only data from browsers/mixed feeders are utilised as they are considered most reliable. All samples derived from Kenyan sites. The fossil material was sampled from Allia Bay (3.9 Ma), Koobi Fora (older than 1.9 Ma) and Olorgesailie (0.7 Ma), while the comparative material was collected from the Masai Mara; the latter site is generally considered to exhibit similar climatic conditions to these fossil localities, particularly those at Allia Bay. In general, East Africa experiences a bimodal rainfall pattern.

The following taxa were sampled: *Taurotragus oryx*, *Giraffa camelopardalus*, *Deinotherium bozasi*, *Hexaprotodon protoamphibius*, *Tragelaphus* sp., *Hippopotamus* sp., *Theropithecus oswaldi*. Histological sections were prepared and the spacing between accentuated striae of Retzius were counted, whereby an algorithm was developed to filter out randomly occurring accentuated striae, i.e. noise (see Macho *et al.*, 2003 for sample preparation and more detail about methods). Within and between each locality the probability of finding comparable recurrent patterns with three lines per year is 0.001%. The stress patterns thus established form the basis for the present analysis, although only the spacings between the assumed onsets of successive rainy seasons are reported here. This period, which is shorter than the dry season, is marked by increased overall humidity and, hence, constitutes a time of relative abundance of food sources. By focusing on this period over time, it is possible to indirectly also assess whether the proportional length of the dry season has changed during the Pleistocene.

Results and discussion

Mammalian evolution and climate are intricately linked (e.g. Vrba *et al.*, 1995; Potts 1998a, b; Bromage and Schrenk, 1999; deMenocal 2004). For hominins, there exists a broad correspondence between hominin species turnover and global climatic changes (deMenocal, 2004), as well as directional morphological changes in some lineages as a result of the environment becoming harsher (e.g. Suwa *et al.*, 1997). Despite these overall patterns, however, the causes for key stages in hominin evolution, specifically bipedalism and brain evolution, are still not understood. In order to resolve some of these issues, recent studies have called for an integration of large-scale climatic changes and locally restricted environmental information within a coherent framework (e.g. Feibel, 1999; Wynn, 2004). The present study aims to contribute to this approach and to introduce a potentially useful tool for the reconstruction of past environments. By analysing recurrent stress lines in mammalian teeth the timing of seasonal rainfall of three hominin fossil-bearing sites from different time periods is inferred. Limitations of the present study need to be borne in mind, however. First, to fully appraise the validity of the approach taken here, more controlled studies are required on the effects of environment on stress lines across diverse groups of mammals (Dirks *et al.*, 2002). Second, sample sizes need to be increased. At present, the small sample only allows for tentative conclusions to be drawn, although it is noteworthy that the results presented

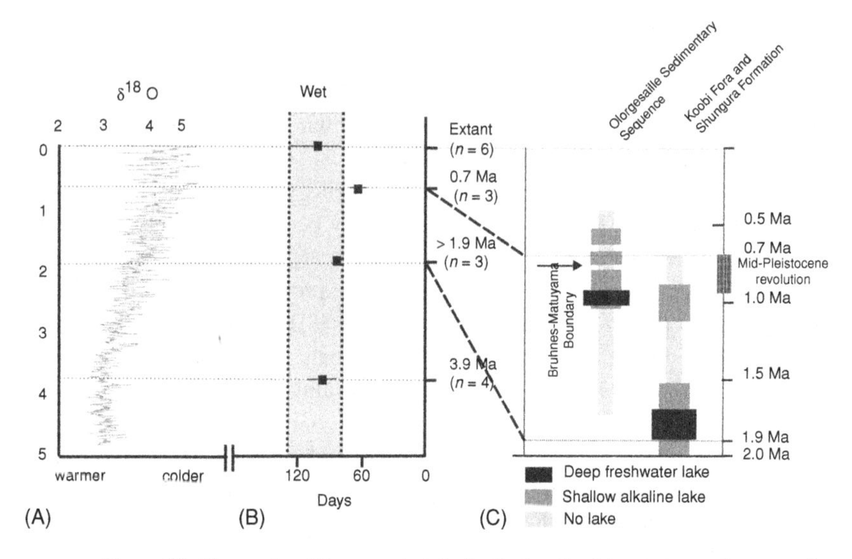

Figure 6.3 The results of the present study for the length of time between the onset of the rainy seasons (i.e. relative moisture during the annual cycle) (B) are contrasted with isotope data (A) and the moisture inferred from diatoms for the same geological deposits (C) (modified from Trauth *et al.*, 2005).

here are statistically significant. Perhaps most importantly, despite the shortcomings of the sample, trends are apparent which can be interpreted within the context of evolutionary patterns.

Figure 6.3 shows the pattern of rainfall seasonality across the time periods sampled vis-à-vis the isotope records (http://131.111.44.196/coredata/v677846.html) and some indication of moisture based on diatom analyses provided by Trauth and colleagues (2005). The results indicate no major climatic fluctuations for the lifetime of *Australopithecus anamensis* between 3.9 Ma and 4.2 Ma. Apparently, *A. anamensis* lived and died in a mixed habitat consisting of a gallery forest, floodplain grasslands and marginal dry bushland (Brown and Feibel, 1991; Wynn, 2000). There is also good evidence for seasonal availability of water, with mean annual precipitation having been estimated at 350–600 mm (Wynn, 2000), or, more recently, at about 620 ± 100 mm per year (Wynn, 2004). In addition, the present results would indicate that the pattern of seasonality was apparently similar to conditions found at the Masai Mara today (Figure 6.3); the onset of each rainy season and, hence, the duration of the dry season, fall within the modern range. Importantly, the standard deviation found for the sample from Allia Bay is smaller than it is for the modern sample, implying that the climate may have been more stable and predictable. This inference, although speculative, is supported by more detailed analyses of the frequency of stress lines found within teeth of mammals from Allia Bay (Macho *et al.*, 2003); on average, the fossil specimens showed fewer stress lines than their extant counterparts (Macho *et al.*, 2003). Hence, on the basis of previous research, as well as present analyses, there is little indication that *A. anamensis* was faced with major climatic changes during its lifetime, nor can it be inferred that the short-term abiotic conditions were stressful and unpredictable. Relating the short life-span of the species (Leakey *et al.*, 1995, 1998) to environmental factors would therefore be inappropriate at present. However, this may not be the case for the time preceding the emergence of *A. anamensis*. Isotope data suggest substantial large-scale climatic fluctuations prior to the origin of the species (Figure 6.3), which may have led to natural selection forces triggering its origin, although not the evolution of bipedalism or brain expansion.

Geological evidence shows a marked decrease in precipitation from about 600 mm to 380 ± 240 mm per year after 2.5 Ma, although conditions apparently returned to normal at about 1.9 Ma (Wynn, 2004); diatom analyses similarly provide evidence for sufficient moisture in East Africa after 1.9 Ma (Trauth *et al.*, 2005; Figure 6.3). Yet, despite these considerable fluctuations in climatic conditions, the impact on hominin speciation, extinction and morphological change has been relatively minor. Like *A. afarensis* (Bonnefille *et al.*, 2004), hominins living in East Africa during this period were probably capable of coping with the environmental changes, perhaps through behavioural plasticity. This is not manifest as cultural advances, however. Unfortunately, the data presented here for seasonal pattern of rainfall add only little to the debate as they are ambiguous for a number of reasons. The sample analysed comes from horizons which may fall just outside the crucial period of climatic fluctuations (Wynn, 2004; Trauth *et al.*, 2005), and important information may thus have been missed (as in the case for *A. anamensis*). Furthermore, sample size is extremely small, calling for caution in interpretation. However, given the low probability of finding exactly the same stress pattern by chance, it is remarkable to find that the two different individuals show comparable, relatively short timing between the rainy seasons and, hence, a somewhat prolonged dry season. Although the data from Koobi Fora are, overall, difficult to interpret, this latter finding could mark the beginning of an overall trend towards prolongation of the dry season, which becomes exaggerated during the Pleistocene (Figure 6.3). This shift in rainfall pattern, together with major global changes in climate, may have crucially affected hominin evolution during the Pleistocene.

During the mid-Pleistocene transition climatic fluctuations became more prevalent, global temperatures dropped and rhythmic development of global ice volume commenced. The lowered sea surface temperature after 0.9 Ma reduced the tropical evaporation and atmospheric moisture, leading to large-scale aridification and persistent C_4 grasslands (Schefuß *et al.*, 2003). Even the interglacials during the mid Pleistocene differed, such that the ice sheets in the northern hemisphere remained, and ocean waters stayed colder than they are today (Shackleton, 1987). If the results of the present study reflect reality, it would appear that the overall aridification was further exacerbated by a prolongation of the dry season beyond the range encountered today (Figure 6.3). The resulting habitat fragmentation and environmental unpredictability must have been considerable and are likely to underlie the demographic and behavioural changes inferred for middle Pleistocene hominins.

The intrinsic rate of increase of a population is dependent on mortality (Ross and Jones, 1999), which, in turn, depends largely on external climatic conditions. As a case in point, even relatively minor fluctuations in seasonal rainfall were recently found to affect the survival of infants of older mothers in lemurs (King *et al.*, 2005). Given that modern humans are similarly susceptible to environmental stressors (e.g. Arudo *et al.*, 2003; Figure 6.2) it is reasonable to infer that during the middle Pleistocene the decreased temperatures and aridification of Africa, with the establishment of C_4 grasslands, together with the prolongation of the dry season, influenced the reproductive rate of hominins, resulting in population bottlenecks (Foley and Lahr, 1998; Templeton, 2002). While these demographic changes in and by themselves are important for evolutionary processes, it is proposed that the deterioration of the habitat conceivably also caused the evolution of large-brained mid-Pleistocene hominins, i.e. *Homo heidelbergensis*.

Large brains are generally considered to confer behavioural flexibility which is particularly advantageous in variable, unpredictable ecological situations. The hypotheses underlying the evolution of this flexibility in cognitive abilities are traditionally either social in nature (e.g. Byrne and Whiten, 1988) or are related to foraging strategies (e.g. Milton, 1988). However, while the correlation between these variables and increased brain size is undisputed, partitioning the effects of various factors has been found difficult. Doing so is perhaps irrelevant, however, as all these cognitive abilities are associated with one common behavioural trait: innovation. Sol and colleagues (2005) convincingly demonstrated for birds that the key to survival and successful migration is the ability to find novel solutions to problems. In other words, fluctuating and unpredictable environments demand behavioural flexibility if the individuals/groups (species?) are to survive and to successfully invade new territories. Studies on primates similarly arrived at this conclusion (Reader and Laland, 2002). The ability to innovate and, by proxy, brain size, may therefore be selected for in variable environments. Given the severe environmental climatic changes during the mid-Pleistocene transition outlined above, and their effects on habitat fragmentation and population size, the evolutionary changes in brain size observed within the *Homo* lineage may thus be unsurprising (Ruff *et al.*, 1997; Rightmire, 2004). Indirect support for this suggestion is provided by the fact that advances in material culture also seem to have gone hand in hand with these environmental and morphological changes (Foley and Lahr, 1998; Klein, 2000), as did the subsequent migration of hominins out of Africa (Templeton, 2002). The timing of the migration of hominins between 0.42 Ma and 0.84 Ma, given by Templeton (2002), may be too broad, however. Based solely on the climatic data presented elsewhere, northward migration during isotope stage 16–18 seems improbable, but can be envisaged during the subsequent interglacials when substantial amounts of ice were still locked up at the poles (Vrba, 1995). Hence, a proposed movement after about 620 kyr would not only seem probable, but is also in line with evolutionary trends in brain size (Ruff *et al.*, 1997; Rightmire, 2004) and cultural innovations documented in Europe (Klein, 2000).

Conclusion

Although abiotic factors underlie evolutionary processes, the relative importance of global and local conditions on hominin evolution remains unclear. Over the last decade it has become evident that global climatic fluctuations do not unequivocally account for, or correlate with, evolutionary pathways among hominins. For example, recent studies have demonstrated that *A. afarensis* lived in a changing environment, yet there is little evidence for morphological and/or behavioural evolution in this species. Similarly, hominins living in East Africa around 2 Ma encountered drastic changes in the ecological set-up, but these cannot be readily correlated with what is known about hominin palaeobiology. The present study further investigated the relationship between environmental factors and palaeoanthropology, and postulated that local conditions may be of similar or even greater importance than global climate changes in determining the species' evolutionary fate. In particular, unpredictability of the environment may put (groups of) individuals under stress, which may affect their reproductive success. Based on neontological studies interested in cognition, it is further hypothesised that this unpredictability exerts selection pressure on the evolving brain. Using indicators of smaller-scale climatic fluctuations, i.e. seasonal patterns of rainfall, these predictions seem to be borne out. Although sample sizes are small, the tentative conclusion can be drawn that evolutionary changes among hominins can only be correlated with climatic data when both large-scale global and small-scale local data vary simultaneously. This seems to lend support to suggestions that only a combined approach will ultimately elucidate the influence of climate on hominin evolution.

Acknowledgements

Meave Leakey supported the study throughout and provided the fossil samples. The comparative material was collected by Louise Leakey and Thure Cerling. Technical help was provided by Pam Walton, Don Reid and Daisy Williamson. Financial support was given by the Leverhulme Trust. Finally, I wish to thank Holger Schutkowski for inviting me to contribute this work to the book.

References

Arudo, J., Gimnig, J. E., Ter Kuile, F. O. *et al.* (2003). Comparison of government statistics and demographic surveillance to monitor mortality in children less than five years old in rural Western Kenya. *American Journal of Tropical Medicine and Hygiene*, **68**, 30–37.

Berry, W. B. N. and Barker, R. M. (1968). Fossil bivalve shells indicate longer month and year in Cretaceous than present. *Nature*, **217**, 938–9.

Bonnefille, R., Potts, R., Chalié, F., Jolly, D. and Peyron, O. (2004). High-resolution vegetation and climate change associated with Pliocene *Australopithecus afarensis*. *Proceedings of the National Academy of Sciences*, **101**, 12 125–9.

Bromage, T. G. and Schrenk, F. (1999). *African Biography. Climate Change & Human Evolution*. New York: Oxford University Press.

Bronikowski, A. M., Alberts, S. C., Altmann, J. *et al.* (2002). The aging baboon: Comparative demography in a non-human primate. *Proceedings of the National Academy of Sciences*, **99**, 9591–5.

Brown, F. H. and Feibel, S. S. (1991). Stratigraphy, depositional environments, and palaeogeography of the Koobi Fora Formation. In *Koobi Fora Research Project. Vol. 3, The Fossil Ungulates: Geology, Fossil Artiodactyls and Palaeoenvironments*. Oxford: Clarendon, pp. 1–30.

Byrne, R. W. and Whiten, A. (1988). *Machiavellian Intelligence: Social Expertise and the Evolution of Intellect in Monkeys, Apes, and Humans*. Oxford: Oxford University Press.

Charnov, E. L. (1993). *Life History Invariants*. Oxford: Oxford University Press.

Dirks, W., Reid, D. J., Jolly, C. J., Phillips-Conroy, J. E. and Brett, F. L. (2002). Out of the mouths of baboons: stress, life history, and dental development in the Awash National Park hybrid zone, Ethiopia. *American Journal of Physical Anthropology*, **118**, 239–52.

Feibel, C. S. (1999). Basin evolution, sedimentary dynamics and hominid habitats in East Africa. In *African Biography. Climate Change & Human Evolution*, eds. T. G. Bromage and F. Schrenk. New York: Oxford University Press, pp. 276–81.

Foley, R. and Lahr, M. M. (1998). Towards a theory of modern human origins: geography, demography, and diversity in recent human evolution. *Yearbook of Physical Anthropology*, **41**, 137–76.

Guatelli-Steinberg, D. and Skinner, M. (2000). Prevalence and etiology of linear enamel hypoplasia in monkeys and apes from Asia and Africa. *Folia Primatologica*, **71**, 115–32.

Hannibal D. L. and Guatelli-Steinberg, D. (2005). Linear enamel hypoplasia in the great apes: analysis by genus and locality. *American Journal of Physical Anthropology*, **127**, 13–25.

King, S. J., Arrigo-Nelson, S. J., Pochron, S. T. *et al.* (2005). Dental senescence in a long-lived primate links infant survival to rainfall. *Proceedings of the National Academy of Sciences*, **102**, 16 579–83.

Klein, R. G. (2000). Archeology and the evolution of human behaviour. *Evolutionary Anthropology*, **9**, 17–36.

Leakey, M. G., Feibel, C. S., McDougall, I. and Walker, A. (1995). New four-million-year-old hominid species from Kanapoi and Allia Bay, Kenya. *Nature*, **376**, 565–71.

Leakey, M. G., Feibel, C. S., McDougall, I., Ward, C. and Walker, A. (1998). New specimens and confirmation of an early age for *Australopithecus anamensis*. *Nature* **393**, 62–6.

Macho, G. A. and Williamson, D. K. (2002). The effects of ecology on life history strategies and metabolic disturbances during development: an example from the African bovids. *Biological Journal of the Linnean Society*, **75**, 271–9.

Macho, G. A., Reid, D. J., Leakey, M. G., Jablonski, N. G. and Beynon, A. D. (1996). Climatic effects on dental development of *Theropithecus oswaldi* from Koobi Fora and Olorgesailie. *Journal of Human Evolution*, **30**, 57–70.

Macho, G. A., Leakey, M. G., Williamson, D. K. and Jiang, Y. (2003). Palaeoenvironmental reconstruction: evidence for seasonality at Allia Bay, Kenya, at 3.9 million years. *Palaeogeography, Palaeoclimatology, Palaeoecology* **199**, 17–30.

deMenocal, P. B. (2004). African climate change and faunal evolution during the Pliocene–Pleistocene. *Earth and Planetary Science Letters*, **220**, 3–24.

Milton, K. (1988). Foraging behavior and the evolution of primate cognition. In *Machiavellian Intelligence: Social Expertise and the Evolution of Intellect in Monkeys, Apes and Humans*, eds. R. W. Byrne and A. Whiten. Oxford: Oxford University Press, pp. 285–306.

Partridge, L. and Harvey, P. H. (1988). The ecological context of life history evolution. *Science*, **241**, 1449–55.

Potts, R. (1998a). Environmental hypotheses of hominin evolution. *Yearbook of Physical Anthropology* **41**, 93–136.

Potts, R. (1998b). Variability selection in hominid evolution. *Evolutionary Anthropology* **7**, 81–96.

Ramakrishnan, U., Hadly, E. A. and Mountain, J. L. (2005). Detecting past population bottlenecks using temporal genetic data. *Molecular Ecology*, **14**, 2915–22.

Reader, S. M. and Laland, K. N. (2002). Social intelligence, innovation, and enhanced brain size in primates. *Proceedings of the National Academy of Sciences*, **99**, 4436–41.

Reader, S. M. and MacDonald, K. (2003). Environmental variability and primate behavioural flexibility. In *Animal Innovation*, eds. S. M. Reader and K. N. Laland. Oxford: Oxford University Press, pp. 83–116.

Rendell, L. and Whitehead, H. (2001). Culture in whales and dolphins. *Behavioral and Brain Sciences*, **24**, 309–92.

Rightmire, G. P. (2004). Brain size and encephalization in Early to mid-Pleistocene *Homo*. *American Journal of Physical Anthropology*, **124**, 109–23.

Rose, J. C. (1977). Defective enamel histology of prehistoric teeth from Illinois. *American Journal of Physical Anthropology*, **46**, 439–46.

Ross, C. (1992). Environmental correlates of the intrinsic rate of natural increase in primates. *Oecologia*, **90**, 383–90.

Ross, C. and Jones, K. E. (1999). Socioecology and the evolution of primate reproductive rates. In *Comparative Primate Socioecology*, ed. P. C. Lee. Cambridge: Cambridge University Press, pp. 73–110.

Rudney, J. D. (1983). Dental indicators of growth disturbance in a series of ancient lower Nubian populations: changes over time. *American Journal of Physical Anthropology*, **60**, 463–70.

Ruff, C. B., Trinkaus, E. and Holliday, T. W. (1997). Body mass and encephalization in Pleistocene *Homo*. *Nature*, **387**, 173–6.

Schefuβ, E., Schouten, S., Jansen, J. H. F. and Damsté, J. S. S. (2003). African vegetation controlled by tropical sea surface temperatures in the mid-Pleistocene period. *Nature*, **422**, 418–21.

Shackleton, N. J. (1987). Oxygen isotopes, ice volume and sea levels. *Quarternary Science Reviews*, **6**, 183–90.

Skinner, M. F. (1986). Enamel hypoplasia in sympatric chimpanzee and gorilla. *Human Evolution*, **1**, 289–312.

Sol, D., Duncan, R. P., Blackburn, T. M., Cassey, P. and Lefebvre, L. (2005). Big brains, enhanced cognition, and response of birds to novel environments. *Proceedings of the National Academy of Sciences*, **102**, 5460–5.

Stearns, S. C. (1992). *The Evolution of Life Histories*. Oxford: Oxford University Press.

Suwa, G., Asfaw, B., Beyene, Y. *et al.* (1997). The first skull of *Australopithecus boisei*. *Nature*, **389**, 489–92.

Templeton, A. R. (2002). Out of Africa again and again. *Nature*, **416**, 45–51.

Trauth, M. H., Maslin, M. A., Deino, A. and Strecker, M. R. (2005). Late Cenozoic moisture history of East Africa. *Science*, **309**, 2051–3.

Vrba, E. S. (1995). On the connections between palaeoclimate and evolution. In *Paleoclimate and Evolution, with Emphasis on Human Origins*, eds. E. S. Vrba, G. H. Denton, T. C. Partidge and L. H. Burckle, New Haven: Yale University Press, pp. 24–45.

Vrba, E. S., Denton, G. H., Partridge, T. C. and Burckle, L. H. (eds.) (1995). *Paleoclimate and Evolution, with Emphasis on Human Origins*. New Haven: Yale University Press.

Wynn, J. G. (2000). Paleosols, stable carbon isotopes, and paleoenvironmental interpretation of Kanapoi, Northern Kenya. *Journal of Human Evolution*, **39**, 411–32.

Wynn, J. G. (2004). Influence of Plio-Pleistocene aridification on human evolution: evidence from paleosols of the Turkana Basin, Kenya. *Journal of Physical Anthropology*, **123**, 106–18.

7 *Thoughts for food: evidence and meaning of past dietary habits*

HOLGER SCHUTKOWSKI

Introduction

Arguably, humans are the only animals that have taken food beyond its immediate function of satisfying caloric and nutritional needs. Cuisine is distinctively human; it is the cultural transformation of diet into meaning. This notwithstanding, food is essential to facilitate the basic needs of growth and physiological maintenance. Thus, humans uniquely combine two aspects of eating, by sharing with all other organisms of their habitats the task of consuming food for nutritional requirement, while at the same time using food, or rather culturally acquired dietary habits, as a signifier of individual, group, or even national preferences and identities.

Human nutrition, therefore, has to be viewed in a conceptual framework, which covers constitutional and environmental factors (Vorster and Hautvast, 2002). The health aspect relates to the physiological and biochemical level, i.e. dietary allowances, nutrient metabolism and body maintenance. These issues also define clinical concepts about the adequacy of nutritional value and intake, resulting in optimal, over-, under- or malnutrition. The influence of the environment covers a much wider range of diverse, immediately tangible factors, including food procurement and production strategies, material culture, socioeconomic circumstances and household economics. Food and nutrition, however, encompasses another component, one that relates to the immaterial, to cosmology, belief systems, ritual, and inclusiveness. Thus, food has multiple meanings and roles, and it permeates all levels of human existence. For humans, it entails not only the fulfilment of a basic need, but is a truly biocultural entity.

Humans are unspecialised omnivores with little nutritional demand beyond what is required for growth and maintenance. They are capable of not only exploiting a wide range of foodstuffs but also of extracting nutrients and providing a diet from the widest range of biomes and ecological settings. Humans are nutritional generalists. Their opportunistic feeding behaviour generally puts little restriction on habitat requirement, and it is this trait that provided humans

Between Biology and Culture, ed. Holger Schutkowski. Published by Cambridge University Press.

with an evolutionary selective advantage, and equipped them superbly to inhabit virtually all known terrestrial ecotopes and every continent except Antarctica. Humans occupy an unusually broad ecological niche, or rather an unusually diversified one. It is part of their ecological profession (*sensu* Odum) to intentionally alter environments in order to implement subsistence strategies, and to form landscapes, far beyond anything known from animals, for the purpose of food procurement. Unlike all other organisms this does not stop at converting plant and animal biomass directly into nutritional value. While this does happen, e.g. by picking berries or nuts, humans typically process food. Humans convert food into diet.

In this sense food relates to any edible substance that contains nutrients, such as proteins, fats, carbohydrates, minerals or vitamins. Food facilitates nutrition, i.e. the uptake of nutrients, which are ingested and metabolised into energy for tissue growth and maintenance. Diet, in contrast, rather refers to a behavioural category. It denotes the regular consumption of foodstuffs and drinks, and thus constitutes the customary selection from food and drink options available to an individual or a human community. These basic relations apply universally and are thus pertinent to human communities today as well as in the past.

Ecological aspects of human nutrition

Despite being so dominant in almost every habitat in which they live, humans are nonetheless tied into the same basic ecological principles that apply to other organisms within a habitat. Humans are higher-order consumers who derive their food from producers (plant biomass) and lower-order consumers, i.e. other heterotrophic organisms. Such trophic relationships form complex links of organisms interconnected through the processes of eating and being eaten, i.e. a food web. Dietary relationships are linked through flows of energy and materials in space and time within an ecosystem or a habitat. Matter is distributed through the system as a result of biogeochemical processes, comprising slow cycles constantly fed from the abiotic reservoir of the environment, and accelerated processes, in which organisms immediately interact with their environments through the recycling of materials from the biotic realm. The latter provides the macro- and micronutrients required for metabolic action of growth, maintenance and thermoregulation. On average, human diet is composed of *c*. 80 % plant foods and *c*. 20 % animal foods, but there is considerable geographical, seasonal and cultural variation. The consumed foodstuffs provide humans with the energy supply necessary to keep up live functions. Nutrition has to meet the requirements of energy expenditure; the fixed expenditure necessary to maintain basic physiological function, and the variable expenditure which provides energy for growth, activity or pregnancy and lactation, to name a few (Eastwood, 2003, p. 52). In addition, requirements vary with age, sex, genetic disposition, health status and socio-economic circumstances. There is considerable individual variation, but recommended dietary allowances (or recommended daily amounts) give an indication about levels of intake for different classes of nutrients to avoid development of deficiency symptoms.

The building blocks differ with regard to their energetic efficiency. Carbohydrates and proteins provide about 4 kcal of energy per gram consumed, while fats yield 9 kcal/g (Goran and Astrup, 2002). Energy requirements vary greatly between individuals and populations, depending on climate/latitude, body mass, age, lifestyle or physical activity levels. According to WHO data, average energy requirements across a wide range of developing and industrialised countries fall between *c.* 1900 and 2400 kcal/capita/day (James and Schofield, 1990, p. 31), with males requiring more than females due to differences in body weight. Recommended dietary allowances indicate the amount of nutrients generally thought to be adequate for a balanced nutrition. An average human adult would thus require about 40 g of protein per day, whereas pregnant or lactating women should take in an extra six to eleven grams, respectively. In a balanced diet the fat content should be 30–35 % of the energy content, with unsaturated fat providing 10 % of the food energy. Starch is the bulk carbohydrate to meet energy requirements and is recommended to amount to *c.* 37 % of the total dietary energy. Non-milk sugar should contribute no more than 10 % of the total daily dietary intake (see Eastwood, 2003 for all recommended intakes). Such values are inevitably crude proxies only, or geared towards Western diets, when in fact they are highly dependent on the availability of certain food components in the diverse range of habitats. Traditional circum-polar societies, for example, face a general scarcity of carbohydrates in their habitats. These building blocks, however, are crucial for the sufficient supply of sugars to the brain. While this supply route is the most immediate and efficient, other indirect and energetically more costly biochemical pathways can be used to meet the glucose requirements, i.e. through gluconeogenesis, the synthesis of sugars from proteins and fat, those building blocks that are abundant in the Inuit diet (Draper, 2000). Other nutritional deficiencies, mostly involving micronutrients such as trace minerals, cannot be compensated so easily. Habitats with iodine or selenium deficiencies are not uninhabitable; yet leave the human population prone to developing dietary deficiency symptoms, such as goitre or Keshan's disease, respectively (Hetzel and Dunn, 1989; Rayman, 2000).

Human dietary variation is influenced by the availability of food items in the habitats, in the first place. The prevailing ecological circumstances, e.g. geomorphology, climate, seasonality or soil quality, impinge on the spectrum of plant and animal species that can be exploited for food. Different habitats offer specific sets of dietary or nutritional opportunities, but also pose challenges, most notably through variation in the abundance of food resources or different levels of biological productivity. But what is eventually eaten may both be guided by choice and availability, and this impacts on the organisation of food procurement. In the neotropics of Amazonia, for example, societies with access to productive aquatic resources have developed a more sedentary lifestyle due to the stationary nature of the resource, and this facilitates the formation of horticultural production. Where such access is not possible, hunting is the main means of acquiring animal protein, which has to be supported by a mobile, foraging lifestyle (Ross, 1987).

How many individuals can be sustained in a given habitat depends on the severity of limiting factors. Resources known to follow Liebig's Law of the Minimum govern the distribution, size and density of human populations. Arid environments with high amounts of evapotranspiration and erratic precipitation constrain people through fluctuations in the availability of water; temperate environments and their limited vegetation periods provide certain foodstuffs only during the growing season, thus creating times of leanness and plenty in the course of the yearly cycle; and the humid tropics are characterised by soils with thin humus layers whose nutrients are quickly depleted, thus allowing cultivation in one location only for a very limited amount of time, with long fallow periods in between (e.g. Moran, 2000). Humans combat these constraints largely by cultural means. The development of subsistence modes and the invention of technological implements are cultural responses to overcome, moderate or compensate for such shortfalls, and to facilitate survival in a given habitat. Such cultural regulatory adjustments to the conditions of the natural habitat are geared towards the management of available resources by, first of all, participating in, but certainly also by intervening in and controlling, flows of energy and matter in the habitat, i.e. by actively steering the availability of energy and material resources. The considerable success, which human communities show in accomplishing survival in the most diverse array of habitats, has led to a hallmark of their population ecology.

Typically, humans are end-members of the trophic relations they participate in. Under normal circumstances, and owing to the fact that the amount of usable energy decreases by 80–90 % from one trophic level to the next, high-order consumers can only be sustained in small numbers. The longer or more complex the food chain, the less effective the energy yield. Humans, however, distinctly deviate from this pattern by displaying unexpectedly large population densities and sizes despite being at the top end of the trophic pyramid. This testifies to the advantages of being an opportunistic and generalised feeder, which enables a flexible extraction of food energy from diverse biotic components. More importantly, it is the result of unmatched sophistication in the establishment of food procurement strategies, irrespective of the subsistence mode pursued.

The crucial aspect appears to be that these strategies are firmly tied into social arrangements that govern and regulate their practical implementation, e.g. by organising the tasks of food production and division of labour, professional specialisation, or the exchange, barter and trade of goods between individuals or communities. Ecosystems shaped by human activity thus show a variability that is largely characterised by adaptations in subsistence activities. These serve several purposes towards long-term habitation and survival, from securing sufficient food energy supply and an adequate provision with essential nutrients, to tackling food-related limiting habitat factors through the co-ordination of food procurement within the social and political structures of a community. The accomplishment of these tasks lends itself to categories based on differences in patterns of societal organisation.

Food and the organisation of human societies

Food procurement and the ecological conditions, i.e. eventually the available subsistence options of a habitat, are connected through resource management at varying levels of complexity in socio-political representation and involvement (e.g. Danforth, 1999). A convincing model has been developed by Johnson and Earle (2000), which suggests that there are three levels of social differentiation in the course of socio-cultural evolution from foraging groups to agrarian states (see Schutkowski, 2006, pp. 154 ff.). (i) Family-Level Groups are generally organised into networks of familial association and bilateral descent lines. Leadership functions develop on an ad hoc basis, and land use is regulated through customary rights. Subsistence relies on the natural abundance of wild species which at times can lead to unpredictable resource supplies, especially when societies have been pushed into marginal ecosystems, for example the !Kung in southwestern Africa. However, there are human populations at the family level of social organisation that have developed domestication and food production, e.g. horticulturalists in the Amazonian rain forest, without evolving into the higher complexities of social and economic systems. Food procurement is divided by age and sex groups, yet without formalised division of labour. The provision of food is individualised or involves the group members and, in times of hardship, relies on reciprocal distribution so as to minimise risk through interfamilial networks. (ii) Local Groups represent a wide spectrum of socio-cultural organisation, from clan-based foragers and horticulturalists to big-man societies of collectors and pastoralists. They are characterised by corporate structures, often defined by genealogical descent, and extending into rights of possession and defended territories and resources. Representation often involves temporarily formalised leadership based on skill and merit. The modes of subsistence mostly rely on domesticated species, and seasonal food supplies are stored.

Resource management implies both private ownership of, for example, technological investment, and collective use of the implement, for example a fish weir. Compared to the organisation of family groups, the extent of economic and social integration is higher, which is in part reflected by the distribution of resource surplus through political channels, i.e. a corporate responsibility for dispersing resource risk. (iii) Regional Polities span across the entire range of chiefdoms, archaic states and agrarian peasant economies. They are able to sustain high population densities. Their socio-political organisation displays stratification and regional institutions with established leadership. Even though the mode of production is still tied to rich and highly abundant resources, considerable technological investment is necessary to produce food on a large scale from entirely domesticated species, which aims to secure resource supplies through intensive agriculture, long-distance trade and market orientation. The subsistence system is characterised by loss of diversity and food production is reliant on few, highly controlled food items. There is a trend towards reducing the food chains to make production energetically less costly. The generation of surplus is crucial, and empirical evidence shows that no state-level society can be sustained in a habitat that does not allow the accumulation of foodstuffs beyond subsistence level (Danforth, 1999). Regional polities support the formation of regional centres, towns and cities, accompanied by institutionalised ownership and state-guaranteed rights of possession. This, in turn, facilitates differential control of resources and thus the long-term possibility for a concentration of political power and centralised resource control.

This general trend in the evolution of social systems is reflected in an increasing amount of manipulation of the environment, accompanied by specialisation and change within the ecosystem in order to gain better and more efficient control over flows of energy and matter. While a predominantly foraging lifestyle can afford little alteration to what is essentially a pristine habitat, horticulture and basic animal husbandry require regular human ecosystem interference. Fully developed agriculture, pastoralism and any industrialised food production are only sustainable with constant human intervention and the creation of managed environments (e.g. Ellen, 1982). For a discussion of associated trends in population growth see e.g. Livi-Bacci (1997), Weisdorf (2005) or Schutkowski (2006).

Food as cultured biology

Even though it has been argued that the development of civilisation and what we would today perceive as civilised mannerisms is profoundly linked with the refinement of food and table manners (Visser, 1992; Elias, 1997), all humans are distinctive from animals in their ability to habitually alter the raw state of food into items that are considered more digestible and more palatable, irrespective of a society's mode of subsistence or social organisation. The capability to master fire had certainly favourably influenced this behavioural trait already early on in hominin evolution (e.g. Gowlett, 2006), but sophistication has gone far beyond heating and roasting. While being a generalised omnivore is advantageous in that it enables humans to sample from the widest variety of edible items in the habitat, there are two main obstacles that need to be overcome: avoiding toxic food, and ensuring an overall balanced nutrition. Both are essential for longer-term survival in a habitat, and both have created ingenious solutions of cuisine, i.e. the typical preparation and processing of food items. Manioc (*Manihot* sp.), for example, a tropical starchy root staple with high yields, is highly toxic in its native state due to large amounts of cyanogenic glycosides. Only after applying a complicated set of detoxifying steps is this rich source of carbohydrates ready for consumption. Maize (*Zea mays*), on the other hand, is deficient in certain amino acids and when eaten as staple cereal can cause deficiency disease, unless it is biochemically mitigated by cooking maize in lime water (Katz *et al.*, 1974; FAO, 1992).

Cuisine requires cultural modification of food, yet the acquisition of cuisine traits is said to combine biological and cultural mechanisms (Rozin, 2000). Cultural acquisition of nutritional habits has been described as often representing a two-tier process. First an individual behaviour, the 'invention' of a cuisine trait, is adopted and maintained by the group, and thus accepted by the community and embedded into the socio-cultural environment. This purposeful exposition to the trait then becomes internalised through personal motivation; now already emancipated from the socio-cultural environment it is the anticipatory thrill of a certain taste that triggers acculturation and the eventual learning of food preferences. Food choice becomes the result of interplay between biological factors, cultural influences and individual experiences.[1] The life practical advantage of lime water that makes tortilla dough malleable is much easier to understand, because the result is readily acceptable without having to comprehend the complex biochemistry that improves the nutrient value of maize. In the case of manioc, it is difficult to imagine, though, how exactly the invention of transforming a toxic tuber into a basic foodstuff was realised – considering the likely intoxication of those who first tried to establish a procedure and the fact that humans have an inherited aversion to foodstuffs that taste or smell unpleasant; but once recognised, every single step of the recipe has a very high chance of cultural transmission. Despite the fact that toxic substances often have a bitter taste, and should therefore be avoided as part of an evolutionary adaptive behavioural mechanism, many cultures know at least one food item that should be inherently unpalatable, yet is an important or regular part of the cuisine,

Figure 7.1 Stained glass window depicting a communal meal of the butcher's guild. Endris Dittwerdt, 1586. Heimatmuseum Reutlingen.

e.g. beer, bitter lemon drink, chicory, or Brussels sprouts. Cultural modification and sublimation is obviously able to even reverse inherited nutritional or taste preferences, which otherwise form a psychological category of taste rejection.

Cuisine has been characterised as containing three essential components: the staple ingredients, i.e. mainly the major food components available in a habitat; the flavour principles, i.e. the combination of spices and herbs that make food distinct (e.g. 'Indian' or 'Chinese'); and the method of processing (Rozin, 1982). Cuisine therefore denotes ethnic or geographical provenance, and nutritional preferences or habits. Food becomes a multi-layered expression of identity, and eating carries meaning (e.g. Scholliers, 2001; Ashley *et al.*, 2004). French cuisine has become synonymous with an at least three-course meal; roast turkey represents Thanksgiving; sausages, chips, sushi all have immediate connotations with regional or national food traditions. But food and eating are above all social entities. Sharing food, be it in immediate terms, e.g. meat sharing among foragers, or more indirectly in terms of food aid programs, creates a sense of inclusiveness. Sharing a meal is often the formal conclusion of a successful business deal or other celebratory occasions, and not only today. Medieval stained glass windows show guild members gathered around a table sharing their sense of social belonging through a joint meal (Zischka *et al.*, 1993; Figure 7.1); Christian ritual makes reference to the most iconic of communal meals. On the other hand, a lack of meaning is deplored today, when the fragmentation of social cohesion in post-modern society is being partly blamed on the loss of the family meal as the focal point of core social reassurance – whatever reasons and circumstances may lead to this situation.

Apart from ethnic distinction, the cultural connotations of food have created sub-layers within society that qualify as group identifiers. A study carried out at Arizona State University in the 1980s produced revealing correlations between nutritional habits and personality self-evaluations, to an extent that food preferences almost became the alter ego of individual lifestyle indicators (Sadalla and Burroughs, 1981). 'The gourmet', the 'vegetarian' and 'the health food enthusiast' suggested traits that would not only characterise themselves in very specific ways, but these traits would also be recognised by external subjects as typical of people with distinct food preferences. At a certain level, such categorisations create a set of small-scale cosmologies, where cultural traits, beliefs and attitudes co-vary with food preferences. The question is to what extent perceptions that have become part of everyday culture and social attitude transcend to the wider context of social organisation and worldviews that constitute belief systems beyond the level of fast-moving fashion. How closely related is food with conceptions of the world?

Ever since Lévi-Strauss offered a structuralist view on food categories (e.g. Lévi-Strauss, 1964; 1965), the idea that food should not only be good to eat but, more importantly, good to think, has become a widespread notion. Indeed, food preferences, expressed through acceptance and rejection of food items, do contain a strong element of identity. Böhmer-Bauer (1990) has presented a fascinating case in point; the culinary triangle of the Maasai. The Maasai are pastoralists who live in the steppe habitats north and south of the Kenyan–Tanzanian border. As in all pastoralist societies, the distinct lifestyle provides a strong sense of pastoralist identity and serves to delimitate them from neighbouring agriculturalist (Bantu) or forager (Dorobo) societies with whom, however, they have close social and economic relations. Three important food items; milk, blood and meat, are directly derived from the most eminent subsistence resource, the cattle, and are perceived as ideal foodstuffs, despite the fact that horticultural and other produce are consumed as well. Milk, blood and meat epitomise distinct sociocultural areas, linking, for example, roast meat with herdsmen and pasture; raw blood with bush land and hunting, or milk with women, settlement and crops (Figure 7.2). By combining food with form and location of consumption, association with a social group and allocation of an activity or mode of production, the Maasai define their self-perception. Mingling the ideal food categories would create impurity and would threaten the cultural and social integrity of the society. The body of the animal symbolises the unity of the society. Its distribution forms part of ritual behaviour that creates meaning, the confirmation of social order and

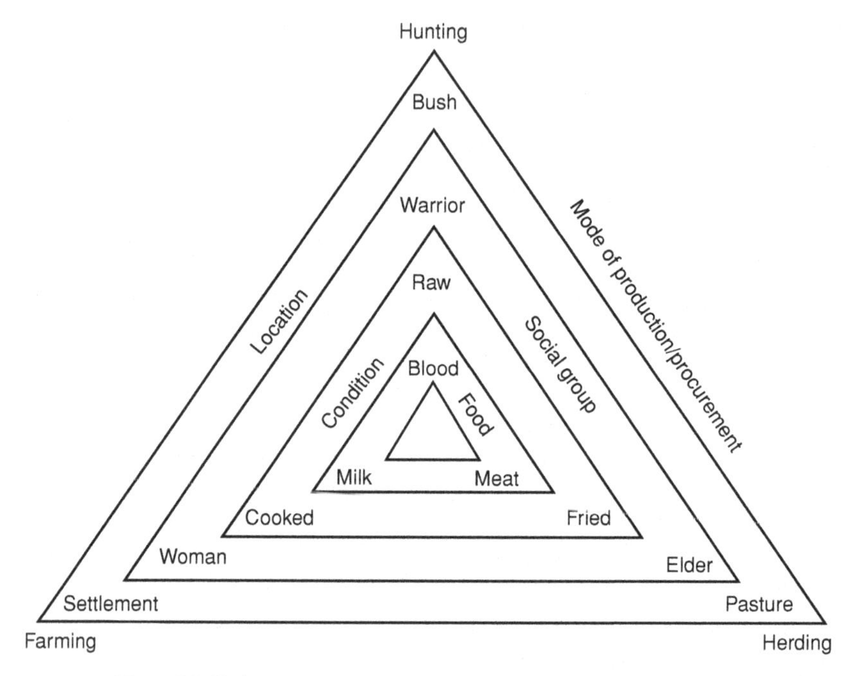

Figure 7.2 Culinary triangle of the Maasai. Modified after Böhmer-Bauer 1990.

the renewal of connectedness with the divine. Food in its ideal manifestations stands for important values of Maasai society, it is good to think. It reinforces group identities, even though they do not necessarily have to match up with reality.

Perhaps somewhat unexpectedly, very similar notions can be found almost at our doorstep. In a Breton fishing community, food has been found to symbolise the division of the male and female realm along idealised, socio-cultural partitioning lines (Chapman, 1990). Drinking considerable quantities of strong red wine and smoking strong cigarettes reinforce toughness, strength and endurance in males, whilst milk is considered weak and is associated with women and children, and thought to be unhealthy and indigestible for males. Pork meat on the one side, and tarts and cakes on the other, complement the stereotypical notion. The selection of food is connected with the idea of appropriateness and inappropriateness. Food categories reflect cultural categories.

These examples demonstrate that human communities have not only, through trial and error, found ingenious ways to appropriate food into cuisine. They have, probably more importantly, sublimated food as the supplier for basic metabolic needs into a meaningful expression of identity. It is not unreasonable to assume that very similar mechanisms were in place in past societies and that food or diet was one of many ways to elaborate difference and distinction within and between communities.

Past diets: classes of evidence and principles of assessment

Methodological approaches to the reconstruction of diet or dietary circumstances in past populations broadly fall into three categories: the analysis of faunal and floral assemblages, the analysis of human skeletal remains for signs of pathological alterations indicative of metabolic and nutrition-related disease, and the chemical analysis of artefacts and mineralised tissues.

The analysis of ecofacts, i.e. animal bones and plant macro remains, provides information about the menu humans would have been able to choose from. The nature of ecofact assemblages usually allows no quantification or direct inference of diet. Faunal material is usually abundant in the archaeological record across all time periods, from the late Pleistocene (e.g. Hockett and Haws, 2005) to medieval and post-medieval times. It allows clear understanding of the variety provided by the menu, whether it was diverse or rather monotonous, nutritious in principle or deficient. Morphological and metric analyses of animal bones give insights into hunting strategies in foragers, or into domestication, husbandry and culling practices in farmers and pastoralists, and thus provide information beyond the range of available foodstuffs by allowing inferences about modes of subsistence and political economy (e.g. Mainland and Halstead, 2005). Floral remains suffer from reduced chances of survival, usually only facilitated by dry, cold or waterlogged conditions and through charring. Like animal remains, they complement the reconstruction of a menu of plant staples or supplements that would have formed regular parts of a diet in past populations. Especially since the Neolithic, plant remains have broadened our understanding of plant domestication, the advent of new cultivars and the development of horti- and agricultural skills and techniques.

While ecofacts, despite their inability to help reconstruct past diets, provide positive evidence of food use, bone pathological alterations display, as it were, the negative side of nutritional conditions, i.e. usually signs of dietary deficiency. Nevertheless, these skeletal vestiges of under-nourishment allow inferences about socio-economic status, well-being and ecological circumstances, as they are a direct reflection of ambient living conditions (see also Chapter 5). Diseases associated with dietary supply can be separated into those related to episodes of stress, which occur localised, and those with a more systemic impact. Disruption in nutrient supply can, for example, lead to impairment or failure of mineralisation. Dental enamel hypoplasia, usually lines or pits as a result of malformation during tooth crown development, is frequently observed in human skeletal remains. While the result of irregular deposition of calcium phosphates into the mineralising matrix is evident, the exact reasons for undersupply may vary. Cribra orbitalia refers to porotic pitting of the orbital roof and is related to persistent anaemic conditions. The body reacts by enlarging areas of haematopoietic activity in the skeleton in an attempt to compensate for the deficiency. Again, the formation of cribra orbitalia may have multiple causes, and not all of them are nutrition-related, e.g. infestation with intestinal parasites. Yet, the majority of possible reasons are linked to deficient supply with certain nutrients, be it iron, folic acid or, in the case of scurvy, vitamin C. So-called Harris lines provide evidence of arrested growth, notably in long bones. Stress situations of varying kinds cause a failure of bone formation at the epiphyseal plate. Mineralisation continues, but growth is halted. Only after the stress situation is overcome will normal growth resume and this is marked by a line of increased mineral density visible in a radiograph.

Rickets, or osteomalacia as the equivalent in adult individuals, develops when a lack of vitamin D leads to failure of mineralisation of the osteoid, the bone precursor ground substance. Strictly speaking, this condition is not nutrient-related, but the effects of rickets can be remedied or prevented by dietary supplements or supplies in cases where synthesis of vitamin D from the pro-vitamin is impaired. For example, the consumption by Inuit of food rich in vitamin D prevents the development of rickets or osteomalacia which, given the lack of sunshine in the ambient natural environment, would be prevalent. Scurvy, on the contrary, is directly related to a lack of supply with vitamin C. The vitamin is also involved in the synthesis of collagen Type I, which forms the organic matrix of bone and can co-occur with iron deficiency anaemia. The deficiency leads to haemorrhages that may occur anywhere in the body, including the gingiva, the joint capsule or bone tissue beneath the periosteum.

In contrast to these macroscopic approaches, studies at the molecular level utilise chemical signatures that allow the detection of food components and the reconstruction of diets. The analysis of organic residues associated with archaeological artefacts, such as lipids or resins, using various types of gas chromatography/mass spectrometry (GC/MS) or high-performance liquid chromatography (HPLC) greatly helps in the identification of foodstuffs, such as dairy products or fish (Brown and Heron, 2003; Craig *et al.*, 2005), but also contributes to a better understanding of trade routes and relations with food commodities. Analyses of residues from amphorae reveal the various provenances of food components (Stern *et al.*, 2008). The examples demonstrate how important these investigations are to decipher the menu of past diets.

At present, however, the only approach that provides direct information about the diet of past individuals and populations is through chemical analysis of human remains. It is based on the principle that the chemical composition of foodstuffs is passed on to consumers and, subject to known systematic biochemical alteration, reflects both the trophic level and relative proportions of food derived from plants and animals. The most commonly employed method is the mass spectrometric analysis of stable carbon and nitrogen isotopes of collagen extracted from calcified (bones, teeth) or, to a lesser extent, keratinised tissues (hair, nails), due to the good prospects of collagen preservation in these substances. For details on other approaches, e.g. the use of isotopes from other elements, isotopic ratios derived from the mineral portion, or trace element analyses see e.g. Sealy (2001) or Pollard *et al.* (2007).

The vast majority of studies that aim to reconstruct past diets use bulk analyses of light stable isotope ratios measured in bone collagen, whilst the analysis of compound-specific isotopic ratios, for example from single amino or fatty acids, only recently opened up a new field of inquiry (Copley *et al.*, 2003). Generally, what is actually detected is the protein intake metabolised from different foodstuffs, as the body uses these building blocks to synthesise collagen molecules, and the ratio denotes the relative amount of the isotopes ^{12}C to ^{13}C, and ^{14}N to ^{15}N, respectively. When food is digested and broken down into its components, the minute mass difference between the isotopes is sufficient to cause discrimination against the heavier isotope. This leads to a change in the isotopic ratio of the food and that detected in the tissue of consumers, a process called fractionation. Since fractionation occurs in the processes that involve protein metabolism, the value of the isotopic ratio indicates trophic position relative to other individuals and those organisms that supply the food web. Carbon isotope ratios are measured as fractionation in per mil ($\delta^{13}C$) against the standard isotopic ratio of carbon in the PeeDee Belemnite formation, while nitrogen ratios ($\delta^{15}N$) vary against nitrogen in air (Ambient Inhalable Reservoir). Typically, there is a systematic shift of 1–3 ‰ between trophic levels for carbon, whereas the trophic level difference for nitrogen ratios is 3–5 ‰.

Stable isotope ratios of carbon differentiate between plant sources depending on photosynthetic pathway. The vast majority of plants used for cultivation and consumption fall into either C_3 or C_4 plant groups. C_3 plants follow the Calvin-Benson cycle, which utilises building blocks consisting of three carbon atoms, while C_4 plants follow the Hatch-Slack cycle involving strings of four carbon atoms. The first group consists of cultivars from temperate climates, including common staples such as wheat, rye or barley and most vegetables; the second group includes plants from more-arid zones with maize and the millets as most prominent examples. A further group of plants, including cacti and amaranth, use a Crassulacean acid metabolism (CAM), characterised by switching between C_3 and C_4 pathways, but they are of little relevance as dietary sources in temperate environments. Carbon isotopic ratios thus allow identification of the relative amount of major plant groups contributing to the diet. This can be of particular relevance if the introduction of a new, isotopically different staple crop occurs, for example in cases of major subsistence change such as the widespread adoption of maize, a C_4 plant, into C_3-based economies of prehistoric Meso- and North America. Carbon derived from animal protein introduces a further fractionation step and as a result, human $\delta^{13}C$ values will be less negative when domestic or wild animals are contributing to the diet in measurable quantities. In the marine biotope, carbon isotopic ratios are generally more enriched and therefore allow the detection of seawater sources, especially when the terrestrial food components are C_3-based. Carbon isotopic ratios measured from bone collagen reflect the regular and average diet of the last five to ten years before death. There is no significant difference in the ratios between different bones of the same skeleton or between tissues and tissue products. Therefore, milk is isotopically indistinguishable from bulk bone collagen. Dietary regimes that refer to longer periods can be detected from carbon in the mineral fraction, because not only proteins, but also lipids and carbohydrates contribute to the isotopic ratio. Especially bones with extended turnover times, such as long bones, provide a good source of approximate lifetime diets, but caution needs to be exercised to check for diagenetic alterations of the mineral. Dental enamel, a tissue much less prone to taphonomic change, is a recommended alternative material, however, with the restriction that it only reflects dietary intake during the mineralisation of the tooth crown. But since the formation of enamel is an incremental process, it opens up the possibility of detecting the ontogenetic development of dietary habits.

The isotopic ratios of nitrogen are solely based on protein intake and are therefore a direct reflection of the nitrogen source. Even though a distinction between legumes and non-leguminous plants is possible, nitrogen ratios best describe protein intake from animal sources due to their much higher protein content, reflecting, for example, trophic level effects caused by meat consumption. As with carbon, fractionation occurs in the marine biotope. In arid environments more positive nitrogen isotopic ratios can be the result of water stress.

Diet as signifier

Even though our knowledge about contemporary food ways is detailed and quite exhaustive, and enables us to view human dietary behaviour within its immediate economic, social and political context (e.g. Harris and Ross, 1987; Ungar and Teaford, 2002), it is reasonable to assume that dietary information gleaned from the skeletal record through isotope analysis can equally help elucidate the circumstances and consequences of dietary habits in the past.

After all, it is this context, rather than the dietary reconstruction itself, that informs the research question. Light stable isotope studies have made invaluable contributions to our understanding of diet and subsistence in the past, both on a large scale involving trophic level spacing, and in detecting the more subtle differences of dietary options and preferences within populations. The purpose of this section is not to provide an extensive overview of the various facets of past human dietary behaviour that have been investigated (for this aspect see e.g. Ambrose and Katzenberg, 2000), but to present a selection of cases in point where diet and modes of subsistence have become distinctive and detectable features of socio-cultural expression, and which can be related to the general remarks on human dietary patterns and habits made above. The following case studies provide examples both of dietary transitions that result in large-scale dietary differences, and of smaller-scale differences expressed by groups within societies.

Changing subsistence modes

Transitions in subsistence modes are known to be caused or at least facilitated by changes in socio-economic conditions. Cross-cultural analyses have identified numerous examples from the relatively recent past (e.g. Bradley *et al.*, 1990) where colonial rule and the expansion of market-orientated economies have entailed substantial, if not complete, change. The transformations brought about by the effects of globalisation today are expansions on this theme and provide salient cases much closer to our immediate perception. Subsistence change in the more distant past is perhaps less obvious on the smaller scale, but certainly a prominent feature of big events such as the rapid incorporation of maize as a staple crop into Native American economies (see Smith, 1995, for an overview) or the spread of the Neolithic into northwestern Europe (Bogucki, 2000).

The adoption of a Neolithic lifestyle by Mesolithic populations in Southern Scandinavia and the British Isles essentially equates to the implementation of a fully developed subsistence technique and the concomitant shift in dietary patterns. In both cases the extensive exploitation of marine food resources, characteristic of late Mesolithic populations along the Atlantic and North Sea coasts (Arias, 1999), was replaced by the consumption and cultivation of terrestrial food – or more precisely food items that produce a terrestrial isotopic signature. Likewise, in both cases the replacement was wholesale, dramatic and a matter of a few generations only. The specific conditions, though, would have been distinctly different.

Late Mesolithic societies of southern Scandinavia had long-established contacts with the farming communities of the Linearbandkeramik (LBK) further south. They quite obviously knew about the cultivation of domesticated plants and animal husbandry. Yet, for about one thousand years after the arrival of the LBK on the north-German plain they continued to live a life as foragers, in sedentary and socially diversified communities (see above, Local Groups). The affluence of available food resources did not necessitate a change in food procurement strategies, as long as the existing social system could be sustained and the natural resources continued to be abundant. It was probably a combination of sea-level change and an increase in social differentiation that eventually led to the adoption of agriculture, and it has been argued that especially the latter would have required more clearly defined corporate structures that allowed rights of possession, resource control and management as well as private ownership (Price, 2000), requirements that could be met by a new culture kit (see also Rowley-Conwy, 2004). Such intrinsic motives are powerful drivers and stable isotope analyses (Richards *et al.*, 2003a), supported by the archaeological and faunal record, indeed demonstrate that the transition from foraging to farming was almost precipitous (in prehistoric terms) (but see Price *et al.*, 2007). Ecologically, i.e. in terms of the management of material and energy flows, this makes sense as well (see Begon *et al.*, 2003, p. 858). Modelling the dynamics of subsistence change reveals that only a rapid transition allows the system to recover quickly from the disturbance of change and to settle in a new stability domain (Schutkowski, 2006, p. 260), and thus to make the shift successful and lasting. The best pre-condition for that is a long availability phase during which the new subsistence mode can be explored and its implementation mentally prepared, i.e. exactly the situation so characteristic of the late adoption of the Neolithic in the north.

For the arrival of the Neolithic in Britain the same ecological rationale applies and again is borne out by the isotopic evidence (Richards *et al.*, 2003b). Contrary to traditional views of a rather gradual mode of change,[2] the sudden shift in isotopic signatures (Figure 7.3), indicating a sharp decline in marine resources to a diet entirely reliant on terrestrial (and estuarine) food, is best explained by the export of the complete Neolithic tool kit from the Continent, obviously without ample previous opportunity for Mesolithic populations to become acquainted with this lifestyle. The effect must have been equally dramatic and quite possibly disruptive. Yet, the fact that after *c.* 4000 cal BC there is fully established agriculture, reflected in typically terrestrial – or rather non-marine – isotopic signatures, demonstrates the successful transition and a quick consolidation of new biocultural equilibrium. Interestingly enough, this appears to be so extraordinary, it has been argued (Thomas, 2003) that such an event was internally supported – or mentally facilitated – by powerful food taboos against marine resources. This may or may not have been the case, but is not unknown as the examples outlined above show (see above, Food as cultured biology), and food

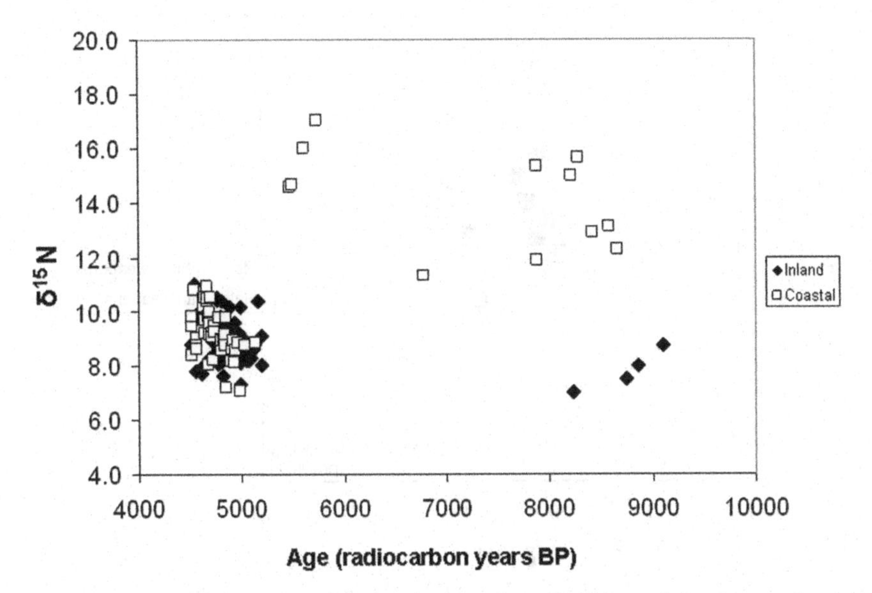

Figure 7.3 Sharp drop of nitrogen isotope values of British populations at around 4000 BC, indicating a wholesale dietary shift. Data courtesy of M. P. Richards, (Richards, 2005).

taboos, or rather socio-culturally engrained food preferences or ascriptions (see Grivetti, 2000), may very well have accompanied and stabilised the transition by helping provide and implement the new identity.

Variation within communities

Whilst dietary transitions connected with social-political change often take the form of trophic level spacing, there is evidence for isotopically more subtle differences in dietary behaviour within societies. Perhaps the most obvious ones are those that reflect control over, and differential access to, food resources. The evolution of societal complexity is accompanied by a development of socio-political structures that allow for increasingly sophisticated forms of resource management and distribution. Therefore, in societies that can be classified as Local Groups or Regional Polities, i.e. that show distinct social differentiation and often-institutionalised leadership, dietary differences are likely associated with and caused by differences in power and influence. The detection of social differentiation in the archaeological record is usually based on evidence from the mortuary context, for example the level of sophistication displayed in the burial arrangement and grave inclusions, indicative of a social rank attained

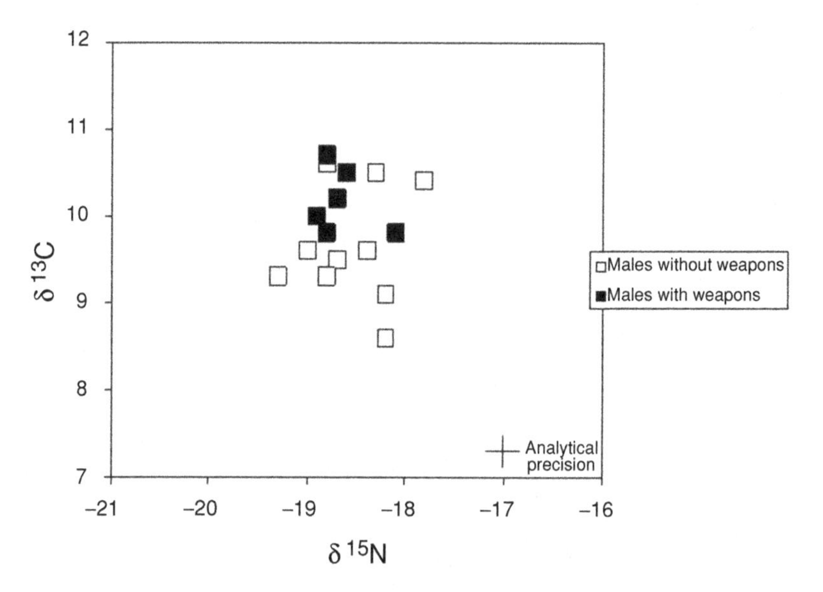

Figure 7.4 Differences in consumption of animal protein between elite (warrior) and common males in communities of Tišice and Makotrasy, Iron Age Bohemia. Data from Le Huray *et al*., 2006.

or the perception of an individual's importance by the community (e.g. Parker Pearson, 1999, pp. 72 ff.). Where such connections between status and diet have been identified through isotope analysis they either point towards better access to animal protein and concomitant differences in consumption patterns, or the ability to afford a diet derived from sources that require the input of more than average energy, material and socio-economic control. For example, in the context of societies from the La Tène period in Bohemia, males attributed to belong to a high-status warrior elite consumed a diet that contained significantly more animal protein than that of non-elite males (Kutná Hora-Karlov and Radovesice, Le Huray and Schutkowski, 2005; see also Le Huray *et al.*, 2006; Figure 7.4). Similar status-related differences, this time identified for women, were found in Continental early medieval (Kirchheim, Schutkowski, 2006, pp. 170 ff.) and prehistoric North American (Cahokia, Ambrose *et al.*, 2003) contexts. Here, social variation appears to translate directly into dietary quality, again suggesting better access to animal protein for women of higher social standing.

At early medieval Weingarten (Schutkowski, 2006) and medieval Berinsfield (Privat *et al.*, 2002), however, there was altogether significantly less dietary variation in high-status individuals compared with their common counterparts. It has been suggested that their diet would have been almost exclusively derived from resource-intensive husbandry or farming practices that require more input of resources, rather than from low-maintenance animals. It can thus be argued that, by implication, this would mean that those individuals or groups within a stratified society who have the privilege of choice through power and control also have the means to depart from a largely opportunistic dietary behaviour, and rather than having to rely on the breadth of available options, can instead be more selective, as it were as an expression of the possibilities of resource use ensuing from status variation. The result must not necessarily be a difference in nutritional quality, but is likely to be a matter of identity and social display, instead. After all, to provide a modern analogy, the protein content per edible portion of haddock and lobster may be equivalent (16.8 and 16.9 g, respectively; Southgate, 2000), but what a difference in statement it makes to be able and seen to eat the crustacean delicacy.

Conclusion

The last twenty years or so have seen fascinating progress in the reliable detection of past dietary habits. It seems, though, that we are just at the beginning of exploring the full potential that chemical analysis of human remains can hold. This is perhaps not surprising considering that the majority of bulk analyses aim to establish the enormous range of, for example, isotopic variation, to better understand the dietary meaning of mass spectrometry data output. It would appear that only a multidisciplinary approach, where scientific archaeology is enlisting the abundant contributions from ecology, climatology, sedimentology or hydrology, is able to address the complexities associated with the generation of isotopic fractionation, i.e. to establish the ecological framework of dietary signatures (see Fry, 2006). The analysis of (past) human dietary behaviour, however, takes a step beyond understanding ecological circumstances of the natural environment and extends the base of enquiry to explicitly include the socio-cultural environment as an expression of the human ecological niche (see above, Introduction) and as an explanation for dietary variation.

If human ecology can be broadly defined as the biological outcomes of cultural strategies (Schutkowski, 2006, p. 23) then the task would be to reconcile the artefactual evidence and the archaeological record with dietary signatures retrieved from human remains and organic residues, to begin to understand how this information is reflected by and translated into the socio-cultural realm.[3] As numerous ethnographic examples demonstrate, however, the cultural differences between and within societies may be substantial and obvious to the contemporary and retrospective observer. The extent to which these become manifest in chemical information recorded in calcified or keratinised tissues, i.e. to what extent they become accessible as a biologically detectable outcome, can be a different matter entirely. Even the salient examples of broad shifts in food availability or preference connected with subsistence transitions mentioned above only show up in the isotopic signatures because resource exploitation strategies happened to rely on a switch to foodstuffs that represent a different trophic level, for example from largely marine to largely terrestrial. But already a distinction between the consumption of meat or milk and milk products, known to be an ascribed dietary social signifier (the Breton and Maasai examples), would go unnoticed even on a large scale because there is no fractionation between body tissue and fluid and thus no obvious difference in isotopic representation. This, however, is not to say that the fine and subtle changes would escape detection. In fact, within populations one would hardly expect to find dietary differences equating to trophic level spacing. Variation of human diet develops within the confines of omnivorous food procurement, which means that dietary differentiation, if biochemically distinct, usually occurs across a band of 3–5 ‰ variation. This specifically calls for interpretations where socio-cultural evidence can independently corroborate the dietary variation as expressed through the isotopic signature. Dietary differences between the genders in general would seem to be particularly promising, but have only just begun to be explored more widely. What little is known, however (e.g. Dürrwächter *et al.*, 2006; Schutkowski and Richards, 2008), suggests that in socially diversified societies there is consistent variation pointing towards a dietary expression of difference. In cases of known social differentiation within a society the archaeological record is particularly useful, e.g. through the distinctiveness and quality of grave offerings, to support indications of small-scale differences in resource supplies and availabilities. It is at this level of resolution, where chemical information can help establish the fundamental significance of diet and its transformation into self-expression and group codification, where food becomes a signifier, and diet an item of identity. It is only just the beginning.

Acknowledgement

I thank Gabriele Macho for thoughtful comments on a previous version of this chapter.

Notes

1. Whether in this sense the invention by Japanese macaques on Koshima Islet to take sweet potatoes they were provisioned with and wash them in salt water is the beginning of cuisine in non-human primates adds a challenging note.

2. The ongoing heated debate over timing and rate of this change (see Milner *et al.*, 2004; Richards and Schulting, 2006) demonstrates the necessity to analyse the evidence for paradigm shifts in archaeological interpretation in a synthetic way that includes modelling and theoretical assumptions.
3. This would eventually encompass the significance of cooking as a cultural, yet biologically relevant, achievement (see Wrangham and Conklin-Brittain, 2003).

References

Ambrose, S. H. and Katzenberg, M. A. (eds.) (2000). *Geochemical Approaches to Palaeodietary Analysis*. London: Kluwer/Plenum.

Ambrose, S. H., Buikstra, J. and Krueger, H. W. (2003). Status and gender differences in diet at mound 72, Cahokia, revealed by isotopic analysis of bone. *Journal of Anthropological Archaeology*, **22**, 217–26.

Arias, P. (1999). The origins of the Neolithic along the Atlantic coast of continental Europe. *Journal of World Prehistory*, **13**, 403–64.

Ashley, B., Hollows, J., Jones, S. and Taylor, B. (2004). *Food and Cultural Studies*. London: Routledge.

Begon, M., Harper, J. L. and Townsend, L. (2003). *Ecology: Individuals, Populations and Communities*. 3rd edn. Oxford: Blackwell Scientific.

Bogucki, P. (2000). How agriculture came to north-central Europe. In *Europe's First Farmers*, ed. T. D. Price. Cambridge: Cambridge University Press, pp. 179–218.

Böhmer-Bauer, K. (1990). *Nahrung, Weltbild und Gesellschaft. Ernährung und Nahrungsregeln der Massai als Spiegel der gesellschaftlichen Ordnung*. Saarbrücken: Breitenbach.

Bradley, C., Moore, C. C., Burton, M. L. and White, D. R. (1990). A cross-cultural historical analysis of subsistence change. *American Anthropologist*, **92**, 447–57.

Brown, L. D. and Heron, C. (2003). Boiling oil: The potential role of ceramics in recognising direct evidence for the exploitation of fish. In *Prehistoric Pottery: People, Pattern and Purpose*, ed. Gibson, A., BAR International Series 1156, Oxford, pp. 35–41.

Chapman, M. (1990). The social definition of want. In *Food for Humanity*, eds. M. Chapman and H. Macbeth, Oxford: Oxford Polytechnic, pp. 26–33.

Copley, M. S., Berstan, R., Dudd, S. N. *et al.* (2003). Direct chemical evidence for widespread dairying in prehistoric Britain. *Proceedings of the National Academy of Sciences*, **100**, 1524–9.

Craig, O. E., Chapman, J., Heron, C. *et al.* (2005). Did the first farmers of central and eastern Europe produce dairy foods? *Antiquity*, **79**, 882–94.

Danforth, M. E. (1999). Nutrition and politics in prehistory. *Annual Review of Anthropology*, **28**, 1–25.

Draper, H. H. (2000). Human nutritional adaptation: biological and cultural aspects. In *The Cambridge World History of Food*, eds. K. F. Kiple and K. C. Ornelas. Cambridge: Cambridge University Press, pp. 1466–76.

Dürrwächter, C., Craig, O. E., Collins, M. J., Burger, J. and Alt, K. (2006). Beyond the grave: variability in Neolithic diets in Southern Germany? *Journal of Archaeological Science*, **33**, 39–48.

Eastwood, M. (2003). *Principles of Human Nutrition*. 2nd edn. Oxford: Blackwell Science.

Elias, N. (1997). *Über den Prozess der Zivilisation. Soziogenetische und psychogenetische Untersuchungen*. Frankfurt am Main: Suhrkamp.

Ellen, R. (1982). *Environment, Subsistence and System. The Ecology of Small-Scale Social Formations*. Cambridge: Cambridge University Press.

FAO (1992). *Maize in Human Nutrition*. Rome: Food and Agricultural Organization.

Fry, B. (2006). *Stable Isotope Ecology*. London: Springer.

Goran, M. I. and Astrup, A. V. (2002). *Introduction to Human Nutrition*. Oxford: Blackwell Science.

Gowlett, J. A. J. (2006). The early settlement of northern Europe: Fire history in the context of climate change and the social brain. *Comptes Rendus Palevol*, **5**, 299–310.

Grivetti, L. E. (2000). Food prejudices and taboos. In *The Cambridge World History of Food*, eds. K. F. Kiple and K. C. Ornelas. Cambridge: Cambridge University Press, pp. 1495–1513.

Harris, M. and Ross, E. B. (eds.) (1987). *Food and Evolution. Toward a Theory of Human Food Habits*. Philadelphia: Temple University Press.

Hetzel, B. S. and Dunn, J. T. (1989). The iodine deficiency disorders. Their nature and prevention. *Annual Review of Nutrition*, **9**, 21–38.

Hockett, B. and Haws, J. A. (2005). Nutritional ecology and the human demography of Neandertal extinction. *Quaternary International*, **137**, 21–34.

James, W. P. T and Schofield, E. C. (1990). *Human Energy Requirements*. Oxford: Oxford University Press.

Johnson, A. W. and Earle, T. (2000). *The Evolution of Human Societies. From Foraging Group to Agrarian State*. 2nd edn. Stanford: Stanford University Press.

Katz, S. H., Hediger, L. and Valleroy, L. A. (1974). Traditional maize processing techniques in the New World. *Science*, **184**, 765–73.

Le Huray, J. D. and Schutkowski, H. (2005). Diet and social status during the La Tène period in Bohemia – carbon and nitrogen stable isotope analysis of bone collagen from Kutná Hora-Karlov and Radovesice. *Journal of Anthropological Archaeology*, **24**, 135–47.

Le Huray, J. D., Schutkowski, H. and Richards, M. P. (2006). Le Tène dietary variation in Central Europe: A stable isotope study of human skeletal remains from Bohemia. In *Social Archaeology of Funerary Remains*, eds. R. Gowland and C. Knüsel. Oxford: Oxbow, pp. 99–116.

Lévi-Strauss, C. (1964). *Mythologiques I: Le Cru et le Cuit*. Paris: Plon.

Lévi-Strauss, C. (1965). Le triangle culinaire. *L'Arc*, **26**, 19–29.

Livi-Bacci, M. (1997). *A Concise History of World Population*. 2nd edn. Oxford: Blackwell.

Mainland, I. and Halstead, P. (2005). The economics of sheep and goat husbandry in Norse Greenland. *Arctic Anthropology*, **43**, 103–20.

Milner, N., Craig, O. E., Bailey, G. N., Pedersen K. and Andersen, S. H. (2004). Something fishy in the Neolithic? A re-evaluation of stable isotope analysis of Mesolithic and Neolithic coastal populations. *Antiquity*, **78**, 9–22.

Moran, E. F. (2000). *Human Adaptability. An Introduction to Ecological Anthropology*. 2nd edn. Boulder: Westview.

Parker Pearson, M. (1999). *The Archaeology of Death and Burial*. Stroud: Sutton.

Pollard, A. M., Batt, C. M., Stern, B. and Young, S. M. M. (2007). *Analytical Chemistry in Archaeology*. Cambridge: Cambridge University Press.

Price, T. D. (2000). The introduction of farming in northern Europe. In *Europe's First Farmers*, ed. T. D. Price. Cambridge: Cambridge University Press, pp. 260–300.

Price, T. D., Ambrose, S. H., Bennike, P. *et al.* (2007). New information on the Stone Age graves at Dragsholm, Denmark. *Acta Archaeologica*, **78**(2), 193–219.

Privat, K. L., O'Connell, T. C. and Richards, M. P. (2002). Stable isotope analysis of human and faunal remains from the Anglo-Saxon cemetery at Berinsfield, Oxfordshire: dietary and social implications. *Journal of Archaeological Science*, **29**, 779–90.

Rayman, M. P. (2000). The importance of selenium to human health. *Lancet*, **356**, 233–41.

Richards, M. P. (2005). *Diet Shifts Across the Mesolithic-Neolithic Transition in North-West Europe. Going Over: the Mesolithic-Neolithic Transition in North-West Europe*. Cardiff, Wales: Cardiff University.

Richards, M. P. and Schulting, R. J. (2006). Against the grain? A response to Milner *et al.* (2004). *Antiquity*, **80**, 444–58.

Richards, M. P., Price, T. D. and Koch, E. (2003a). Mesolithic and Neolithic subsistence in Denmark: new stable isotope data. *Current Anthropology*, **44**, 288–95.

Richards, M. P., Schulting, R. J. and Hedges, R. E. M. (2003b). Sharp shift in diet at onset of Neolithic. *Nature*, **425**, 366.

Ross, E. B. (1987). An overview of trends in dietary variation from hunter-gatherer to modern capitalist societies. In *Food and Evolution. Toward a Theory of Human Food Habits*, eds. M. Harris and E. B. Ross. Philadelphia: Temple University Press, pp. 7–55.

Rowley-Conwy, P. (2004). How the West was lost. A reconsideration of agricultural origins in Britain, Ireland and Southern Scandinavia. *Current Anthropology*, **45** [Supplement], S83–S113.

Rozin, E. (1982). The structure of cuisine. In *The Psychobiology of Human Food Selection*, ed. L. M. Barker. Westport: AVI Publishing, pp. 189–203.

Rozin, P. (2000). The psychology of food and food choice. In *The Cambridge World History of Food*, eds. K. F. Kiple and K. C. Ornelas. Cambridge: Cambridge University Press, pp. 1476–86.

Sadalla, E. and Burroughs, J. (1981). Profiles in eating. Sexy vegetarians and other diet-based social stereotypes. *Psychology Today*, Oct 1981, 51–7.

Scholliers, P. (2001). Meals, food narratives, and sentiments of belonging in past and present. In *Food, Drink and Identity*, ed. P. Scholliers. Oxford: Berg, pp. 3–22.

Schutkowski, H. (2006). *Human Ecology. Biocultural Adaptations in Human Communities*. Berlin: Springer.

Schutkowski, H. and Richards, M. P. (2008). Diet and subsistence during the Middle Bronze Age at Sidon, Lebanon – first isotopic evidence of coastal Levantine food ways. In *Paleonutrition and Food Practices in the Ancient Near East: Towards a Multidisciplinary Approach*. History of the Ancient Near East/Monographs IX, ed. L. Milano. Padova: SARGON.

Sealy, J. (2001). Body tissue chemistry and palaeodiet. In *Handbook of Archaeological Science*, eds. D. R. Brothwell and A. M. Pollard. Chichester: Wiley, pp. 269–79.

Smith, B. D. (1995). *The Emergence of Agriculture*. New York: Scientific American Library.

Southgate, D. A. T. (2000). Meat, fish, eggs and novel proteins. In *Human Nutrition and Dietetics*, eds. J. S. Garrow, W. P. T. James and A. Ralph. London: Churchill Livingstone, pp. 363–74.

Stern, B., Lampert Moore, C. D., Heron, C. and Pollard, A. M. (2008). Bulk stable light isotope ratios in recent and archaeological resins: Towards detecting the transport of resins in antiquity? *Archaeometry*, **50**(3), 351–70.

Thomas, J. (2003). Thoughts on the 'repacked' Neolithic revolution. *Antiquity*, **77**, 67–74.

Ungar, P. S. and Teaford, M. F. (eds.) (2002). *Human Diet: its Origin and Evolution*. Westport: Bergin & Garvey.

Visser, M. (1992). *The Rituals of Dinner*. London: Penguin.

Vorster, H. H. and Hautvast, J. (2002). Introduction to human nutrition: a global perspective on food and nutrition. In *Introduction to Human Nutrition*, eds. M. J. Gibney, H. H. Vorster and F. J. Kok. Oxford: Blackwell Science, pp. 1–11.

Weisdorf, J. L. (2005). From foraging to farming: explaining the Neolithic revolution. *Journal of Economic Surveys*, **19**, 561–86.

Wrangham, R. and Conklin-Brittain, N. (2003). Cooking as a biological trait. *Comparative Biochemistry and Physiology Part A*, **136**, 35–46.

Zischka, U., Ottomeyer, H. and Bäumler, S. (eds.) (1993). *Die anständige Lust. Von Esskultur und Tafelsitten*. Munich: Edition Spangenberg.

8 *Ancient proteins: what remains to be detected?*

M. J. COLLINS, E. CAPPELLINI, M. BUCKLEY,
K. E. H. PENKMAN, R. C. GRIFFIN AND H. E. C. KOON

Can an understanding of taphonomic processes help inform our choice of research questions and methods? Living in York one is all too aware of the importance of molecular degradation, as its foundations and economic success are built upon slowly decaying Medieval rubbish. Arguably, the foundation of biomolecular archaeology is an understanding of molecular diagenesis, but all too often the importance of this has been overlooked. Perhaps this is because it is all too obvious; the same factors that preserved foods – (semi)sterile, dry or cold conditions – are those which intuitively will enhance the preservation of ancient biomolecules.

Having identified cultural, biological and chemical processes which enhance preservation of human remains, the selection of analytical techniques becomes less daunting. Early postmortem decay will limit the ability to obtain larger intact biomolecules and lead to considerable overprinting. Increased porosity of the bone caused by microbial decay will enhance secondary contamination. The abundance of degradation products may actually mask other biomolecules. Having avoided these taphonomic pitfalls, the two major analytical challenges are separation of complex mixtures and increased sensitivity of detection. Advances in both of these areas has in recent years begun to expand the range of analytical options open to anthropologists and therefore the range of potential research questions and the time period over which this research can be undertaken.

Introduction

This is an essay about proteins, ancient proteins to be precise, lying in a volume entitled ‘Between Biology and Culture’. Proteins – the engines and architecture of the cell – lie entrapped within or sorbed upon archaeological artefacts. In the same way as the field archaeologist seeks for artefacts within the remains of abandoned and decayed structures, ‘molecular archaeologists’ attempt to seek out and interpret the molecules themselves. However, biomolecular archaeology

Between Biology and Culture, ed. Holger Schutkowski. Published by Cambridge University Press.

has not become engaged in the sometimes polarised debates in Archaeology between Modernist divides (subject | object, structure | agency, nature | culture, etc.) that has characterised the material analysis of artefacts. In the discussion surrounding Jones' suggestion that 'materiality' is a route to rapprochement within archaeology between science and theory (Jones, 2004 and comments in *Archaeometry* **467**, 1), only Gosden (2005) mentions biomolecular data.

Unlike the chemical and physical analysis of artefacts, biomolecular analysis is primarily used to establish identity. This can be the identity of a material from an artefact (e.g. the source of fuel in lamps, Evershed *et al.*, 1997; Copley *et al.*, 2005d), the material from which an artefact is made (Burger *et al.*, 2000), the identity of biological remains (Burger *et al.*, 2001), the presence of organisms where no remains survive (Hofreiter *et al.*, 2003) and, in the near future, phenotypes (such as cattle coat colour). In addition to identity, stable isotopes are used to estimate the contribution of different sources to the formation of proteins (Schulting and Richards, 2002), whilst DNA has been used to investigate phylogenies (Lalueza-Fox *et al.*, 2002; Burger *et al.*, 2004) and patterns of exchange or contact between wild populations (Barnes *et al.*, 2002; Shapiro *et al.*, 2004) or domestic animals (Anderung *et al.*, 2005).

In most of these cases, the types of data obtained do not engage with the *nature* | *culture* debate because there are no opportunities for selection by human agency. This is not always the case; dietary choices (such as those at the time of the transition to farming, Richards *et al.*, 2003), and selection of specific materials for mummification (Buckley *et al.*, 2004), or specific phenotypes are all examples in which 'materiality' matters.

In this chapter, we seek to highlight examples in which the study of ancient proteins and their remains may contribute to the study of culture. Specifically, the ways in which the taphonomy, diagenesis and analysis of proteins entrapped in skeletons have the potential to expand our interpretation of skeletal remains.

Ancient proteins, a brief primer

In order to consider what remains we must first examine the original structures. **Amino acids**, the building blocks of proteins, have been identified as widespread in ancient materials since the pioneering work of Abelson (1954). As Abelson discovered, if the system is well defined, the **compositional pattern** of amino acids can **identify** the dominant protein, well illustrated by the findings of ancient collagen-based 'glue' from the Nahal Hemar cave by Arie Nissenbaum (1997). Not long after Abelson's first investigations, Hare and Abelson (1968) appreciated that the rate of decay of proteins, specifically the inter-conversion of the L (or *laevo*) form found in proteins to the chiral D (*dextro*) form amino acids (**racemisation**) could be used as a **clock** to assess the relative ages of sediments from the same region. The larger the proportion of D-amino acids entrapped in a sample, the more degraded and hence the older a sample was.

Information encoded by the genome is expressed as protein sequence, and attempts have been made to detect protein sequence in order to both **identify** samples and to assess **phylogenetic** relationships. Although some attempts were made to directly **sequence** proteins and protein fragments (Huq *et al.*, 1990) or to examine distributions of peptides following **proteolysis** (Armstrong *et al.*, 1983), both approaches were compromised because for optimal application they require extremely well-preserved protein. **Immunology**, which can detect small regions or structures (epitopes), proved to be the most simple and reliable method both to detect unknown protein **residues** (Cattaneo *et al.*, 1990) and to infer phylogenetic distance (Lowenstein, 1981).

Protein mass spectrometry was first applied to fossil proteins by Ostrom *et al.*, (2000) and holds great promise because peptides can be identified by direct sequencing, thus in theory both **identity** and **sequence analysis** are possible, as is deriving data on patterns of **deterioration**.

What remains: decay and transformation

Decaying bodies

The Archaeological Data Service based in the Department of Archaeology at the University of York holds (as of September, 2005) almost one million records of ancient monuments. So many records, so many sites! With limited resources the prioritisation of ancient monuments becomes a challenge to heritage managers and politicians alike. In the same way, at a single site we may be faced with the possibility of studying thousands of bones and, with new technologies, numerous options for analysis. So the question is the same, which samples to select for destructive analysis. In order to do this we must understand decay; in other words, *what remains*?

Bone deterioration, from acid rain to mummification and perinatal mortality

What remains of organic remains? Normally not much, often nothing at all. The *Scotland's First Settlers* project has found little organic remains from a comprehensive survey of Mesolithic sites, even bone is rarely well preserved and then only when buried in shell middens. The paucity of information in Scotland is in stark contrast to waterlogged sites in the Baltic which have revealed spectacular evidence of fishing technology (e.g. Price *et al.*, 2001). The surprising insights into Neolithic-Copper Age technology which have arisen following the discovery of a frozen body from the Italian Tyrol (Hess *et al.*, 1998; Rollo and Marota, 1999; Cano *et al.*, 2000; Rollo *et al.*, 2000; Makristathis *et al.*, 2002; Muller *et al.*, 2003) only serve to remind us how little normally remains.

In the case of the Scotland's First Settlers project, the organic remains which did survive amongst the shellfish middens were largely bone. Bone is the most common material to bridge the biology–culture divide; but bone, although a tough material, only rarely survives, and survival can vary over short distances (e.g. Stiner *et al.*, 2001).

The absence of bone in predominately acidic, well-drained soils, which cover much of north and western England, indicates that soil conditions are the primary control on bone preservation, and this was found to be the case in a large European-wide investigation (Nord *et al.*, 2005). Perhaps more surprisingly, where bone did survive, cultural factors appeared to be significant.

The most revealing insights into survival were given by measurements of pore size distribution, by intruding mercury into vacuum-dried samples under increasing pressure. Porosimetry revealed that most pores in archaeological bone were sub-micron in size, giving the appearance of a spongiform quality in back scatter scanning electron microscopy (BSEM, Turner-Walker *et al.*, 2002). Under the microscope there is a honeycomb texture of holes and dense apatite 'collars' (Collins *et al.*, 2002). Over the range visualised microscopically, the pore-size distributions were consistent between mercury intrusion and BSEM. We were unable to visualise with sufficiently high resolution below 100 nm, and so frustratingly could not understand why the major small pore dimensions which emerged were centred on about 30 nm, when it was believed that collagen fibre dimensions lay closer to 80–100 nm (Weiner and Traub, 1986). Recently, elegant ion-beam thinning analyses of bone sections has revealed details of the mineral-collagen interaction at this scale, which appears to show a major pore-dimension at approximately 30 nm (Cressey and Cressey, 2003).

The porosity data (taken from compact, predominately limb, bones) revealed apparent differences between animals and humans. Jans *et al.* (2004) noted that the major difference was a preference for human bone to be altered by 'spongiform' (bacterial?) porosity, whereas animal bone was more commonly affected by long penetrating tunnels, believed to be caused by fungi. The propensity for microbial attack of corpses could be explained by the tendency of buried intact corpses to putrefy. Gut bacteria can then travel via the circulatory system into the bone, an idea first suggested by Lynne Bell and colleagues (1996). Fungi, by contrast, will colonise processed bone, or when in the soil, mycorrhizal networks will mine bone phosphate (Ness and Vlek, 2000). Assuming this theory is correct, intact animal burials should behave like human burials and display the same pattern of spongiform porosity – and Jans *et al.* (2004) found that they did.

These findings were used to help understand an unusual burial found at Cladh Hallan, in the Outer Hebrides of Scotland. A crouch burial was discovered carefully positioned in the based of an Iron Age round house, but buried several hundred years after death. The pattern of postmortem alteration was unusual in that extensive microbial attack was restricted to two narrow zones underlying the inner and outer edges. Using Small Angle X-Ray Scattering (SAXS) (Wess *et al.*, 2002; Hiller *et al.*, 2004), Jen Hiller was able to map changes in crystal dimension across a bone shaft and detect increased mineral thickness, even in those inner-regions which had not suffered from evident microbial attack. This suggested that mineral change had occurred in some regions of the bone without attendant microbial decay, suggesting that it may have been exposed to acidity (before or after burial, it is not possible to say) leading to crystal modification. More work needs to be undertaken to establish the nature and extent of this practice, but one possibility is that these findings suggest evidence of mummification occurring in later prehistory in Scotland. If this is the case, it changes our views of attitudes to the dead in these societies (Parker Pearson *et al.*, 2005). It would be interesting to conduct similar investigations of known mummified remains to establish if a similar pattern of intense rapid decay is seen in other bones and if this is a clue to early, arrested putrefaction; indeed, we still remain ignorant of the speed of initial microbial alteration of bone (Bell *et al.*, 1996).

If it is indeed correct that microbial attack of bone is linked to putrefaction then this could be used to provide an insight into changing cultural traditions in the treatment of perinatal infants. Attitudes to infant death changed in Europe in the thirteenth century with the general acceptance that children had souls. After the Fourth Lateran Council in 1215, baptism of the newborn became the only sacrament that could be performed by a layperson, even a woman, and priests were instructed to teach laypeople emergency baptism formulae. S*ectio in mortua* became increasingly common, not in order to save the life of the fetus, but to enable it to be 'born' out of its dead mother so that it could be baptised and its soul saved (Schäfer, 1999, p. 301). The pattern of microbial decay of bones could potentially be used to discriminate between those children who acquired a gut flora and those who died with an aseptic gut (stillbirths and s*ectio in mortua*).

Decay is easy

Under ideal circumstances, in which the burial environment is benign and neither causes the mineral to dissolve, nor encourages microbial assault, what then would be the fate of bone? The collagen will still deteriorate, driven by a largely irresistible chemical clock, collapsing to form soluble gelatine. The gelatine, being soluble, would be lost from the bone and, if the weakened structure was not disturbed, the resulting pore space could then be replaced by mineral, resulting in a fossil with excellent histological preservation. The rate of collagen deterioration is predictable and is exponentially related to temperature, so it is possible to assess the approximate time required to remove the collagen under ideal burial conditions in different environments. The advantage of such an approach is that it could identify limits to the persistence of collagen and make some predictions about the nature of fossilisation.

Using estimates of effective burial temperature from ISLSCP data (Meeson *et al.*, 1995), Colin Smith estimated the rate of collagen deterioration in Africa and found it to be suspiciously related to the distribution of hominin sites and faunal remains. We tested the idea by comparing the distribution of sites that recorded lithics and those that recorded bones. However, despite the initial premise, there was no significant match between those regions in which collagen would persist (and hence the bones remain tough) and the sites which yielded bones (Holmes *et al.*, 2005). The idea that only bones spared microbial attack would persist into the fossil record did appear to be correct (Trueman and Martill, 2002), although such samples, being phosphate rich, would be susceptible to phosphate mining at any time.

We reasoned that as collagen was thermally limited, we could use this knowledge to establish the limit to DNA survival and consequently identify those bones likely to yield ancient sequences. This was a time when museum curators were being approached for rare Neanderthal fossils in order to attempt to extract and amplify ancient DNA, following the first successful amplification, by Svante Pääbo's group (Krings *et al.*, 1997). Our analysis revealed that most were beyond the thermal limit for recovery (Smith *et al.*, 2001; 2003). This limit is an empirical one – we understand little of the rate and mechanism of DNA degradation and the limit will move back in time as extraction and analytical methods improve. Samples from coprolites have thermal ages greater than this estimated limit for bone (Poinar *et al.*, 2003), which may reflect different mechanisms or environments of survival. However, a recent publication by Salamon *et al.*, (2005) related successful amplification of human DNA from intra-crystalline aggregates of a more than 6000-year-old sample from Wadi Makuch. This sample has an estimated thermal age three times that of the limit in bone (similar to the value for coprolites), suggesting either the virtue of extreme aridity, or that improved extraction strategies may improve the yield of ancient DNA.

Deterioration as cooking and cooking as culture

The moment and meaning of eating or consumption is a corporeal blending of economics, politics, sociality, and aesthetics, entwined in life history and cultural memory (James, 2004, p. 2).

It has been argued that culture is often manifested through the production of food, therefore it should follow that bone chemistry can offer some cultural insights. Cultural information can certainly be derived from the stable isotopic composition of bone, which is linked to diet (see Chapter 7), but what about the process of cooking itself? Upon heating, collagen undergoes a dramatic thermal transition to gelatine, and therefore bone collagen should also record cooking events. If cooking is a key component of culture (e.g Swinbank, 2002), then the decomposition of bone collagen has a direct role to play in culture – has it not?

What appears to be a simple problem, a direct contribution from taphonomy to culture, proves more difficult than first expected. Burnt bone can be readily detected (e.g. Shipman *et al.*, 1984; Nicholson, 1993), however burnt bone does not identify cooked bone. During roasting, meat helps to insulate bone and the meat is only palatable if it remains moist (below 100 °C). Boiling (at a constant temperature of 100 °C) is a more aggressive heat treatment, but prolonged boiling is required to convert bone collagen to gelatine (Roberts *et al.*, 2002). The reason for this lies in the multi-level, structural arrangement of collagen.

The collagen polypeptides (α-chains) form a triple helix which self-organises into fibrils, via a quarter stagger alignment, which are further aggregated to form fibres. Following prolonged heating this structural organisation will ultimately collapse to yield denatured polypeptides (gelatine). However the exquisite packaging is remarkably resistant to heat degradation, even more so when the fibrils are mineralised, as they are in bone (Kronick and Cooke, 1996). The prolonged heating required to release gelatine from bone makes bone an unattractive substrate for gelatine production compared to skin and hooves. This thermal stability also makes the detection of cooking difficult. Preparation of meat rarely exposes bone to high or prolonged heat and consequently, despite the profound importance of this process to bone taphonomy, cooking (unlike burning) has been difficult to detect in the archaeological record. Richter (1986) used transmission electron microscopy to detect damage to the fibril organisation as a means of detecting cooking in fish bone. Koon *et al.*, (2003) used a modified version of the same approach to detect cooking in animal bone, although as Roberts *et al.* (2002) had reported previously, changes observed during cooking also mimicked those occurring by diagenesis.

What remains to be discovered: molecular evidence

Crystal cages

Salamon *et al.*, (2005) used bleach treatment to remove organic matter from bone in order to remove contamination and damaged molecules. The concept of intra-crystalline organic matter was first proposed as a source of entrapped organic matter by Ken Towe (1980), but developed by the Weiner group (e.g. Berman *et al.*, 1988; DeNiro and Weiner, 1988). Sadly in the case of bone, individual crystallites both appear too small to retain macromolecules, and also undergo significant post-diagenesis modification (Hiller *et al.*, 2004; Smith *et al.*, 2005). In an elegant study, Trueman *et al.*, (2004) directly measured the size of bone crystallites as they underwent diagenetic modification. The bone apatite crystallites initially had mean lengths of *c*. 40 nm and widths of *c*. 22 nm, although these grew larger during diagenetic modification. Examining the deterioration of osteocalcin, McNulty *et al.* (2002) observed that the most common site for cleavage was between Asp_{14} and Pro_{15}, the region between the flexible N-terminus and the mineral-bound mid-region of the molecule (Hoang *et al.*, 2003). Mineral binding protects highly unstable Gla residues of the mid-region (Ostrom *et al.*, 2000; Nielsen-Marsh *et al.*, 2002; Nielsen-Marsh *et al.*, 2005). It has commonly been suggested that DNA survival in bone is enhanced by binding to apatite. The maximum length of naked double-stranded DNA which could be laid out straight (the most electrostatically favourable arrangement) along a bone crystallite of 40 nm by 22 nm would be less than 140 bp (base pairs) – even if lying diagonally from corner to corner. It is perhaps not surprising that most ancient DNA fragments are less than 200 bp; there are simply too few long stretches of mineral to bind longer fragments. Curiously, diagenesis will generate larger crystallites, which could potentially preserve longer fragments of (allochthonous) DNA.

It would appear that in the case of vertebrate remains, only enamel, the most highly mineralised tissue of all, is a realistic source of biomolecular remains which will survive into deep time.

Amino acid racemisation

Protein decay is more readily tractable than decay of DNA (but see Chapter 10). Predicting decay of proteins is difficult, but specific reactions are more tractable, one of which is the rate of a simple change in the building blocks of the proteins, the inter-conversion of the chiral L form to a racemic mixture of 50 % L and 50 % D. The increase in D-amino acids is used as a method of estimating the age of samples. The method has proved controversial in bone (Marshall, 1990), but has yielded useful results in bird eggshell (Brooks *et al.*, 1990; Miller *et al.*, 1992; 2000). In the case of mollusc shells, the approach has been used to estimate the age of terrestrial Quaternary deposits (Bowen *et al.*, 1989), although some of the age estimates, notably of sites with hominin remains, such as Boxgrove (Bowen and Sykes, 1994; Roberts, 1994), have proved controversial (Stringer and Hublin, 1999). Sykes *et al.*, (1995) observed that an inter-crystalline fraction gave more reliable measures of protein decomposition than analysis of the whole shell. The idea was further developed by Collins and Riley (2000), who argued that only by using an intra-crystalline fraction could reliable estimates be made of the extent of protein degradation, and hence age. This approach to protein dating, an extension of conventional amino acid racemisation geochronology, using both an intra-crystalline fraction and analysis of multiple amino acids (Kaufman and Manley, 1998), appears to hold promise; it has recently been used to help establish the age of the earliest humans in Northern Europe (Parfitt *et al.*, 2005).

Improved dating methods would help address the chronology of 'cultures' as represented by the lithic traditions (but see McNabb *et al.*, 2004) seen in the British Lower/Middle Palaeolithic. Bridgland (2000) suggests that the first Levallois flint processing (a core and flake technology associated with Neandertals) appears in deposits correlated with Oxygen Isotope Stage 8. Prior to this the picture is much more confusing and the original distinction between more 'primitive' Clactonian flakes and more 'advanced' Acheulian bifaces (King and Oakley, 1936) has been reassessed in the light of new finds and improved chronology (Ohel, 1979; Ashton *et al.*, 1994; McNabb, 1996), which reveals 'advanced' knapping in Cromerian deposits. White (2000) argues that the distinction between Acheulian and Clactonian should be retained and that the timing of the reappearance of the two traditions in Britain at the end of OIS 12 and OIS 10 should be examined more closely. High-resolution chronology will be required to resolve this issue, and it may be that the proteins steadily decomposing within closed organic systems offer an unusual bridge between biology and culture.

Aspartic acid and age at death

Racemisation has been used with less controversy in forensic science, where careful estimates of the extent of aspartic acid racemisation in dentine proteins were used to estimate the age at death (Ritz *et al.*, 1990; Ohtani and Yamamoto, 1991). We had suspected, based upon the ideas of Paul Cloos, a doctoral student in Copenhagen, that there was no racemisation in the collagen triple helix, but this absence was hard to prove experimentally, as isolating the helix induced racemisation, and heating samples to accelerate racemisation denatured the helix. Adri van Duin was able to demonstrate that due to the constraints upon the helix the cyclic succinimide, believed to pre-phase racemisation, would not readily form in the helix (van Duin and Collins, 1998). With this knowledge to hand we were then able to model racemisation of collagen, recognising the importance of primary structure (sequence), conformation and the respective ease with which asparaginyl and aspartyl residues can form succinimides (Collins *et al.*, 1999). The modelling has then helped to explain some misconceptions caused by the data generated by this complex set of competing factors (Waite and Collins, 2000).

Residues on pots

One area in which the link between biology and culture is more immediate is the detection of residues found on archaeological artefacts, such as pottery and stone tools.

One of the most intriguing investigations has been that of Tom Loy, who had detected blood residues on lithic artefacts (Loy, 1983). The research had proved controversial (Custer *et al.*, 1988; Gurfinkel and Franklin, 1988; Smith and Wilson, 1992; Cattaneo *et al.*, 1993; Tuross *et al.*, 1996). We attempted to use a similar method to detect residues on pottery (Craig and Collins, 2002), but, like earlier investigators using laboratory-prepared samples, we found that recovery was difficult (Cattaneo *et al.*, 1993; Tuross *et al.*, 1996). Laboratory studies and experimental archaeological approaches, complete with hazardous attempts to fire pots in wood-filled pits, and the usual Neolithic cook-up, highlighted a near irreversible sorption of the milk proteins to the ceramics. After numerous examinations of conventional extraction techniques it was decided that if the protein would not come off the pot, then the pot would have to come off the protein.

Dissolving the ceramic with hydrofluoric acid and the simultaneous capture and protection of the released proteins onto a plastic finally resulted in us being able to see, by immunology, protein which we knew was still there (Craig and Collins, 2000). The method saw rapid application to resolve an argument surrounding the zooarchaeological data at a site in the Outer Hebrides (Cladh Hallan). Here a large number of juvenile cattle bones either suggested a cull of young males to focus effort on a dairy herd, or slaughter of animals in the autumn due to the difficulty of overwintering in a marginal environment. The analysis revealed a high percentage of the pottery contained cows' milk, supporting the view that dairying was indeed practised, although the extent could not be established from the small number of ceramics used (Craig *et al.*, 2000). This method has now been coupled with an elegant and more robust lipid stable isotope approach developed by Dudd and Evershed (1998). Unlike the lipid-based method, which has been used to survey large numbers of sherds (Copley *et al.*, 2005a, b, c), the immunological method is slow and the proteins appear to survive less well. However the two methods have been combined to investigate early evidence of dairying in Copper Age Europe (Craig *et al.*, 2003). The combination of independent methods is an exciting approach which offers advantages over single methods used in isolation.

Protein sequences

The use of antibodies, whilst precise and sensitive, does have one overriding problem: non-specific reactions, which generate false-positive results. These may be due to the very low levels of epitopes surviving, which mean that small increases in reaction, caused for example by humic substances (a catch-all term for the unidentified products of protein degradation), may yield false positives. Another idea is that decomposition and condensation of organic matter at mineral surfaces may lead to a degree of self-organisation to produce decomposition products with charge distributions similar to those of some of the original macromolecules involved in biomineralisation (Collins *et al.*, 1992). Immunology is therefore a useful screening tool, but is not unequivocal evidence of the presence of a protein; protein sequence is the gold standard for authentication of ancient proteins. Although conventional sequencing using Edman degradation proved difficult, partial sequences were reported in well-preserved *Moa* bone (Huq *et al.*, 1990) and foraminifera (Robbins and Brew, 1990), but the methods required relatively large samples of purified target molecules, which proves difficult in the case of most archaeological remains.

Improvements in soft-ionisation and mass-resolution of mass spectrometers have opened up the possibility of resolving large fragile molecules such as proteins, as considered by Robbins *et al.* (1993). The first successful application was the identification of the MH^+ ion of intact osteocalcin in old bones (Ostrom *et al.*, 2000). Peter Hauschka (1980) argued that osteocalcin would be a useful protein for archaeometry. Immunological and amino-acid data (osteocalcin contains an unusual amino acid, γ-carboxy glutamic acid or Gla) suggested that osteocalcin did persist remarkably well in fossils (Ulrich *et al.*, 1987; Muyzer *et al.*, 1992), a finding supported by laboratory investigation (Collins *et al.*, 2000). However, attempts to use it to date low-collagen bones have not proved as useful as hoped (Ajie *et al.*, 1992; Burky *et al.*, 1998).

The small size and high concentration of this protein in bone has meant that it has been used to pioneer mass-spectrometric techniques in fossils. Osteocalcin has been identified and partially sequenced from bison bones from the permafrost (Ostrom *et al.*, 2000; Nielsen-Marsh *et al.*, 2002), and from a Neanderthal (Nielsen-Marsh *et al.*, 2005). Intriguingly in the latter case the mid region of the molecule (regions which are mineral-associated and possess secondary structure) survived, whilst the unstructured amino-terminus did not survive. A complete osteocalcin sequence with an intact di-sulphide bridge was obtained from a 42 ka horse from a temperate cave (Ostrom *et al.*, 2006), correcting an erroneous published sequence obtained from modern material.

Mass spectrometry is well suited to repetitive analysis, and it may be possible to use osteocalcin to discriminate between genera. Unlike DNA-based methods discrimination is very crude; for example it is not possible to discriminate between aurochs and modern *Bos*, but it is possible to differentiate between the main domesticates, sheep, goat, pig, cow, etc. The methods require very small sample size, which, allied to automated methods of sample extraction, could give us an alternative approach to zooarchaeology; Zooarchaeology by Mass Spectrometry or ZooMS for short. With ZooMS it would be possible to identify the genus of processed bone fragments, such as those of the mid-shaft femur – important in improving models of resource exploitation (Marean *et al.*, 2000). In addition, because protein analysis, unlike DNA-based methods, measures what is there, rather than what can be amplified, a bone-rich sample of an occupation layer could be sub-sampled to reveal the percentage of bone in a layer and the distribution (by mass of contributing genera).

Shotgun or sniper?

Much molecular archaeological work has been that of the sniper, targeting specific compounds with high resolution. However such targeted strategies, using monoclonal antibodies, targeted DNA primers, or purified bone proteins, have limited application. In most cultural studies, be this occupation floors, residues in pots or the state of health of an individual, a whole-system approach has much greater potential. Schmidt-Schultz and Schultz (2004) observe, by 2D electrophoresis, a much wider range of proteins surviving in bone than osteocalcin and collagen alone. In soils and dissolved organic matter, complex patterns of protein presence have been documented (Schulze *et al.*, 2005). The latter investigation is interesting because it uses a 'shotgun approach', attempting to identify all the proteins in a sample. A similar approach has also been used for DNA, to identify unseen plants and animals in sediments (Hofreiter *et al.*, 2003; Willerslev *et al.*, 2003) and to amplify as much DNA as possible from fossil bone (Noonan *et al.*, 2005). These shotgun approaches, which have arisen as powerful analytical technologies and are allied with computational methods able to sift large datasets to identify patterns of sequence, will soon be attempted upon archaeological materials.

What remains to be detected?

The title is not ours; it was given to us by Holger Schutkowski as a challenge to speak to, for the opening of the Biological Anthropology Research Centre at Bradford. An answer to the question '*What remains?*' would seem to be '*More than we anticipated.*' when processes such as mineral sorption slow the decay of macromolecules. However, in relatively constant environments, such as the intra-crystalline proteins in carbonate fossils, decay is so consistent that with appropriate sampling, by 500 000 years in Britain all the shells are equally deteriorated, with more than 60 % of the amino acids in their free form and their D:L ratios all remarkably consistent. Conversely in samples which are thermally almost equivalent from Shanidar, a few sequences of polypeptides are preserved. The difference between the two cases would seem to be binding to mineral, which helps reduce the rate of hydrolysis and deterioration of residues in relaxed states on the surface. Bones themselves will decay, but in this case it is the early postmortem changes which in large part seem to govern survival. '*What remains to be detected?*' is an altogether different question. The ability of technology to now obtain direct sequence on ancient proteins offers breathtaking opportunities for archaeology, in terms of enhanced residue analysis and investigation of disease, phylogeny and identification in bones and residues.

Biomolecular archaeology can therefore offer new insights into past artefacts, just as the artefacts themselves can help to shed light upon culture. Yet as studies from Terra del Fuego (Terradas *et al.*, 1999) remind us, even in hunter/gatherer groups the complexity of material culture is bewildering. Molecules can offer new insights into biology and cultural remains; it is difficult to span the divide between artefact and culture, and yet the cultures which may help us are disappearing much faster than ancient biomolecules.

Acknowledgements

The preceding article contains dangerously high levels of speculation. Any innovative ideas have been developed in concert with the following: Geoff Bailey, Ian Barnes, Don Bothwell, Joachim Burger, Andrew Chamberlain, Alan Cooper, Oliver Craig, Adri van Duin, Karen Hardy, Robert Hedges, Jen Hiller, Miranda Jans, Henk Kars, Nicky Milner, Jacqui Mulville, Gerard Muyzer, Terry O'Connor, Mike Parker-Pearson, Alastair Pike, Colin Smith, Gordon Turner-Walker, Noreen Tuross, Steve Weiner, Tim Wess, Peter Westbroek and Mark White; needless to say, any errors are ours.

References

Abelson, P. H. (1954). Organic constituents of fossils. *Carnegie Institute of Washington YearBook*, **53**, 97–101.

Ajie, H. O., Kaplan, I. R., Hauschka, P. V. *et al.* (1992). Radiocarbon dating of bone osteocalcin – isolating and characterizing a noncollagen protein. *Radiocarbon*, **34**(3), 296–305.

Anderung, C., Bouwman, A., Persson, P. *et al.* (2005). Prehistoric contacts over the Straits of Gibraltar indicated by genetic analysis of Iberian Bronze Age cattle. *Proceedings of the National Academy of Sciences*, **102**(24), 8431–5.

Armstrong, W. G., Halstead, L. B., Reed, F. B. and Wood, L. (1983). Fossil proteins in vertebrate calcified tissues. *Philosophical Transactions of the Royal Society of London Series B Biological Sciences*, **301**(1106), 301–43.

Ashton, N., McNabb, J., Irving, B., Lewis, S. and Parfitt, S. (1994). Contemporaneity of Clactonian and Acheulean flint industries at Barnham, Suffolk. *Antiquity*, **68**(260), 585–9.

Barnes, I., Matheus, P., Shapiro, B., Jensen, D. and Cooper, A. (2002). Dynamics of Pleistocene population extinctions in Beringian brown bears. *Science*, **295**(5563), 2267–70.

Bell, L. S., Skinner, M. F. and Jones, S. J. (1996). The speed of postmortem change to the human skeleton and its taphonomic significance. *Forensic Science International*, **82**(2), 129–40.

Berman, A., Addadi, L. and Weiner, S. (1988). Interactions of sea-urchin skeleton macromolecules with growing calcite crystals: a study of intracrystalline proteins. *Nature*, **331**, 546–8.

Bowen, D. Q. and Sykes, G. A. (1994). How old is Boxgrove man? *Nature*, **371**(6500), 751.

Bowen, D. Q., Hughes, S., Sykes, G. A. and Miller, G. H. (1989). Land-sea correlations in the Pleistocene based on isoleucine epimerization in non-marine mollusks. *Nature*, **340**, 49–51.

Bridgland, D. R. (2000). River terrace systems in north-west Europe: an archive of environmental change, uplift and early human occupation. *Quaternary Science Reviews*, **19**(13), 1293–1303.

Brooks, A. S., Hare, P. E., Kokis, J. E. *et al.* (1990). Dating Pleistocene archaeological sites by protein diagenesis in ostrich eggshell. *Science*, **248**(4951), 60–4.

Buckley, S. A., Clark, K. A. and Evershed, R. P. (2004). Complex organic chemical balms of Pharaonic animal mummies. *Nature*, **431**(7006), 294–9.

Burger, J., Hummel, S. and Herrmann, B. (2000). Palaeogenetics and cultural heritage. Species determination and STR-genotyping from ancient DNA in art and artefacts. *Thermochimica Acta*, **365**(1–2), 141–6.

Burger, J., Schoon, R., Zeike, B., Hummel, S. and Herrmann, B. (2001). Species determination using species-discriminating PCR-RFLP of ancient DNA from prehistoric skeletal remains. *Ancient Biomolecules*, **4**(1), 19–23.

Burger, J., Rosendahl, W., Loreille, O. *et al.* (2004). Molecular phylogeny of the extinct cave lion Panthera leo spelaea. *Molecular Phylogenetics and Evolution*, **30**(3), 841–9.

Burky, R. R., Kirner, D. L., Taylor, R. E., Hare, P. E. and Southon, J. R. (1998). C-14 dating of bone using gamma-carboxyglutamic acid and alpha-carboxyglycine (aminomalonate). *Radiocarbon*, **40**(1), 11–20.

Cano, R. J., Tiefenbrunner, F., Ubaldi, M. *et al.* (2000). Sequence analysis of bacterial DNA in the colon and stomach of the Tyrolean Iceman. *American Journal of Physical Anthropology*, **112**(3), 297–309.

Cattaneo, C., Gelsthorpe, K., Phillips, P. and Sokol, R. J. (1990). Blood in acient human bone. *Nature*, **347**, 339.

Cattaneo, C., Gelsthorpe, K., Phillips, P. and Sokol, R. J. (1993). Blood residues on stone tools – indoor and outdoor experiments. *World Archaeology*, **25**(1), 29–43.

Collins, M. J. and Riley, M. (2000). Amino acid racemization in biominerals, the impact of protein degradation and loss. In *Perspectives in Amino Acid and Protein Geochemistry*, eds. G. A. Goodfriend, M. J. Collins, M. L. Fogel, S. A. Macko and J. F. Wehmiller. New York: Oxford University Press, pp. 120–41.

Collins, M. J., Westbroek, P., Muyzer, G. and deLeeuw, J. W. (1992). Experimental evidence for condensation reactions between sugars and proteins in carbonate skeletons. *Geochimica et Cosmochimica Acta*, **56**, 1539–44.

Collins, M. J., Waite, E. R. and van Duin, A. C. T. (1999). Predicting protein decomposition: the case of aspartic-acid racemization kinetics. *Philosophical Transactions of the Royal Society of London Series B-Biological Sciences*, **354**(1379), 51–64.

Collins, M. J., Gernaey, A. M., Nielsen-Marsh, C. M., Vermeer, C. and Westbroek, P. (2000). Slow rates of degradation of osteocalcin: Green light for fossil bone protein? *Geology*, **28**(12), 1139–42.

Collins, M. J., Nielsen-Marsh, C. M., Hiller, J. *et al.* (2002). The survival of organic matter in bone: A review. *Archaeometry*, **44**, 383–94.

Copley, M. S., Berstan, R., Dudd, S. N. *et al.* (2005a). Dairying in antiquity. I. Evidence from absorbed lipid residues dating to the British Iron Age. *Journal of Archaeological Science* **32**(4), 485–503.

Copley, M. S., Berstan, R., Mukherjee, A. J. *et al.* (2005b). Dairying in antiquity. III. Evidence from absorbed lipid residues dating to the British Neolithic. *Journal of Archaeological Science*, **32**(4), 523–46.

Copley, M. S., Berstan, R., Straker, V., Payne, S. and Evershed, R. P. (2005c). Dairying in antiquity. II. Evidence from absorbed lipid residues dating to the British Bronze Age. *Journal of Archaeological Science*, **32**(4), 505–21.

Copley, M. S., Bland, H. A., Rose, P., Horton, M. and Evershed, R. P. (2005d). Gas chromatographic, mass spectrometric and stable carbon isotopic investigations of organic residues of plant oils and animal fats employed as illuminants in archaeological lamps from Egypt. *Analyst*, **130**(6), 860–71.

Craig, O. E. and Collins, M. J. (2000). An improved method for the immunological detection of mineral bound protein using hydrofluoric acid and direct capture. *Journal of Immunological Methods*, **236**(1–2), 89–97.

Craig, O. E. and Collins, M. J. (2002). The removal of protein from mineral surfaces: Implications for residue analysis of archaeological materials. *Journal of Archaeological Science*, **29**(10), 1077–82.

Craig, O., Mulville, J., Parker Pearson, M. *et al.* (2000). Archaeology – detecting milk proteins in ancient pots. *Nature*, **408**(6810), 312.

Craig, O. E., Chapman, J., Figler A. *et al.* (2003). 'Milk jugs' and other myths of the Copper age of central Europe. *European Journal of Archaeology*, **6**, 251–65.

Cressey, B. A. and Cressey, G. (2003). A model for the composite nanostructure of bone suggested by high-resolution transmission electron microscopy. *Mineralogical Magazine*, **67**(6), 1171–82.

Custer, J. F., Ilgenfritz, J. and Doms, K. R. A. (1988). Cautionary note on the use of Chemstrips for detection of blood residues on prehistoric stone tools. *Journal of Archaeological Science*, **15**, 343–5.

DeNiro, M. J. and Weiner, S. (1988). Organic matter within crystalline aggregates of hydroxyapatite: A new substrate for stable isotopic and possibly other biogeochemical analyses of bone. *Geochimica et Cosmochimica Acta*, **52**, 2415–23.

Dudd, S. N. and Evershed, R. P. (1998). Direct demonstration of milk as an element of archaeological economies. *Science*, **282**(5393), 1478–81.

Evershed, R. P., Vaughan, S. J., Dudd, S. N. and Soles, J. S. (1997). Fuel for thought? Beeswax in lamps and conical cups from late Minoan Crete. *Antiquity*, **71**(274), 979–85.

Gosden, C. (2005). Comments III: Is science a foreign country? *Archaeometry*, **47**(1), 182–5.

Gurfinkel, D. and Franklin, U. M. (1988). A study of the feasibility of detecting blood residue on artifacts. *Journal of Archaeological Science*, **15**, 83–97.

Hare, P. E. and Abelson, P. H. (1968). Racemization of amino acids in fossil shells. *Carnegie Institute of Washington Yearbook*, **66**, 526–8.

Hauschka, P. V. (1980). Osteocalcin: a specific protein of bone with potential for fossil dating. In *Biogeochemistry of Amino Acids*, eds. P. E. Hare, T. C. Hoering and K. King. New York: Wiley, pp. 75–82.

Hess, M. W., Klima, G., Pfaller, K., Kunzel, K. H. and Gaber, O. (1998). Histological investigations on the Tyrolean Ice Man. *American Journal of Physical Anthropology*, **106**(4), 521–32.

Hiller, J. C., Collins, M. J., Chamberlain, A. T. and Wess, T. J. (2004). Small-angle X-ray scattering: a high-throughput technique for investigating archaeological bone preservation. *Journal of Archaeological Science*, **31**(10), 1349–59.

Hoang, Q. Q., Sicheri, F., Howard, A. J. and Yang, D. S. C. (2003). Bone recognition mechanism of porcine osteocalcin from crystal structure. *Nature*, **425**(6961), 977–80.

Hofreiter, M., Mead, J. I., Martin, P. and Poinar, H. N. (2003). Molecular caving. *Current Biology*, **13**(18), R693–R695.

Holmes, K. M., Robson Brown, K. A., Oates, W. P. and Collins, M. J. (2005). Assessing the distribution of African Palaeolithic sites: a predictive model of collagen degradation. *Journal of Archaeological Science*, **32**(2), 157–66.

Huq, N. L., Tseng, A. and Chapman, G. E. (1990). Partial amino acid sequence of osteocalcin from an extinct species of ratite bird. *Biochemistry International*, **21**, 491–6.

James, R. (2004). Introduction. Halal pizza: food and culture in a busy world. *The Australian Journal of Anthropology*, **15**(1), 1–11.

Jans, M. M. E., Nielsen-Marsh, C. M., Smith, C. I., Collins, M. J. and Kars, H. (2004). Characterisation of microbial attack on archaeological bone. *Journal of Archaeological Science*, **31**(1), 87–95.

Jones, A. (2004). Archaeometry and materiality: materials-based analysis in theory and practice. *Archaeometry*, **46**(3), 327–38.

Kaufman, D. S. and Manley, W. F. (1998). A new procedure for determining DL amino acid ratios in fossils using reverse phase liquid chromatography. *Quaternary Science Reviews*, **17**(11), 987–1000.

King, W. B. R. and Oakley, K. P. (1936). The Pleistocene succession in the lower part of the Thames Valley. *Proceedings of the Prehistoric Society*, **1**, 52–76.

Koon, H. E. C., Nicholson, R. A. and Collins, M. J. (2003). A practical approach to the identification of low temperature heated bone using TEM. *Journal of Archaeological Science*, **30**(11), 1393–9.

Krings, M., Stone, A., Schmitz, R. W. *et al.* (1997). Neanderthal DNA sequences and the origin of modern humans. *Cell*, **90**, 19–30.

Kronick, P. L. and Cooke, P. (1996). Thermal stablization of collagen fibres by calcification. *Connective Tissue Research*, **33**, 275–82.

Lalueza-Fox, C., Shapiro, B., Bover, P., Alcover, J. A. and Bertranpetit, J. (2002). Molecular phylogeny and evolution of the extinct bovid Myotragus balearicus. *Molecular Phylogenetics and Evolution*, **25**(3), 501–10.

Lowenstein, J. M. (1981). Immunological reactions from fossil material. *Philosophical Transactions of the Royal Society of London*, series B, **292**, 143–9.

Loy, T. H. (1983). Prehistoric blood residues: detection on tool surfaces and identification of species origin. *Science*, **220**, 1269–71.

Makristathis, A., Schwarzmeier, J., Mader, R. M. *et al.* (2002). Fatty acid composition and preservation of the Tyrolean Iceman and other mummies. *Journal of Lipid Research*, **43**(12), 2056–61.

Marean, C. W., Abe, Y., Frey, C. J. and Randall, R. C. (2000). Zooarchaeological and taphonomic analysis of the Die Kelders Cave 1 Layers 10 and 11 Middle Stone Age larger mammal fauna. *Journal of Human Evolution*, **38**(1), 197–233.

Marshall, E. (1990). Racemization dating: great expectations. *Science*, **247**, 799.

McNabb, J. (1996). More from the cutting edge: further discoveries of Clactonian bifaces. *Antiquity*, **70**, 428–36.

McNabb, J., Binyon, F. and Hazelwood, L. (2004). The large cutting tools from the South African Acheulean and the question of social traditions. *Current Anthropology*, **45**(5), 653–77.

McNulty, T., Calkins, A., Ostrom, P. *et al.* (2002). Stable isotope values of bone organic matter: artificial diagenesis experiments and paleoecology of Natural Trap Cave, Wyoming. *Palaios*, **17**, 36–49.

Meeson, B. W., Corprew, F. E., McManus, J. M. P. *et al.* (1995). *ISLSCP Initiative I: Global Data Sets for Land-Atmosphere Models, 1987–1988. Washington: National Aeronautics and Space Administration.* Published on (USA_NASA_GDAAC_ISLSCP_001-USA_NASA_GDDAC_ISLSCP_005).

Miller, G. H., Beaumont, P. B., Jull, A. J. T. and Johnson, B. J. (1992). Pleistocene geochronology and palaeothermometry from protein diagenesis in ostrich eggshells: implications for the evolution of modern humans. *Philosophical*

Transactions of the Royal Society of London Series B Biological Sciences, **337**, 149–57.

Miller, G. H., Hart, C. P., Roark, E. B. and Johnson, B. J. (2000). Isoleucine epimerization in eggshells of the flightless Australian birds Genyornis and Dromaius. In *Perspectives in Amino Acid and Protein Geochemistry*, eds. G. A. Goodfriend, M. J. Collins, M. L. Fogel, S. A. Macko and J. F. Wehmiller. New York: Oxford University Press, pp. 161–81.

Muller, W., Fricke, H., Halliday, A. N., McCulloch, M. T. and Wartho, J. A. (2003). Origin and migration of the Alpine Iceman. *Science*, **302**(5646), 862–6.

Muyzer, G., Sandberg, P., Knapen, M. H. J. *et al.* (1992). Preservation of the bone protein osteocalcin in dinosaurs. *Geology*, **20**, 871–4.

Ness, R. L. L. and Vlek, P. L. G. (2000). Mechanism of calcium and phosphate release from hydroxy-apatite by mycorrhizal hyphae. *Soil Science Society of America Journal*, **64**(3), 949–55.

Nicholson, R. A. (1993). A morphological investigation of burnt animal bone and an evaluation of its utility in archaeology. *Journal of Archaeological Science*, **20**(4), 411–28.

Nielsen-Marsh, C. M., Ostrom, P. H., Gandi, H. *et al.* (2002). Exceptional preservation of bison bones older than 55 ka as demonstrated by protein and DNA sequences. *Geology*, **30**(12), 1099–1102.

Nielsen-Marsh, C. M., Richards, M. P., Hauschka, P. V. *et al.* (2005). Osteocalcin protein sequences of Neanderthals and modern primates. *Proceedings of the National Academy of Sciences*, **102**, 4409–13. Published online, pnas. 0500450102.

Nissenbaum, A. (1997). 8000 years collagen from Nahal Hemar cave. (In Hebrew.) *Archaeology and Natural Sciences*, **5**, 5–9.

Noonan, J. P., Hofreiter, M., Smith, D. *et al.* (2005). Genomic sequencing of Pleistocene cave bears. *Science*, **309**(5734), 597–9.

Nord, A. G., Kars, H., Ullén, I., Tronner, K. and Kars, E. (2005). Deterioration of archaeological bone – a statistical approach. *Journal of Nordic Archaeological Science*, **15**, 77–86.

Ohel, M. Y. (1979). Clactonian – Independent complex or an integral-part of the Acheulean. *Current Anthropology*, **20**(4), 685–726.

Ohtani, S. and Yamamoto, K. (1991). Age estimation using the racemization of amino-acid in human dentin. *Journal of Forensic Sciences*, **36**(3), 792–800.

Ostrom, P. H., Schall, M., Gandhi, H. *et al.* (2000). New strategies for characterizing ancient proteins using matrix-assisted laser desorption ionization mass spectrometry. *Geochimica et Cosmochimica Acta*, **64**(6), 1043–50.

Ostrom, P. H., Gandhi, H., Strahler, J. R. *et al.* (2006). Unraveling the sequence and structure of the protein osteocalcin from a 42 ka fossil horse. *Geochimica et Cosmochimica Acta*, **70**(8), 2034–44.

Parfitt, S. A., Barendregt, R. W., Breda, M. *et al.* (2005). The earliest humans in Northern Europe: artefacts from the Cromer Forest-bed Formation at Pakefield, Suffolk, UK. *Nature*, **438**, 1008–12.

Parker Pearson, M., Chamberlain, A., Craig, O. *et al.* (2005). Evidence for mummification in Bronze Age Britain. *Antiquity*, **79**(305), 529–46.

Poinar, H., Kuch, M., McDonald, G., Martin, P. and Paabo, S. (2003). Nuclear gene sequences from a late Pleistocene sloth coprolite. *Current Biology*, **13**(13), 1150–2.

Price, T. D., Gebauer, A. B., Hede, S. U. *et al.* (2001). Smakkerup Huse: A Mesolithic settlement in NW Zealand, Denmark. *Journal of Field Archaeology*, **28**(1–2), 45–67.

Richards, M. P., Schulting, R. J. and Hedges, R. E. M. (2003). Sharp shift in diet at onset of Neolithic. *Nature*, **425**(6956), 366.

Richter, J. (1986). Experimental study of heat induced morphological changes in fish bone collagen. *Journal of Archaeological Science*, **13**, 477–81.

Ritz, S., Schutz, H. W. and Schwarzer, B. (1990). The extent of aspartic-acid racemization in dentin – a possible method for a more accurate determination of age at death. *Zeitschrift für Rechtsmedizin–Journal of Legal Medicine*, **103**(6), 457–62.

Robbins, L. L. and Brew, K. (1990). Proteins from the organic matrix of core-top and fossil planktonic foraminifera. *Geochimica et Cosmochimica Acta*, **54**, 2285–92.

Robbins, L. L., Muyzer, G. and Brew, K. (1993). Macromolecules from living and fossil biominerals; Implications for the establishment of molecuar phylogenics. In *Organic Geochemistry*, eds. M. H. Engle and S. A. Macko. New York: Plenum, pp. 799–816.

Roberts, M. B. (1994). How old is 'Boxgrove man'? *Nature*, **371**(6500), 751.

Roberts, S. J., Smith, C. I., Millard, A. R. and Collins, M. J. (2002). The taphonomy of cooked bone: characterising boiling and its physico-chemical effects. *Archaeometry*, **44**(4), 485–94.

Rollo, F. and Marota, I. (1999). How microbial ancient DNA, found in association with human remains, can be interpreted. *Philosophical Transactions of the Royal Society of London Series B Biological Sciences*, **354**(1379), 111–19.

Rollo, F., Luciani, S., Canapa, A. and Marota, I. (2000). Analysis of bacterial DNA in skin and muscle of the Tyrolean Iceman offers new insight into the mummification process. *American Journal of Physical Anthropology*, **111**(2), 211–19.

Salamon, M., Tuross, N., Arensburg, B. and Weiner, S. (2005). Relatively well preserved DNA is present in the crystal aggregates of fossil bones. *Proceedings of the National Academy of Sciences*, **102**(39), 13 783–8. Published online, pnas.0503718102.

Schäfer, D. (1999). *Geburt aus dem Tod: Der Kaiserschnitt an Verstorbenen in der abendländischen Kultur.* Hürtgenwald: Guido Pressle.

Schmidt-Schultz, T. H. and Schultz, M. (2004). Bone protects proteins over thousands of years: Extraction, analysis, and interpretation of extracellular matrix proteins in archeological skeletal remains. *American Journal of Physical Anthropology*, **123**(1), 30–9.

Schulting, R. J. and Richards, M. P. (2002). The wet, the wild and the domesticated: the Mesolithic-Neolithic transition on the west coast of Scotland. *European Journal of Archaeology*, **5**(2), 147–89.

Schulze, W. X., Gleixner, G., Kaiser, K. *et al.* (2005). A proteomic fingerprint of dissolved organic carbon and of soil particles. *Oecologia*, **142**(3), 335–43.

Shapiro, B., Drummond, A. J., Rambaut, A. *et al.* (2004). Rise and fall of the Beringian steppe bison. *Science*, **306**(5701), 1561–5.

Shipman, P., Foster, G. and Schoeninger, M. (1984). Burnt bones and teeth: an experimental study of color, morphology, crystal structure and shrinkage. *Journal of Archaeological Science*, **11**, 307–25.

Smith, C. I., Chamberlain, A. T., Riley, M. S. *et al.* (2001). Ancient DNA. Not just old, but old and cold. *Nature*, **410**, 771–2.

Smith, C. I., Chamberlain, A. T., Riley, M. S., Stringer, C. and Collins, M. J. (2003). The thermal history of human fossils and the likelihood of successful DNA amplification. *Journal of Human Evolution*, **45**(3), 203–17.

Smith, C. I., Craig, O. E., Prigodich, R. V. *et al.* (2005). Diagenesis and survival of osteocalcin in archaeological bone. *Journal of Archaeological Science*, **32**(1), 105–13.

Smith, P. R. and Wilson, M. T. (1992). Blood residues on ancient tool surfaces: a cautionary note. *Journal of Archaeological Science*, **19**, 237–41.

Stiner, M. C., Kuhn, S. L., Surovell, T. A. *et al.* (2001). Bone preservation in Hayonim Cave (Israel): A macroscopic and mineralogical study. *Journal of Archaeological Science*, **28**(6), 643–59.

Stringer, C. B. and Hublin, J.-J. (1999). New age estimates for the Swanscombe hominid, and their significance for human evolution. *Journal of Human Evolution*, **37**(6), 873–7.

Swinbank, V. A. (2002). The sexual politics of cooking: A feminist analysis of culinary hierarchy in western culture. *Journal of Historical Sociology*, **15**(4), 464–94.

Sykes, G. A., Collins, M. J. and Walton, D. I. (1995). The significance of a geochemically isolated (intra-crystalline) organic fraction within biominerals. *Organic Geochemistry*, **23**, 1059–66.

Terradas, X., Clemente, I., Vila, A. Y. and Mansur, M. E. (1999). Ethno-neglect or the contradiction between ethnohistorical sources and the archaeological record. The case of stone tools of the Yamana People (Tierra del Fuego, Argentina). In *Ethno-Analogy and the Reconstruction of Prehistoric Artefact Use and Production*, eds. L. Owen and M. Porr. Tübingen: University of Tübingen, pp. 103–15.

Towe, K. M. (1980). Preserved organic ultrastructure: An unreliable indicator for Paleozoic amino acid biogeochemistry. In *Biogeochemistry of Amino Acids*, eds. P. E. Hare, T. C. Hoering and K. King. New York: Wiley, pp. 65–74.

Trueman, C. N. and Martill, D. M. (2002). The long-term survival of bone: the role of bioerosion. *Archaeometry*, **44**(3), 371–82.

Trueman, C. N. G., Behrensmeyer, A. K., Tuross, N. and Weiner, S. (2004). Mineralogical and compositional changes in bones exposed on soil surfaces in Amboseli National Park, Kenya: diagenetic mechanisms and the role of sediment pore fluids. *Journal of Archaeological Science*, **31**(6), 721–39.

Turner-Walker, G., Nielsen-Marsh, C. M., Syversen, U., Kars, H. and Collins, M. J. (2002). Sub-micron spongiform porosity is the major ultra-structural alteration occurring in archaeological bone. *International Journal of Osteoarchaeology*, **12**, 407–14.

Tuross, N., Barnes, I. and Potts, R. (1996). Protein identification of blood residues on experimental stone tools. *Journal of Archaeological Science*, **23**(2), 289–96.

Ulrich, M. M., Perizonius, W. R. K., Spoor, C. F., Sandberg, P. and Vermeer, C. (1987). Extraction of osteocalcin from fossil bones and teeth. *Biochemical and Biophysical Research Communications*, **149**, 712–19.

van Duin, A. C. T. and Collins, M. J. (1998). The effects of conformational constraints on aspartic acid racemization. *Organic Geochemistry*, **29**(5–7), 1227–32.

Waite, E. R. and Collins, M. J. (2000). The interpretation of aspartic acid racemization data. In *Perspectives in Amino Acid and Protein Geochemistry*, eds. G. A. Goodfriend, M. J. Collins, M. L. Fogel, S. A. Macko and J. F. Wehmiller. New York: Oxford University Press, pp. xviff.

Weiner, S. and Traub, W. (1986). Organization of hydroxyapatite crystals within collagen fibrils. *Febs Letters*, **206**(2), 262–6.

Wess, T., Alberts, I., Hiller, J. *et al.* (2002). Microfocus small angle X-ray scattering reveals structural features in archaeological bone samples: Detection of changes in bone mineral habit and size. *Calcified Tissue International*, **70**(2), 103–10.

White, M. J. (2000). The Clactonian question: on the interpretation of core and flake assemblages in the British Lower Paleolithic. *Journal of World Prehistory*, **14**(1), 1–63.

Willerslev, E., Hansen, A. J., Binladen, J. *et al.* (2003). Diverse plant and animal genetic records from Holocene and Pleistocene sediments. *Science*, **300**(5620), 791–5.

9 *Enamel traces of early lifetime events*

LOUISE T. HUMPHREY

Introduction

Nursing practices can have a significant impact on the health and survival of children in a population. The duration of breastfeeding and timing of the weaning process are influenced by cultural norms and access to resources, and vary within and between populations. Reconstructing these processes has become an important focus in studies of the health and population dynamics of past populations. Several indirect types of evidence have been used for this purpose, including skeletal growth curves, the prevalence of enamel hypoplasia and other developmental stress markers, tooth-wear and infant mortality profiles, but each of these approaches can be problematic. More direct evidence of feeding practices in infancy and early childhood can be derived from analyses of the chemical composition of the skeleton and dentition (Katzenberg *et al.*, 1996; Herring *et al.*, 1998).

This review will evaluate the value of human tooth enamel as an archive of information on the weaning process. The dentition has several advantages over the skeleton in terms of its value as a retrospective record of environmental circumstances. Most importantly, since tooth enamel is not remodelled subsequent to its formation, it provides an opportunity to reconstruct specific aspects of the earliest events in a person's life history for as long as the tooth survives. This paper will focus on two different ways in which the tooth enamel has been used to reconstruct nursing behaviour. Linear enamel hypoplasias (LEH) represent a disruption in the normal process of enamel matrix secretion, which affect the contour of the enamel surface at a microscopic or macroscopic level. The age distribution of LEH within a sample can be determined with considerable precision and has been used as indirect evidence of weaning in past populations. Stable isotope and trace element analysis provide direct evidence of diet, but interpretation of this evidence within an appropriate chronological framework represents a greater challenge.

Between Biology and Culture, ed. Holger Schutkowski. Published by Cambridge University Press.

Analysis of each of these parameters requires the tooth to survive in an unmodified state. In practice various factors can contribute to loss of tooth enamel during life, including abrasion and attrition relating to mastication and other uses of the dentition, disease processes, and deliberate cultural modifications such as evulsion or mutilation (Hillson, 1996). The presence of heavy calculus deposits can also prevent observation of enamel surfaces, and would affect the recording of LEH. During life, the chemical composition of the outermost enamel is affected by exchanges within the intra-oral environment. After death, taphonomic factors can affect the preservation and composition of enamel, but generally speaking enamel is less affected than bone, making it an ideal medium for the retrospective interpretation of childhood health and diet.

Weaning

Weaning is a process that involves the gradual introduction of complementary foods and a reduced dependence on breast milk. The process is initiated by the introduction of complementary foods, and ends with the complete cessation of breastfeeding (Katzenberg *et al.*, 1996). The weaning process is determined by the needs and constraints of both mother and infant, and represents an important life-history transition. For the infant, the weaning period is associated with increases in a range of stress factors. Weaning involves the gradual withdrawal of a secure and balanced source of nutrition and loss of immunological support provided by breast milk. Dietary supplementation exposes the infant to new sources of infection associated with contaminated foods, and increases the risk of illness and malnutrition.

For the mother, the age at which complementary foods are introduced and rate at which breast milk is replaced are major determinants of fertility due to their effect on duration of lactational amenorrhea and their relationship to inter-birth interval. Prolonged weaning is costly to the mother, both in terms of her continued energetic investment in the existing infant and also the associated delay in subsequent births. The energetic requirements of a growing individual increase throughout pregnancy and continue to increase postnatally. During pregnancy and exclusive breastfeeding the energetic and nutritional requirements must be met entirely by the mother. As complementary foods are introduced, overall energetic requirements of the infant continue to grow, but the energetic burden on the mother will decrease as the amount of complementary foods increases, offsetting and then exceeding the additional requirements of the developing infant (Lee *et al.*, 1991). The direct maternal contribution to an infant's nutritional requirements therefore peaks at, or very shortly after, the age when complementary foods are introduced, declines during the period of

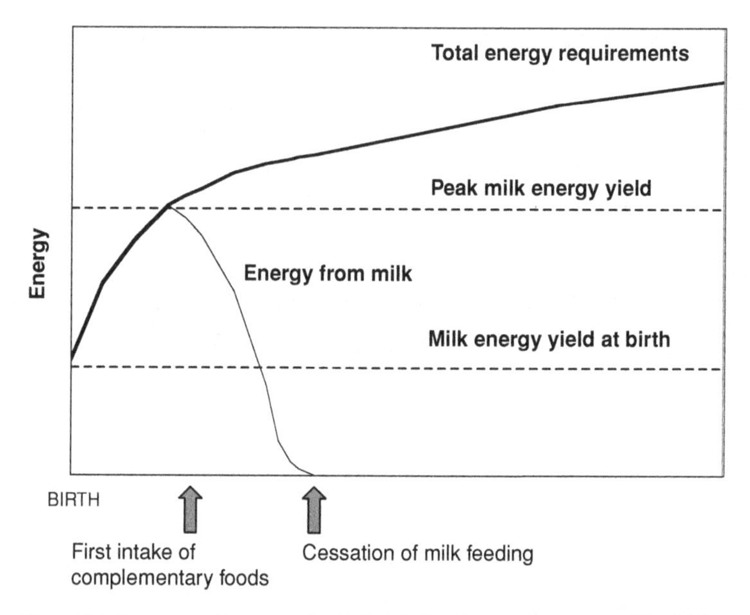

Figure 9.1 Energy requirements of an infant during the weaning process: The thick line illustrates the total energy requirements of the infant and the thin line represents the energy supplied by the mother as breast milk. Age increases along the x-axis. After Lee *et al.* (1991).

mixed feeding and ends at the cessation of breastfeeding (Figure 9.1). During this period the mother's health and nutritional status will have a direct affect on the health and nutritional status of the infant, but arguably these maternal influences will become less important during the weaning process.

The various potential stress factors specifically associated with weaning are introduced at different stages of the weaning process. The introduction of complementary foods marks the onset of the weaning process. In modern humans, the nutritional content of maternal milk is generally sufficient to meet the infants' requirements until about six months after birth (Kramer and Kakuma, 2002). This coincides with the age at which infants acquire peak levels of postnatal adiposity, which may act as a buffer against energy shortfalls during the critical weaning period (Kazawa, 1998). After this age a weaning food of sufficient energy density must be added to the infant's diet to avoid growth faltering. The average reported age for the introduction of complementary foods varies across recent and contemporary non-industrial societies, with a central tendency of between five and six months for the introduction of solid foods (Sellen, 2001). Numerous different factors are likely to affect the age at which different complementary foods are given to a particular infant. These include cultural norms and socio-economic factors, and may involve factors as specific as gender, number of siblings and position in the family, and individual preferences of the child (Katzenberg *et al.*, 1996). These factors will also influence the types of foods given and the rate at which solid foods are introduced.

The introduction of complementary foods is associated with several novel risks for the infant. There is an increased risk of infection associated with the possible use of contaminated foodstuffs, water or feeding implements and leading to diarrhoeal disease and malnutrition. Malnutrition can also result from the use of inappropriate weaning foods, such as those that cannot be easily digested or lack sufficient caloric density. In addition some foods constitute a choking hazard for infants. It is worth noting here that although the age of introduction of complementary foods represents the earliest possible date of exposure to these risks, the risks will continue for some months or even years beyond this transition.

During breastfeeding maternally derived antibodies bolster the infants' immune defences, providing a buffered period in which the infants are able to develop their own specific immune defences (McDade, 2003). Following the cessation of breastfeeding the infant has to rely entirely on its own immune defences. Infectious disease mortality in early childhood is approximately 1/10 of the level during infancy, but remains five times higher than the level in children aged between 5 and 15 years, as newly weaned children continue to encounter novel pathogens (McDade, 2003). During and after weaning, food security for the infant may be compromised. If the weaning process is prolonged, as is common in many human societies today (Sellen, 2001), this will not occur as an abrupt transition, but is nevertheless a vulnerable period. Most newly weaned children are still largely dependent on provisioning from older members of the community and this dependence may be associated with conscious or unconscious neglect (Rousham and Humphrey, 2002).

Not all growth disruptions occurring during infancy and early childhood can be attributed to weaning. Other aspects of the infant's lifestyle relating, for example, to increased infant mobility, social interactions and the reproductive status of the mother can introduce other risks resulting in growth disruptions. Infants regularly suck objects and put them in their mouths, and this may serve the function of calibrating the developing immune system, including antibody production and mucosal immunity, to the local disease ecology (Fessler and Abrams, 2004). The risks associated with this activity are similar to those associated with the introduction of complementary foods, and include choking, poisoning and exposure to pathogens. This behaviour is characteristic of the first two to three years of life but is most intensive in the first year. The risks associated with this behaviour are likely to develop initially in tandem with the infant's grasping abilities and become more frequent as increasing locomotor independence facilitates greater opportunities for independent exploration.

Timing and sequence of tooth crown formation

Human dental development follows a well-documented pattern. Tooth formation takes place as a series of time-successive events, starting with the initiation of crown formation in the earliest forming deciduous teeth and ending with the root apex closure in the latest forming permanent teeth. The earliest forming deciduous tooth crowns initiate enamel secretion around 14–16 weeks after fertilisation and the latest forming deciduous tooth crowns complete their enamel secretion approximately 11 months after birth (Hillson, 1996). Successively developing permanent tooth crowns provide an overlapping and continuous sequence of enamel formation from before birth until mid-childhood (Hillson, 1996). The earliest developing permanent teeth are the first permanent molars, which initiate formation about 30 weeks after fertilisation. Enamel secretion in the second molar continues until approximately 7–8 years after birth. Development of the third molar is highly variable and crown initiation may not take place until after the crowns of the other permanent teeth are complete (Hillson, 1996). As a result it cannot reliably be used to extend an otherwise unbroken sequence of enamel formation.

This sequential development means that different teeth record different parts of the prenatal period, infancy and childhood, and can be studied to address different kinds of research questions. Most studies of enamel defects focus on the permanent dentition since successive loss of the deciduous teeth between the ages of 5 and 11 years limits their value as retrospective stress indicators. The study of permanent tooth crowns provides a retrospective method of evaluating exposure to environmental stress factors during the period of tooth crown formation that can be studied in both older juveniles and adults. The deciduous teeth can be used to investigate the prenatal and early postnatal period and have an important role to play in studies of infant health and dietary history (Fuller *et al.*, 2003; FitzGerald *et al.*, 2006; Humphrey *et al.*, 2007).

Enamel formation

Enamel formation involves the initial production of an organic matrix (secretory stage) and subsequent transformation of the partially mineralised immature enamel into a highly mineralised and durable tissue (maturation stage). Secretion of the organic matrix starts at the tip of the dentine horn and spreads outwards towards the cuspal enamel surface and along the enamel dentine junction towards the cervix. Hydroxyapatite crystals are seeded into the newly produced matrix and elongate in the wake of retreating ameloblasts to produce thin enamel ribbons extending the full thickness of the enamel (Robinson *et al.*, 1997). On reaching the enamel surface, ameloblasts undergo a functional change and enter the maturation phase (Smith, 1998). Maturation involves substantial increase in the width and breadth of the existing apatite crystallites as they grow to fill spaces previously occupied by water and organic material. Complete maturation of the enamel may occur several months or even years after matrix secretion in human teeth.

Incremental markings in enamel are formed during the secretory stage of enamel formation and preserve an accurate and durable record of the enamel growth history (for detailed reviews see Dean, 2000; FitzGerald and Rose, 2000). The neonatal line is an accentuated line that reflects physiological disturbances associated with the birth process (Schour, 1936). The position of the neonatal line reveals the state of development of each tooth at the time of birth, and can be used as a chronological anchor for other developmental markers. In modern humans the neonatal line is present in all deciduous tooth crowns, and in some permanent first molar cusps. Brown striae of Retzius develop with a regular periodicity during normal enamel secretion. They arise from a regular systemic disturbance in enamel matrix formation that affects all ameloblasts involved in enamel secretion at that time. Striae of Retzius mark out successive positions of the forming front of the enamel matrix, thus providing an accurate retrospective record of the external profile of a tooth crown formation at a given point in time. Perikymata are the surface expression of the striae of Retzius, and as such they only appear in the imbricational zone of enamel (i.e. the enamel that is visible at the surface of the crown). Enamel cross striations are caused by cyclical variation in enamel matrix secretion rate, operating with a 24-hour rhythm (FitzGerald, 1998). The number of cross striations between striae is uniform for all teeth within a dentition, but varies between individuals, with an average interval of 7–10 days between striae for modern humans. Corresponding incremental markers occur in dentine but these may be more difficult to interpret (Hillson, 1996).

Enamel microstructures can be used to provide a precise and accurate chronology for developmental events relating to the secretory stage of enamel formation, including systemic growth disruptions. These markers can also be used to determine the age of onset of enamel mineralisation for a given point in the enamel, since this occurs extremely close to the secretory front (Boyde, 1989). During enamel maturation there is a substantial increase in mineral content (Smith, 1998). This results in a time-averaged signal for the chemical composition of the enamel mineral at any given sampling point and complicates the chronological interpretation of isotope and trace element data.

Definition and scoring of linear enamel hypoplasia

Enamel hypoplasias are developmental defects of the enamel surface caused by a temporary disruption in the normal process of enamel matrix secretion. Furrow form defects, commonly referred to as linear enamel hypoplasias, are the most common type of hypoplasia. They occur when a wider than normal band of ameloblasts ceases matrix production early causing a deficiency in enamel thickness (Hillson and Bond, 1997). Linear enamel hypoplasia (LEH) encompasses a continuum of enamel defects from those that would be invisible to the naked eye, to those that would be macroscopically visible. At a microscopic level, LEHs are recognised as a variation in the normal spacing of perikymata grooves (Hillson and Bond, 1997). Specifically an increased spacing is observed in the occlusal wall of an LEH due to the exposure of a wider than normal margin of each brown striae. The spacing of perikymata grooves returns to normal in the floor and cervical wall of the defect. The duration of the disturbance can be determined from the number of perikymata grooves in the occlusal wall, although further research is needed to determine how this relates to the duration and severity of the underlying cause (Hillson and Bond, 1997).

Traditional techniques for estimating the age of occurrence of macroscopic enamel defects involved allocating an LEH to a specific time interval based on the distance between the LEH and the cemento-enamel junction. Such techniques failed to take account of variation in crown height and developmental schedules between individuals and populations (Goodman and Rose, 1990; Reid and Dean, 2000). Hillson and Bond (1997) drew attention to two more fundamental problems. Firstly, traditional methods do not account for the amount of time required for formation of the appositional or cuspal enamel. The growth increments in the appositional enamel are hidden beneath the incisal edge of the tooth and can represent a significant amount of the total crown formation time. Hillson and Bond (1997) note that 15–20 % of growth increments in anterior teeth are appositional compared to 40–50 % of growth increments in the permanent molars. Developmental disruptions that occur during the formation of cuspal enamel cannot be observed in surface enamel. Secondly, in earlier studies the amount of time required for tooth crown formation was equally apportioned between equally spaced enamel segments, thus assuming a constant rate of growth for the entire tooth crown surface. In reality tooth crown extension rates decrease throughout the period of enamel growth (Liversidge *et al.*, 1993). This trend was clearly demonstrated by a recent histological analysis of the anterior teeth (Reid and Dean, 2000). In all teeth examined, the amount of time recorded by successive tenths of tooth crown height increases from the cusp to the cervix of the crown. The earliest forming enamel deciles typically record about 0.2 years of enamel formation whereas the latest forming

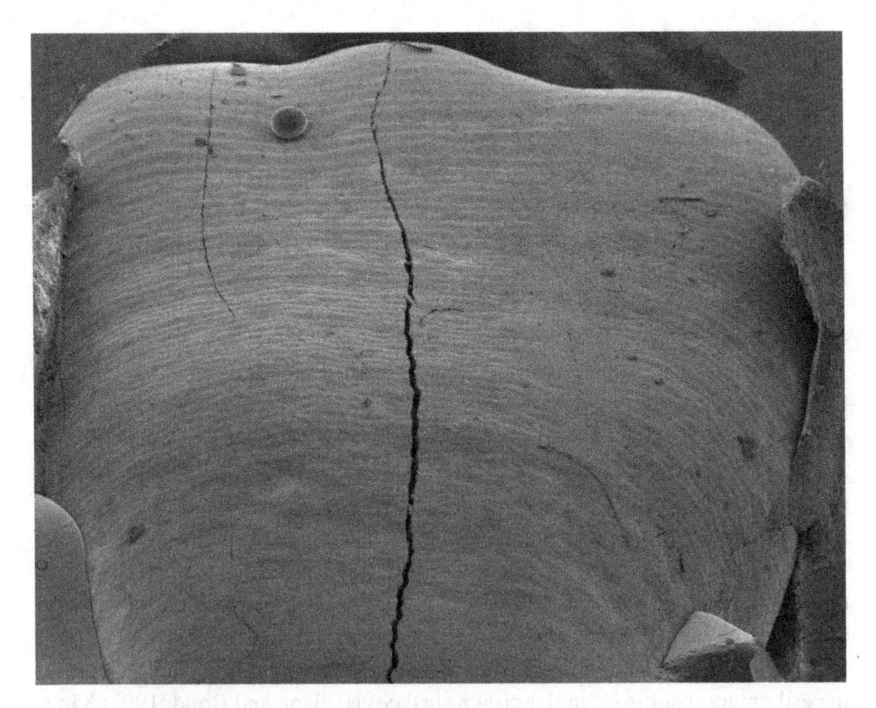

Figure 9.2 Scanning electron microscope image of the labial surface of a lower central incisor from a 12-year-old girl. In this relatively unworn tooth, perikymata can be counted and measured for the entire length of the tooth crown. Image courtesy of Tania King.

deciles in the cervical region of the tooth crown record enamel forming over a period of 0.4 years or longer.

Recent studies of LEH have identified and evaluated defects on the basis of perikymata spacing (King *et al.*, 2002; 2005; Guatelli-Steinberg *et al.*, 2004). King and colleagues (2002) undertook a detailed microscopic evaluation of LEHs in six permanent teeth from a single individual in order to reconstruct a comprehensive individual profile of developmental disruptions occurring during the period of visible enamel formation. Following on from the work of Hillson and Bond (1997), LEHs were recognised as a band of enamel in which there is a greater spacing of perikymata grooves than would be expected in a particular region of the tooth crown (Figure 9.2). Perikymata spacing profiles were used to determine number of defects, number of perikymata involved in each defect, the intervals between defects and the proportion of total enamel formation time affected by growth disruptions. In order to convert counts of perikymata grooves into an actual chronological record of developmental disruptions it was still necessary to estimate some parameters. Firstly, the chronological sequence had to be anchored by estimating the age of completion of appositional enamel formation in the earliest forming tooth, using data from Reid and Dean (2000). Secondly, an average perikymata periodicity had to be used to convert the perikymata counts into estimated time intervals. By estimating these parameters it was possible to determine the age of occurrence of each defect, the duration of each defect and the time interval between defects. In addition, since the date of birth of the individual studied was known it was possible to examine the seasonal distribution of developmental disturbances. The method was subsequently extended to a larger sample of individuals from a post-medieval context (King *et al.*, 2005).

The identification of systemic developmental disruptions involves matching linear enamel hypoplasias across different tooth classes with overlapping developmental schedules (Hillson and Bond, 1997; Malville, 1997; King *et al.*, 2002). The size and prominence of LEH is influenced by the geometry of the tooth crown, spacing of perikymata and tooth wear. When enamel defects are matched across teeth, the appearance of defects resulting from the same developmental disturbance can vary markedly between different teeth forming at the same time (Hillson and Bond, 1997). This reflects the gradient of perikymata spacing on different parts of the tooth crown and on different tooth types. Defects are more difficult to identify on molar teeth due to the steep gradient in perikymata spacing on molar crown surfaces (Hillson and Bond, 1997; King *et al.*, 2005). Defects will appear more prominent and are more readily identified in the midcrown and cervical regions of the teeth since perikymata are more closely spaced and these parts of the tooth are less affected by wear. As a result of these factors, matching of LEHs between all areas of enamel forming at the same time is not always possible (King *et al.*, 2005). Nevertheless, enamel defects should be matched across at least two simultaneously forming teeth in order to ensure that they are not caused by localised factors such as trauma or disease.

Reid and Dean (2000) used incremental markings in enamel to estimate total crown formation times in anterior teeth from Northern European populations. They present data on total crown formation times as well as the times required for the formation of cuspal enamel and of each tenth percentile of visible tooth crown height for upper and lower incisors and canines. Subsequent research (Reid and Dean, 2006) demonstrates that crown formation times for the anterior teeth differ slightly between Southern African and Northern European anterior teeth. Both populations exhibit a gradual trend towards longer enamel formation times between successive deciles of crown length, but this trend is more pronounced in the European sample. As a result, differences between the populations accrue gradually from the cusp to the cervix, and total crown formation times for the anterior teeth were shorter in the Southern African sample.

Population differences in total enamel formation time for the permanent molars were negligible.

The results of Reid and Dean (2000; 2006), and in particular their figures depicting the timing of visible enamel formation, provide a basis for cross-matching LEHs between teeth, and are recommended for this purpose. It is important to note that the deciles presented relate to tooth crown length measured along the enamel surface, and that this distance is often greater than standard measures of tooth crown height particularly in tooth crowns with a more curved labial surface. In some situations it may be unrealistic to compile individual records of enamel defects by reference to perikymata profiles, and in these cases the Reid and Dean charts provide a more realistic means for estimating age of occurrence than traditional measurement-based techniques. The difference in the relative timing of formation of the anterior teeth and molars between Southern African and Northern European populations suggests that schedules for matching defects between teeth may have to be modified for different populations. One way in which to ascertain which schedule would be more suitable for a given sample would be to identify the schedule that results in the lowest frequency of unmatched defects between teeth (Littleton, 2005)

LEH and age at weaning

The term *weaning* has sometimes been used to refer specifically to the age at complete cessation of breastfeeding, but this usage is misleading because it masks a complex transformation (Herring *et al.*, 1998). It also leads implicitly to the understanding that age of weaning is a specific event that can be inferred from the age distribution of dental or osteological developmental disruptions. Following on from this, it has sometimes been assumed that the peak incidence of LEH in an osteological sample corresponds to the age of weaning (i.e. the complete cessation of breastfeeding) within that population. As discussed above, developmental disruptions broadly associated with the weaning process will not necessarily be clustered at a particular stage of that process. Growth disruptions occurring at, or close to, the onset of the weaning period do not lead to the formation of LEH in the permanent dentition because they occur too early. Dietary supplementation is expected to occur at or before the age of six months to prevent growth faltering, and therefore precedes the formation of imbricational enamel in any of the permanent tooth crowns. The complete cessation of breastfeeding would occur within the available age range for the development of LEH in permanent teeth in many populations, but this transition may be both gradual and variable within a population and as a result it may not be easily detected.

Another area that has not received sufficient attention is the underlying age distribution of LEH within the dentition. In evaluating the possible relationship between the weaning process and the expression of LEH, it is necessary to consider what the expected distribution of recordable LEH in the dentition would be in a situation where developmental disruptions were randomly distributed with respect to age. Comparison of the observed age distribution of LEH within a particular sample against the expected or natural LEH distribution would reveal whether LEHs occurred at a higher than expected frequency at any age. As yet no such natural distribution model has been developed, but some suggestions can be made concerning the most relevant parameters and constraints.

- The youngest age at which LEH can be observed is the age of formation of the earliest-forming imbricational enamel. Reid and Dean (2000) found that the earliest permanent teeth to complete their cuspal enamel formation were the lower central and lower lateral incisors at 346 and 358 days respectively. Since very little visible surface enamel is likely to form in any anterior tooth within the first year after birth, developmental disturbances occurring in the first 11 to 12 months after birth will not be recorded as LEH, although they may be detected as accentuated striae or Wilson bands in a histological section.
- Enamel defects should only be recognised as LEH if they can be cross-matched between teeth that overlap in their developmental schedules. If defects must be matched across different tooth types the latest forming defects that can be scored will be those that form at approximately 6 years of age (on the second molar and premolars).
- LEH are more difficult to identify on molar teeth due to the steep gradient in perikymata spacing on molar crown surfaces, and are most easily identified in the anterior teeth. Defects are more readily identified in the midcrown and cervical regions of the teeth, as perikymata are more closely spaced and these regions of the crown are likely to be less affected by wear.

These constraints suggest that most identified LEH will have formed between the ages of 1 and 6 years. In practice this age range may be further constrained to between 1.5 and 5 years as a result of wear of the earliest-forming perikymata on incisors and first molars and because very few studies include the second (or third) permanent molars. In line with this prediction, many studies of enamel defects report a peak frequency of LEH occurring between 2 and 4 years (e.g. Wood, 1996; Malville, 1997; Dittmann and Grupe, 2000; King *et al.*, 2005; Littleton, 2005), but such a distribution does not necessarily reflect a real enhancement in the age-related distribution of developmental disruptions. Tellingly, the observed peak in age of occurrence of LEH in a historically documented sample from Belleville Ontario, which falls within the 2–4 year age range, was inconsistent with demographic, isotopic and historical evidence of the weaning process in that population (Saunders and Keenleyside, 1999).

Finally, evaluation of the health and living conditions of a past population or of groups within that population based on the quantification of LEH is not straightforward (Wood *et al.*, 1992). Three categories of individuals need to be considered. The first group comprises individuals lacking LEH. These individuals were either spared from exposure to the adverse environmental circumstances that cause LEH, or were resistant to these factors so that amelogenesis was not disrupted. The second group comprises individuals exhibiting LEH. The presence of LEH is evidence that the individual experienced one or more periods of developmental disturbance severe enough to cause an interruption in the normal process of enamel matrix secretion, and that the individual subsequently recovered and survived for long enough to resume and complete normal enamel formation. A third group comprises individuals who died during the period of enamel formation as a result of a failure to survive similar or more severe environmental insults. Arguably the first group of individuals, those who were spared from developmental disruptions, represent the most privileged section of society. The second group, those who recovered, show evidence of a less privileged childhood, and perhaps less optimal nursing histories, but still represent a healthier or more advantaged section of society than the third group of individuals, those who failed to survive. Studies of LEH are typically restricted to fully developed tooth crowns. As a result, the third group of non-survivors is excluded and becomes an invisible sub-sample representing an unknown percentage of the population from which the sample was drawn. In view of the relationship between infant mortality and nursing behaviours (Fildes, 1995), there is a risk that this hidden sub-sample would have experienced different and less adequate nursing histories than those who thrived.

Analysis of weaning using chemical techniques

Stable isotope and trace element analysis of ancient bones and teeth provide direct evidence of diet in the past. Studies of weaning in past populations have used nitrogen, oxygen and carbon stable isotopes or strontium/calcium ratios. Since these techniques are of value for understanding different stages of the weaning process and reflect the intake of different kinds of nutrients, many studies have combined evidence from two or more approaches. Although most published studies of infant diet in past populations have been carried out on bone, each of the techniques discussed below could be conducted on samples from enamel and/or dentine. Other physiological and environmental parameters can contribute to measured stable isotope and trace element values, and need to be taken into consideration when data are interpreted.

Stable nitrogen isotopes from collagen in bone and dentine can provide direct evidence of sources of dietary proteins. Experimental studies have demonstrated that there is an increase in $\delta^{15}N$ values of about 2–4‰ between successive trophic levels. Fogel and Tuross (1989) examined $\delta^{15}N$ values in fingernail clippings of lactating women and their infants, and found that breastfed infants showed significant enrichment in $\delta^{15}N$ values when compared to their mother. Following on from this, numerous recent studies have sought to reconstruct nursing histories in earlier human populations using $\delta^{15}N$ values from bone collagen (e.g. Katzenberg *et al.*, 1993; Herring *et al.*, 1998; Dittmann and Grupe, 2000; Dupras *et al.*, 2001; Mays *et al.*, 2002; Richards *et al.*, 2002; Williams *et al.*, 2005; Clayton *et al.*, 2006). The technique is of particular value for identifying the decreasing importance of breast milk as a source of protein during the nursing period. Bone collagen from a newborn infant has a similar $\delta^{15}N$ to that of the mother, reflecting the composition of collagen deposited in utero. During exclusive breastfeeding the infant is effectively one trophic level higher than its mother, such that the $\delta^{15}N$ value in bone collagen is expected to rise during the first few months after birth, stabilising at a level approximately 3‰ above the maternal level. Following the introduction of alternative sources of protein the $\delta^{15}N$ value is expected to decline (Katzenberg *et al.*, 1996; Herring *et al.*, 1998). A concern with this technique relates to uncertainties over collagen turnover rates in infants and small children (Katzenberg *et al.*, 1996). Measured stable isotope values will reflect a time-averaged signal and may draw on collagen formed during different dietary phases. This will introduce a time lag between the change in diet and the reflection of that change in bone collagen (Herring *et al.*, 1998). For example, samples taken from young infants will contain a mix of collagen from bone deposited in utero and bone deposited during the nursing period, and this may differ between skeletal elements depending on age of onset of bone formation, rates of prenatal and postnatal growth and rates of collagen turnover for a particular bone.

The analysis of stable oxygen isotopes can contribute to the interpretation of weaning practices in past populations by identifying the importance of breast milk as a source of water in an infant's diet. Since the liquid consumed by breastfed infants is richer in $\delta^{18}O$ than the liquids consumed by their mothers (and presumably other adults within the population), the replacement of breast milk with other sources of water should be accompanied by a reduction in $\delta^{18}O$ (Wright and Schwarcz, 1998; 1999). In line with this expectation, analysis of $\delta^{18}O$ in enamel from permanent teeth from an archaeological sample from Guatemala revealed a decrease in $\delta^{18}O$ between successively later forming teeth, consistent with a childhood transition from isotopically enriched breastmilk to isotopically lighter rain or stream water as the main source of liquids.

Stable carbon isotopes have been used primarily to distinguish between diets based on C_3 and C_4 plants. In the Americas, $\delta^{13}C$ values have been used to detect consumption of maize, which is the principle cultigen that uses a C_4 photosynthetic pathway. The introduction of maize-based gruels into an infant's diet could lead to an enrichment in $\delta^{13}C$, signalling the onset of complementary feeding (Katzenberg *et al.*, 1993; Wright and Schwarcz, 1998). Additionally, since suckling children have a $\delta^{13}C$ value approximately 1–2 ‰ more enriched than the mother due to a trophic level effect in carbon, it may be possible to use carbon isotopes to interpret wearing behaviour even in the absence of any input from C_4 plants (Fuller *et al.*, 2003). Studies of carbon isotopes based on bone or dentine are often undertaken in conjunction with nitrogen isotope analysis (e.g. Wright and Schwarcz, 1999; Fuller *et al.*, 2003; Clayton *et al.*, 2006).

Physiological levels of strontium and calcium depend on dietary intake and are modified by the preferential transport of calcium across the placental, mammary gland and digestive tract (Comar, 1963). Strontium/calcium (Sr/Ca) ratios are useful for detecting dietary change during infancy and early childhood because human milk has a very low Sr/Ca ratio compared to most solid foods, particularly those given during weaning (Sillen and Smith, 1984). Sr/Ca ratios in bone samples from juvenile skeletons can be used to explore patterns of dietary supplementation in past human populations (Sillen and Smith, 1984; Mays, 2003). Changes in Sr/Ca ratios related to diet must be interpreted against an underlying trend for Sr/Ca ratios to fall during the first decade of postnatal life as a result of the development of intestinal discrimination against strontium (Comar, 1963; Rivera and Harley, 1965). Sr/Ca ratios are expected to fall slightly during the period of exclusive breastfeeding, rise during the introduction of complementary foods and subsequently fall until the intestinal discrimination stabilises at adult levels.

Limitations of cross-sectional weaning studies

Typically studies of dietary change during infancy and early childhood in past populations use a cross-sectional approach, in which a single tissue sample from each skeleton is analysed. This enables the dietary intake in the period leading up to death to be inferred. Thus, most previous chemical analyses of the weaning process have been based on tissue samples from a series of infant skeletons, and changes in diet were inferred from sample trends, on the basis of changing isotopic/elemental values in successive age classes or from interpolation of these values plotted against estimated age at death (e.g. Sillen and Smith, 1984; Herring *et al.* 1998; Dittmann and Grupe, 2000; Dupras *et al.*, 2001; Mays *et al.*, 2002; Mays, 2003).

Cross-sectional studies have revealed useful information about infant diet in past populations, but have several limitations. The sampling strategy is constrained by the number and age distribution of infants and young children in a series, and certain key age ranges may be under-represented. The lack of precision in age-at-death determinations for young infants and children could introduce another source of error, particularly in cases where tooth germs were not recovered. In these cases, age at death must be estimated from skeletal dimensions, and may be underestimated in those children who had suffered chronic growth disruptions related to their nutritional status (Humphrey, 2003). Cross-sectional techniques establish a single averaged result for the entire sample. It is sometimes possible to identify specific individuals who do not conform to the general population trend, but interpretation of what this means in terms of individual dietary histories is not straightforward (Herring *et al.*, 1998). The restriction on sample size means that it is not usually practicable to divide the sample and make comparisons within the group, for example between individuals belonging to different social categories.

Of greater concern is whether a dietary reconstruction based on bone samples from infants and children provides a realistic representation of the population as a whole since osteological assemblages may be heavily biased towards those children who were inadequately nursed. Historical data on the prevalence of breastfeeding and the mode of feeding of infants who died from diarrhoea and other causes suggest that mortality in the past was markedly higher in infants who were bottle-fed than in those who were breastfed (Fildes, 1995). Infants who are given complementary foods prematurely may be at greater risk than other infants (Cunningham, 1995). Ideally, research into the nursing practices of past populations should be undertaken on individuals who are known to have survived the weaning process (Wright and Schwarcz, 1998), or compare those who survived to those who did not (Fuller *et al.*, 2003).

Multiple sampling approaches

Discrete isotopic or trace element analysis of different skeletal and dental tissues that form at different developmental stages can be used to track changes in diet that occurred during the lifetime of a single individual. Over the last decade this type of analytical approach has become increasingly common, and has enormous potential in a variety of archaeological contexts, including the interpretation of residential mobility, seasonality patterns in occupation or subsistence strategies, and infant diet (e.g. Sealy *et al.*, 1995; Wright and Schwarcz, 1998; Balasse *et al.*, 2003). Dental tissues are particularly useful in this respect since their development is well documented and covers the period of infancy and early childhood. Enamel retains elemental and isotopic signals laid down during its formation throughout life and therefore provides the opportunity for a retrospective analysis of childhood diet in individuals who survived the weaning process. Enamel is also the biological apatite most resistant to diagenetic alteration, and less likely than bone or dentine to undergo isotopic and elemental exchange within a burial environment (Koch *et al.*, 1997; Budd *et al.*, 2000). Areas of enamel that are more susceptible to chemical alteration within the intra-oral or burial environment should nevertheless be avoided for dietary analysis. Underlying patterns of heterogeneity in isotope and trace element distributions within enamel also need to be carefully considered when designing multiple sampling strategies, regardless of the size of the analytical samples.

Multiple sampling of human dental tissues has not yet been widely used in the interpretation of dietary shifts in infancy and childhood, but those studies that have been carried out demonstrate the potential value of such an approach. By analysing the stable isotope composition of three tooth crowns that mineralise at different ages for each individual in their sample, Wright and Schwarcz (1998) were able to evaluate diet at three different stages of development. Overall, there was a trend for decreasing $\delta^{18}O$ and increasing $\delta^{13}C$ between teeth developing in infancy and early childhood and those developing in later childhood, but results also implied some variation between individuals within the sample. A subsequent analysis investigated the $\delta^{15}N$ and $\delta^{13}C$ values in dentine collagen in teeth from the same site (Wright and Schwarcz, 1999).

A more recent study compared $\delta^{15}N$ and $\delta^{13}C$ values in dentine and bone in individuals from a medieval osteological series from Wharram Percy (Fuller *et al.*, 2003). Three separate samples were obtained from each tooth root, consisting of dentine enclosed under the tooth crown and the cervical and apical halves of the exposed roots. Since horizontal sectioning was used, the boundaries between samples did not correspond to the developing dentine front, resulting in some chronological overlap between samples. Nevertheless the samples represented a broadly time-successive sequence. Results comparing $\delta^{15}N$ values for deciduous second molar dentine with ribs from the same individuals showed the following pattern of enrichment: crown > cervical half of root > apical half of root > rib. Results were consistent with a gradual reduction of breastfeeding and an increased intake of other sources of protein during the period of dentine growth in this deciduous tooth. A similar depletion pattern was found for $\delta^{13}C$ values in most individuals.

For a truly longitudinal analysis of changing diet during infancy and early childhood, the chronological resolution achieved by published studies will need to be significantly enhanced. Various techniques now exist for analysing the chemical composition of very small quantities of enamel, providing a basis for discrete multiple sampling investigations of variation in the chemical composition of one or more teeth from each individual in a sample (Wurster *et al.*, 1999; Zazzo *et al.*, 2005; Humphrey *et al.*, 2007). The chronological resolution that can be achieved in reconstructing individual nursing histories still has to be resolved for different analytical techniques and tissue types. Small samples of enamel can be analysed from the solid state using laser ablation inductively coupled plasma mass spectrometry (LA-ICP-MS), and the analysis of thin sections allows the position of discrete sampling points to be cross-referenced to incremental growth markers in the tooth (Humphrey *et al.*, 2007). These markers provide a record of the timing of enamel matrix secretion and the onset of mineralisation at any given sampling point, but additional mineral reflecting a later stage of development is added during enamel maturation, resulting in a time-averaged signal. As a result, it will only be possible to generate reliable data on the timing of dietary shifts that occurred during the secretory stage of enamel formation if the chemical changes associated with these shifts are not completely overwhelmed during the maturation stage. Preliminary results have shown that changes in Sr/Ca ratio across the neonatal line in human deciduous teeth are consistent with predictions based on the mode of feeding at birth, suggesting that abrupt dietary transitions such as this leave a permanent signal in enamel (Humphrey *et al.*, 2007). Further work on children of known dietary history will help to clarify these issues.

The successful application of multisampling approaches to the interpretation of weaning histories from stable isotopes and trace elements in enamel and dentine will address many of the limitations of traditional cross-sectional analyses, and enable variation in nursing practices within past populations to be evaluated. This in turn will provide a basis for addressing the possible causal factors associated with such differential treatment (e.g. socio-economic status) and the developmental impact of these differences (e.g. skeletal growth attainment, enamel defects).

Acknowledgements

I thank Holger Schutkowski for the invitation to contribute to this volume. Special thanks to Christopher Dean, Simon Hillson, Teresa Jeffries and Tania King for valuable insights into the processes discussed in this paper.

References

Balasse, M., Smith, A. B., Ambrose, S. H. and Leigh, S. R. (2003). Determining sheep birth seasonality by analysis of tooth enamel oxygen isotope ratios: The late stone age site of Kasteelberg (South Africa). *Journal of Archaeological Science*, **30**, 205–15.

Boyde, A. (1989). Enamel. In *Teeth*, eds. B. K. B. Berkovitz, A. Boyde, R. M. Frank *et al.* Berlin: Springer-Verlag, pp. 309–473.

Budd, P., Montgomery, J., Barreiro, B. and Thomas, R. G. (2000). Differential diagenesis of strontium in archaeological human dental tissues. *Applied Geochemistry* **15**, 687–94.

Clayton, F., Sealy, J. and Pfeiffer, S. (2006). Weaning age among foragers at Matjes River Rock Shelter, South Africa, from stable nitrogen and carbon isotope analyses. *American Journal of Physical Anthropology*, **129**, 311–17.

Comar, C. L. (1963). Some over-all aspects of strontium-calcium discrimination. In *The Transfer of Calcium and Strontium Across Biological Membranes*, ed. R. H. Wasserman. New York: Academic Press, pp. 405–19.

Cunningham, A. S. (1995). Breastfeeding: adaptive behaviour for child health and longevity. In *Breastfeeding: Biocultural Perspectives*, eds. P. Stuart-Macadam and K. A. Dettwyler. New York: Aldine de Gruyter, pp. 243–64.

Dean, C. (2000). Progress in understanding hominoid dental development. *Journal of Anatomy*, **197**, 77–101.

Dittmann, K. and Grupe, G. (2000). Biochemical and palaeopathological investigations on weaning and infant mortality in the early Middle Ages. *Anthropologischer Anzeiger*, **58**, 345–55.

Dupras, T. L., Schwarcz, H. P. and Fairgrieve, S. I. (2001). Infant feeding and weaning practices in Roman Egypt. *American Journal of Physical Anthropology*, **115**, 204–12.

Fessler, D. M. T. and Abrams, E. T. (2004). Infant mouthing behavior: the immunocalibration hypothesis. *Medical Hypotheses*, **63**, 925–32.

Fildes, V. A. (1995). The culture and biology of breastfeeding: an historical review of Western Europe. In *Breastfeeding: Biocultural Perspectives*, eds. P. Stuart-Macadam and K. A. Dettwyler. New York: Aldine de Gruyter, pp. 101–26.

FitzGerald, C. M. (1998). Do enamel microstructures have a regular time dependency? Conclusions from the literature and a large-scale study. *Journal of Human Evolution*, **35**, 371–86.

FitzGerald, C. M. and Rose, J. C. (2000). Reading between the lines: Dental development and subadult age assessment using the microstructural growth markers of teeth. In *Biological Anthropology of the Human Skeleton*, eds. M. A. Katzenberg and S. R. Saunders. New York: Wiley-Liss, pp. 163–86.

FitzGerald, C., Saunders, S., Bondioli, L. and Macchiarelli, R. (2006). Health of infants in an Imperial Roman skeletal sample: Perspective from dental microstructure. *American Journal of Physical Anthropology*, **130**, 179–89.

Fogel, M. L. and Tuross, N. (1989). Nitrogen isotope tracers of human lactation in modern and archaeological populations. *Carnegie Institute of Washington Yearbook*, **88**, 111–17.

Fuller, B. T., Richards, M. P. and Mays, S. A. (2003). Stable carbon and nitrogen isotope variations in tooth dentine serial sections from Wharram Percy. *Journal of Archaeological Science*, **30**, 1673–84.

Goodman, A. H. and Rose, J. C. (1990). Assessment of systemic physiological perturbations from dental enamel hypoplasias and associated histological structures. *Yearbook of Physical Anthropology*, **33**, 59–110.

Guatelli-Steinberg, D., Larsen, C. S. and Hutchinson, D. L. (2004). Prevalence and the duration of linear enamel hypoplasia: a comparative study of Neandertals and Inuit foragers. *Journal of Human Evolution*, **47**, 65–84.

Herring, D. A., Saunders, S. R. and Katzenberg, M. A. (1998). Investigating the weaning process in past populations. *American Journal of Physical Anthropology*, **105**, 425–40.

Hillson, S. (1996). *Dental Anthropology*. Cambridge: Cambridge University Press.

Hillson, S. and Bond, S. (1997). Relationship of enamel hypoplasia to the pattern of tooth crown growth: a discussion. *American Journal of Physical Anthropology*, **104**, 89–103.

Humphrey, L. T. (2003). Linear growth variation in the archaeological record. In *Patterns of Growth Variation in the Genus Homo*, eds. J. L. Thompson, G. E. Krovitz and A. J. Nelson. Cambridge: Cambridge University Press, pp. 144–69.

Humphrey, L. T., Dean, M. C. and Jeffries, T. E. (2007). An evaluation of changes in strontium/calcium ratios across the neonatal line in human deciduous teeth. In *Dental Perspectives on Human Evolution: State-of-the-Art Research in Dental Anthropology*, eds. S. E. Bailey and J.-J. Hublin. Dordrecht: Springer, pp. 301–17.

Katzenberg, M. A., Saunders, S. R. and Fitzgerald, W. R. (1993). Age-differences in stable carbon and nitrogen isotope ratios in a population of prehistoric maize horticulturists. *American Journal of Physical Anthropology*, **90**, 267–81.

Katzenberg, M. A., Herring, A. and Saunders, S. R. (1996). Weaning and infant mortality: evaluating the skeletal evidence. *Yearbook of Physical Anthropology*, **39**, 177–99.

Kazawa, C. W. (1998). Adipose tissue in human infancy and childhood: An evolutionary perspective. *Yearbook of Physical Anthropology*, **41**, 177–209.

King, T., Hillson, S. and Humphrey, L. T. (2002). Developmental stress in the past: A detailed study of enamel hypoplasia in a Post-Medieval adolescent of known age and sex. *Archives of Oral Biology*, **47**, 29–39.

King, T., Humphrey, L. T. and Hillson, S. (2005). Linear enamel hypoplasias as indicators of systemic physiological stress: Evidence from two known age-at-death and sex populations from postmedieval London. *American Journal of Physical Anthropology*, **128**, 547–59.

Koch, P. L., Tuross, N. and Fogel, M. L. (1997). The effects of sample treatment and diagenesis on the isotopic integrity of carbonate in biogenic hydroxylapatite. *Journal of Archaeological Science*, **24**, 417–29.

Kramer, M. and Kakuma, R. (2002). *The Optimal Duration of Exclusive Breastfeeding: A Systematic Review*. Geneva: World Health Organization.

Lee, P. C., Majluf, P. and Gordon, I. J. (1991). Growth, weaning and maternal investment from a comparative perspective. *Journal of the Zoological Society of London*, **225**, 99–114.

Littleton, J. (2005). Invisible impacts but long-term consequences: hypoplasia and contact in Central Australia. *American Journal of Physical Anthropology*, **126**, 295–304.

Liversidge, H., Dean, M. C. and Molleson, T. I. (1993). Increasing human tooth length between birth and 5.4 Years. *American Journal of Physical Anthropology*, **90**, 307–13.

Malville, N. J. (1997). Enamel hypoplasia in ancestral Puebloan populations from southwestern Colorado.1. Permanent dentition. *American Journal of Physical Anthropology*, **102**, 351–67.

Mays, S. (2003). Bone strontium: calcium ratios and duration of breastfeeding in a Mediaeval skeletal population. *Journal of Archaeological Science*, **30**, 731–41.

Mays, S. A., Richards, M. P. and Fuller, B. T. (2002). Bone stable isotope evidence for infant feeding in England. *Antiquity*, **76**, 654–6.

McDade, T. W. (2003). Life history theory and the immune system: Steps toward a human ecological immunology. *Yearbook of Physical Anthropology*, **46**, 100–25

Reid, D. J. and Dean, M. C. (2000). The timing of linear enamel hypoplasias on human anterior teeth. *American Journal of Physical Anthropology*, **113**, 135–9.

Reid, D. J. and Dean, M. C. (2006). Variation in modern human enamel formation times. *Journal of Human Evolution*, **113**, 329–46.

Richards, M. P., Mays, S. and Fuller, B. T. (2002). Stable carbon and nitrogen isotope values of bone and teeth reflect weaning age at the Medieval Wharram Percy site, Yorkshire, UK. *American Journal of Physical Anthropology*, **119**, 205–10.

Rivera, J. and Harley, J. H. (1965). *The HASL Bone Program: 1961–1964*. US Atomic Energy Commission Health and Safety Laboratory Report No. 163.

Robinson, C., Brookes, S. J., Bonass, W. A., Shore, R. C. and Kirkham, J. (1997). Enamel maturation. In *Dental Enamel. Proceedings of the Ciba Foundation Symposium 205*, eds. D. J. Chadwick and G. Cardew. Chichester: John Wiley and Sons Ltd, pp. 118–34.

Rousham, E. K. and Humphrey, L. T. (2002). The dynamics of child survival. In *Human Population Dynamics: Cross-Disciplinary Perspectives*, eds. H. Macbeth and P. Collinson. Cambridge: Cambridge University Press, pp. 124–40.

Saunders, S. R. and Keenleyside, A. (1999). Enamel hypoplasia in a Canadian historic sample. *American Journal of Human Biology*, **11**, 513–24.

Schour, I. (1936). Neonatal line in enamel and dentin of human deciduous teeth and first permanent molar. *Journal of the American Dental Association*, **23**, 1946–55.

Sealy, J., Armstrong, R. and Schrire, C. (1995). Beyond lifetime averages: tracing life histories through isotopic analysis of different calcified tissues from archaeological human skeletons. *Antiquity*, **69**, 290–300.

Sellen, D. W. (2001). Comparison of infant feeding patterns reported for nonindustrial populations with current recommendations. *Journal of Nutrition*, **131**, 2707–15.

Sillen, A. and Smith, P. (1984). Weaning patterns are reflected in strontium-calcium ratios of juvenile skeletons. *Journal of Archaeological Science*, **11**, 237–45.

Smith, C. E. (1998). Cellular and chemical events during enamel maturation. *Critical Reviews in Oral Biology & Medicine*, **9**, 128–61.

Williams, J. S., White, C. D. and Longstaffe, F. J. (2005). Trophic level and macronutrient shift effects associated with the weaning process in the Postclassic Maya. *American Journal of Physical Anthropology*, **128**, 781–90.

Wood, J. W., Milner, G. R., Harpending, H. C. and Weis, K. M. (1992). The osteological paradox: problems of interpreting prehistoric health from skeletal samples. *Current Anthropology*, **33**, 343–70.

Wood, L. (1996). Frequency and chronological distribution of linear enamel hypoplasia in a North American colonial skeletal sample. *American Journal of Physical Anthropology*, **100**, 247–59.

Wright, L. E. and Schwarcz, H. P. (1998). Stable carbon and oxygen isotopes in human tooth enamel: identifying breastfeeding and weaning in prehistory. *American Journal of Physical Anthropology*, **106**, 1–18.

Wright, L. E. and Schwarcz, H. P. (1999).Correspondence between stable carbon, oxygen and nitrogen isotopes in human tooth enamel and dentine: Infant diets at Kaminaljuyú. *Journal of Archaeological Science*, **26**, 1159–70

Wurster, C. M., Patterson, W. P. and Cheatham, M. M. (1999). Advances in micromilling techniques: a new apparatus for acquiring high-resolution oxygen and carbon stable isotope values and major/minor elemental ratios from accretionary carbonate. *Computers & Geosciences*, **25**, 1159–66.

Zazzo, A., Balasse, M., Patterson, W. P. and Patterson, P. (2005). High-resolution delta C-13 intratooth profiles in bovine enamel: Implications for mineralization pattern and isotopic attenuation. *Geochimica et Cosmochimica Acta*, **69**, 3631–42.

10 *Using DNA to investigate the human past*

BETH SHAPIRO,[1] M. THOMAS P. GILBERT AND IAN BARNES

Over the last two decades, biological anthropology has benefited from significant advances in many fields, but perhaps none so much as molecular genetics. Although we progressed from initial applications of heritable molecular markers to the publication of the complete sequence of the human genome, we are still only beginning to make sense of what our DNA can tell us about who we are, where we came from, and why we behave the way we do. In this chapter, we review many facets of human molecular genetic research, focusing on recent developments and successes while exploring some of the potential pitfalls associated with working with human DNA sequences. Our goal is not to present a comprehensive analytical review; for this we refer the interested reader to a recent and thorough treatment of the subject (Jobling *et al.*, 2004). Instead, we aim to enable the interested non-specialist to understand and hopefully to critically evaluate the evidence supplied by DNA studies, so as to be able to employ it in their own work.

We begin the chapter with an overview of the chemical and biological processes involved in changing and shaping our DNA sequences, and how these processes can be modelled and used to reconstruct recent and distant evolutionary history. This section will necessarily provide only the briefest introduction to a complicated subject with a rich history and literature. We will introduce the most commonly utilised sources of genetic information, and discuss how each of these can be applied both to ancient and modern data. Finally, we discuss several phylogenetic and population genetic analysis tools, illustrating how they have been applied to human molecular evolutionary studies.

The development of genetic methods in biological anthropology

The roots of the genetic study of humans and human populations are found in the discovery and widespread characterisation of blood groups in the early twentieth century (e.g. Landsteiner, 1900; Hirzfeld and Hirzfeld, 1919) and the use of these markers as a means to understand human history, which began in

Between Biology and Culture, ed. Holger Schutkowski. Published by Cambridge University Press.

earnest with the work of L. L. Cavalli-Sforza from the mid 1960s (see Cavalli-Sforza *et al.*, 1994). The subsequent application of genetic methods to problems of human history has been intricately tied to technological developments, in particular DNA manipulation techniques.

While classical markers make use of variation in phenotype, DNA markers record differences in the underlying genotype, and as such allow a greater level of discriminatory power. The development of techniques such as the analysis of restriction fragment length polymorphisms (RFLP) in the 1970s, the polymerase chain reaction (PCR) and DNA sequencing in the 1980s and 1990s, and DNA microarrays in the present decade, has resulted in a much-expanded data set for analysis, the ultimate incarnation of which is the recently completed human genome.

A molecular primer

What is DNA?

Genetic information is stored as linear sequences of deoxyribonucleic acid, or DNA, molecules. In every living organism, DNA is made up of four nucleotide 'bases' that pair with each other according to their chemical structures, forming long anti-parallel polymer chains. These chains fold into the characteristic double-helix shape that was first recognised in the 1950s following the pioneering X-ray crystallographic work of Maurice Wilkins and Rosalind Franklin. The human genome is organised into 24 different types of double-stranded DNA molecules, or chromosomes. Of these 22 are autosomes, and 2 (X and Y) are sex chromosomes.

Autosomal DNA and DNA from the sex chromosomes have different mechanisms of inheritance. The Y-chromosome, for example, is inherited uniparentally, passing only from father to son. This is important to human molecular evolutionary studies, because it provides a mechanism for tracing paternal lineages through time, independently of the evolutionary history of the maternal lineage. The autosomal and X-chromosomes are subject to recombination between maternal and paternal lineages, and will therefore show a mixed evolutionary history. If the histories of the two lineages are likely to differ, for example in societies where male philopatry (in which males are likely to remain with their natal group, and females may disperse and join a group in a different population) is the norm, then the analysis of molecular markers on autosomal chromosomes may lead to different conclusions than that of the Y-chromosome.

Sequencing the human genome

The human nuclear genome contains approximately three billion base-pairs (bp), the majority for which no function has yet been identified. Of these three billion bases, 98.7 % have been shown to be identical to that of our closest living relative, the chimpanzee (TCSaAC 2005; Khaitovich *et al.*, 2005). Rival groups published two separate 'rough drafts' of the human genome in February of 2001 (IHGSC, 2001; Venter *et al.*, 2001), and since that time between 20 000 and 25 000 genes (only 1.2 % of the entire genome) have been identified (IHGSC, 2004). A significant proportion of the non-protein-coding human genome comprises repeat regions, for example microsatellites or minisatellites, which are short DNA sequences, often only two or three nucleotides long, that are repeated in tandem arrays. Although the specific function of microsatellites is not well understood, they tend to quickly mutate by gaining or losing whole repeats, making them useful markers in population genetic studies. Other elements, such as endogenous retroviruses, transposable elements and ribosomal RNA genes, are also found in the non-protein coding regions of the human genome.

Mitochondrial versus nuclear DNA

Many of the most well-known human molecular analyses have examined the variation in mitochondrial DNA (mtDNA) rather than nuclear DNA. Mitochondria are membrane-enclosed organelles that are found in the cytoplasm of most eukaryotic cells and whose main function is in the energy-generating process of oxidative phosphorylation.

Mitochondria are believed to have originated as endosymbiotic bacteria, having since lost the ability to survive independently. Mitochondria have a separate, circular genome that in humans is approximately 16.5 kb in length and contains 37 genes and a variable non-coding region, often referred to as the control region, or the displacement- or d-loop.

Like the Y-chromosome, mitochondria are inherited uniparentally, however mitochondria are passed down via the oocyte from mother to child, making it possible to use changes accumulating in the mtDNA sequences to trace the evolutionary history of the maternal lineage.

MtDNA has been particularly important in analyses that incorporate 'ancient DNA' (aDNA) sequences, simply because of numbers. As opposed to autosomal DNA, for which every diploid cell has two copies, there are between two and ten copies of the circular mtDNA in every mitochondrion, and thousands of mitochondria in every cell, depending on that cell's specific energy requirement. As DNA degrades through time, it will be much more likely that a specific fragment of mtDNA than of nuclear DNA will have survived.

Using DNA sequences to investigate the human past

One of the major difficulties in human molecular research is in choosing what populations, and what genes in those populations, will be most appropriate to address a particular evolutionary hypothesis. Where a population study is intended, a suitable group needs to identified, bearing in mind that they must be willing to take part in the study, have a statistically significant number of individuals, and be able to grant appropriate ethical consent. Sampling procedure is also important, as incomplete or biased sampling can affect results. For example, if the analysis requires that samples are independent, care must be taken that samples are not taken from close relatives. It may also be important to ensure a reasonable longevity of association with the sampling locality. Major conurbations, for example, carry the risk of significant amounts of population turnover.

Because of the potential for contamination with exogenous human DNA, care must be taken in the collection and storage of samples. A mouth (or buccal) swab procedure is frequently used for living subjects, which, while simple, often results in relatively low DNA yield. The inverse is true of blood sampling: good DNA yields, but a much more complex procedure, including the need to keep samples cold post-sampling, and a requirement for processing as soon as possible. Blood samples have the potential to be immortalised using Epstein-Barr virus, providing a permanent source of DNA (Ohlin *et al.*, 1992). Both hair and excrement are also sources of DNA, and are most often used in wildlife studies (see Morin *et al.*, 2001). Hair in particular is easy to manipulate and resistant to contamination, although the midshaft contains only mtDNA and roots are required for nuclear DNA (Gilbert *et al.*, 2004; 2006).

Adding ancient specimens to an analysis provides the potential to expand the depth of time during which a population can be analysed. However, the use of ancient specimens also increases the difficulty in thoroughly and appropriately sampling across a population. When sampling from ancient populations, it is often not possible to know if sufficient samples exist, or how these samples might be biased (for example, due to a high degree of relatedness). Cultural hypotheses should be constructed within the terms of population genetics, bearing in mind that relatively few samples may be available, and that relatively little may be known about those that are available. In such instances, it is generally useful to have as much prior information as possible about the present-day genetic landscape.

Dealing with ancient DNA

The problems of experimental design and sample acquisition are generally a secondary consideration when dealing with ancient samples, in comparison to those inherent to aDNA analysis. Fortunately, inhibition, damage and contamination, the three major sources of problems in aDNA research, have been well characterised during the last two decades. Of these, **inhibition** of the polymerase enzyme used in PCR has been perhaps the easiest to address. Current extraction methods use stringent protocols to remove soil and other environmental components that otherwise block the activity of *Taq* polymerase. Most aDNA PCR mixes will also include some form of enzyme-stabilising or inhibitor-binding agent, such as bovine serum albumen (BSA).

Due to chemical reactions between the DNA and other molecules during burial and storage, DNA **damage** plays a much greater part in determining the character of aDNA research. The primary effect of DNA damage is to reduce the average fragment length in ancient extracts to less than 200 bp. To recover longer lengths of sequence, multiple, overlapping amplifications are required, significantly increasing both the financial and time cost of aDNA research. In addition, nucleotide changes may be observed in the aDNA sequences that are the consequence not of evolution, but of incorrect copying of damaged bases (Hofreiter *et al.*, 2001a; Gilbert *et al.*, 2003a). The extent of DNA damage in an ancient extract is largely determined by the temperature regime experienced by the specimen, and the time it has been at that temperature (**thermal history**). It is therefore no surprise that the most consistent data from very ancient samples come from high Arctic permafrost (Hofreiter *et al.*, 2001b; Shapiro *et al.*, 2004). PCR works best with high-quality DNA, and once the concentration of undamaged molecules is sufficiently low, the reaction will preferentially amplify exogenous, **contaminant** DNA. The two primary sources of contaminant DNA are PCR products from previous reactions and (when the target sequence is human) DNA shed by handling of the sample or laboratory materials. Other reported contaminant sequences, such as mouse (from museum storage) or pig (from animal-based glues in conservation), are just as insidious, but considerably easier to detect. The methods required to avoid and detect contamination have been the most heavily discussed aspects of aDNA techniques (e.g. Cooper and Poinar, 2000; Hofreiter *et al.*, 2001b; Pääbo *et al.*, 2004; Gilbert *et al.*, 2005a) and remain contentious yet crucial to the wider application of aDNA data.

Despite the difficulties of generating high-quality data, aDNA provides important and otherwise unattainable insights into human molecular evolution. For example, with modern data alone, population history must be inferred using a model (either explicit or implicit), and as such is vulnerable to any inaccuracies or weaknesses inherent to that model. Ancient DNA can circumvent some of a model's assumptions by providing a genetic signature at a particular point in time. This is particularly important when population history is sufficiently complex that it cannot be confidently inferred, for example in situations where extensive migration and back-migration occur, multiple populations with independent histories co-exist, or, in the most extreme cases, the population or taxon in question is extinct.

5'....CGCATTAC**GAATTC**TTAATAGCA....3'
3'....GCGTAATG**CTTAAG**AATTATCGT....5'

+ EcoRI

5'....CGCATTAC**G** **AATTC**TTAATAGCA....3
3'....GCGTAATG**CTTAA** **G**AATTATCGT....5'

Figure 10.1 The restriction enzyme EcoRI (purified from *Escherichia coli*) recognises the 6-bp DNA motif GAATTC and its complement. When DNA is incubated with EcoRI, the double-stranded DNA is cleaved every time that motif occurs, resulting in a distribution of differently sized fragments.

Methods for extracting and amplifying DNA sequences

Several techniques have been developed to purify DNA from tissue, and these will not be described in detail here. In general, tissues are mechanically disrupted prior to digestion in a 'cocktail' of detergents, digestive enzymes (such as proteinase K) and salts to release the DNA, and then purified using either organic solvents or the DNA-binding properties of silica. In situations where the DNA content of the sample is greatly reduced (e.g. old forensic samples or aDNA) techniques may be used that enable high recovery efficiency (see Cattaneo *et al.*, 1997).

The result of the initial purification step is a mixture of all DNA (mitochondrial, nuclear, contaminant) present in the sample, necessitating an additional step to target specific DNA fragments. Prior to the widespread availability of DNA sequencing platforms, the most common technique was Restriction Fragment Length Polymorphisms (RFLP). Here, the mixture of purified DNA is incubated with naturally occurring **restriction enzymes** that cleave double-stranded DNA at specific nucleotide motifs, resulting in a distribution of fragment sizes (Figure 10.1). Differences between DNA sequences result in different patterns, making comparison between individuals possible. Some examples in which RFLP was used in human molecular evolution include the analysis of variation in human mtDNA (Brown, 1980; Johnson *et al.*, 1983; Cann *et al.*, 1987) and the original DNA fingerprinting methodology (Jeffreys *et al.*, 1985).

Comparison of RFLP patterns provides some information about the genetic differences between individuals; however comparing actual DNA sequences can be much more powerful. Prior to the development of PCR, this was accomplished by a technique called plasmid cloning, in which DNA fragments are individually incorporated into bacterial genomes, which are then grown into colonies and sequenced (Higuchi *et al.*, 1984; Pääbo, 1985a, b, 1986). This process is inefficient, however, as it is impossible to target specific DNA fragments, and repetition is often required before the targeted sequence is identified. It has been estimated that approximately 500 000 bacterial colonies would be needed to generate a complete clone library of a modern mammalian genome (O'Rourke *et al.*, 2000).

The introduction of PCR (Mullis and Faloona, 1987; Saiki *et al.*, 1988) in the late 1980s provided a mechanism for targeting specific DNA fragments, and has been one of the most important technological developments in molecular evolution. PCR is an iterative process in which the two DNA strands are separated by heating, followed by a cooling step, during which specifically designed **primers** locate and bind to the target fragment. A polymerase enzyme then extends the sequence between the primers, resulting in a complementary copy of the targeted sequence. This process is then cyclically repeated, resulting in the exponential amplification of the targeted DNA fragment.

Once DNA has been amplified, it can be used to address different types of hypotheses. For example, RFLP can be used to determine whether a sample has a particular nucleotide at a particular position (termed **single nucleotide polymorphism**, or **SNP**), which can determine the mitochondrial haplotype to which that sample belongs (Figure 10.1). A more direct method of determining DNA sequences is to use Chain Terminator Sequencing (Sanger and Coulson, 1975). The process is similar to PCR, however in addition to using standard nucleotides, a low concentration of labelled chain-terminating nucleotides are added to the reaction mix. Chain-terminating nucleotides are randomly incorporated as the sequence is extended, causing the extension step to stop. At the end of the cycling process, fragments of every length should be present in the reaction, each with a label on the final (terminating) nucleotide. These fragments are then separated by size, and the exact DNA sequence is read from the labels on the terminating nucleotides. Although terminating nucleotides were initially radioactively labelled (Figure 10.2A) dye-terminator sequencing, in which each of the four bases is labelled with a different fluorescent dye molecule, is now more common (Ansorge *et al.*, 1986; Smith *et al.*, 1986; Prober *et al.*, 1987; Figure 10.2B).

In recent years, other methods have been introduced which promise to be of use in anthropological studies, including sequencing by hybridisation (Microarrays) (e.g. Syvanen, 2005), nanopore sequencing (Deamer and Akeson, 2000), and pyrosequencing (Ronaghi *et al.*, 1996; 1998), including a recently described parallel version, capable of sequencing 1.6 million reactions in one run, that promises to be exceedingly powerful (Margulies *et al.*, 2005; Poinar *et al.*, 2006).

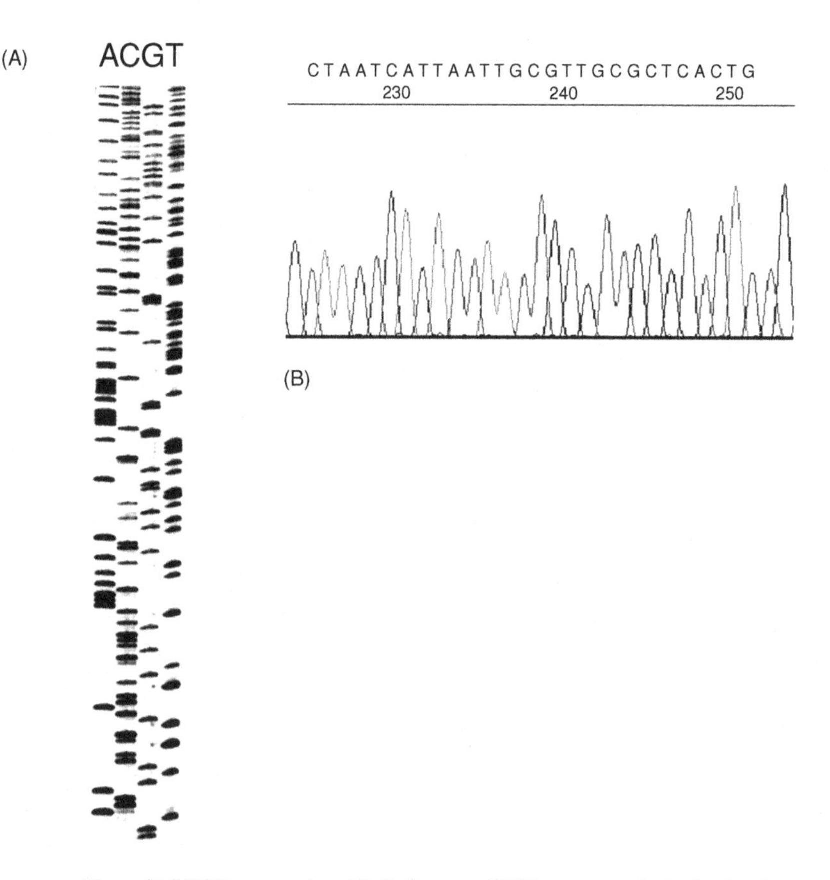

Figure 10.2 **DNA sequencing.** (A) Radiogram of DNA sequence obtained using the original chain terminator method (Sanger and Coulson, 1975). Four parallel DNA sequencing reactions were run for each sample to be sequenced, in each reaction one of the four nucleotides was radiolabelled (in this case from left to right A,C,G,T). Sequenced products were then denatured and separated by gel electrophoresis, before being visualised by exposing the gel to radiation-sensitive film. As the shortest products migrate the most quickly towards the positive cathode, the sequence can be read by identifying the shortest fragment (here a G at the bottom of the figure), then moving up the gel by fragment size (here a second G, then a T, then an A, then 2 C's and so on). Radiogram courtesy of David Heaf and the International Forum for Genetic Engineering (www.ifgene.org). (B) Dye-terminator DNA sequencing. In place of radiolabelling, each nucleotide is labelled with a dye (one colour for each of the four nucleotides) that can be excited by a laser. The colours enable all nucleotides to be incorporated into a single reaction, after which DNA fragments can be separated using gel electrophoresis, or through capillaries. The fluorescence of the dyes is captured and interpreted by sequencing software, generating a sequencing electropherogram. This method of sequencing does not rely on user discrimination between size of fragments, and has the additional benefit of enabling the generation of much longer fragments in a single sequencing run.

Manipulating DNA data

Tracing changes in DNA sequences through time

DNA data are used in a variety of ways to address hypotheses about human evolution.

For example, DNA data can be used to determine how modern populations are related to past populations (e.g. Rudbeck *et al.*, 2005; Töpf *et al.*, 2006), even if past populations are extinct, admixed with immigrant DNA, or simply inaccessible (e.g. Lalueza-Fox *et al.*, 2001; 2004; Endicott *et al.*, 2003; Haak *et al.*, 2005). DNA can also be used to identify specimens, by providing information about familial relationships, or through DNA fingerprinting (discussed in detail below).

The primary force that underlies changes in DNA sequences is mutation. Importantly, mutations can happen in any cell, however only those mutations that occur in germ-line cells have the potential to be passed to the next generation. In most cases, once a mutation arises in a population, it either disappears again, or it eventually becomes a substitution, in which descendent individuals in that population carry the new mutation.

Various statistical methods have been described that rely on specific assumptions about how likely different mutations are to occur through time. These assumptions become an **evolutionary model**, and are used to predict the relationships and genetic distance (which can be extended to distance in time) between DNA sequences. Models of molecular evolution vary from the extremely simple, for example the Jukes-Cantor model (1969), which assumes that all nucleotides, which are present at equal frequency in the genome, are equally likely to change to all other nucleotides; to very complex, for example the general time-reversible model (REV), which allows for different rates of change between all nucleotides and unequal base frequencies. A general description of the different evolutionary models can be found in Swofford *et al.* (1996) or Lió and Goldman (1998).

Phylogenetic methods combine an evolutionary model with statistical techniques used to construct and evaluate **phylogenetic trees**, which are dendrograms describing the relationships between a set of DNA sequences (Figure 10.3). We will not review the methods used for estimating and evaluating phylogenetic trees here, but refer the interested reader to a comprehensive recent review by J. Felsenstein (2003). There are several free or inexpensive tools available for phylogeny estimation; among the most common are PAUP (Swofford, 1999), PHYLIP (Felsenstein, 1989), MEGA (Kumar *et al.*, 2004) and MrBayes (Huelsenbeck and Ronquist, 2001; Ronquist and Huelsenbeck, 2003), and as such the representation of results in the form of phylogenetic trees has become the norm in molecular evolutionary studies. However, it is important for the

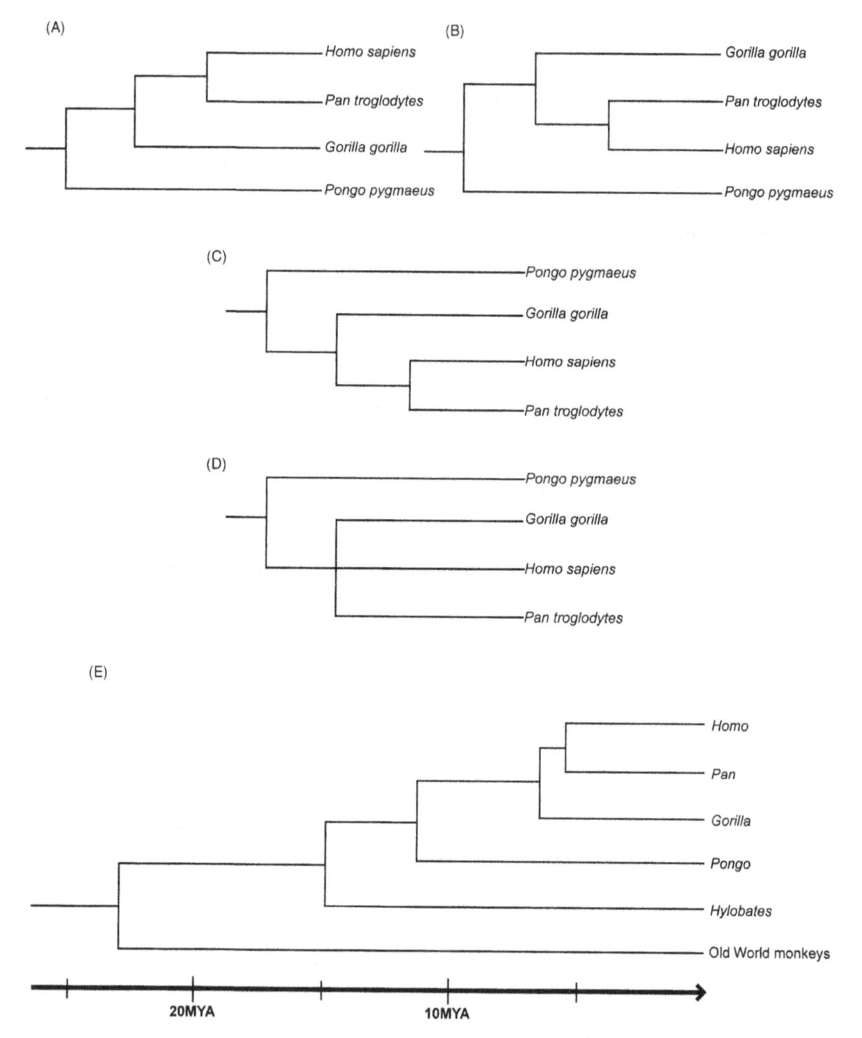

Figure 10.3 **Phylogenetic trees.** Rooted phylogenetic tree for four great apes; humans (*Homo sapiens*), common chimpanzee (*Pan troglodytes*), West African gorilla (*Gorilla gorilla*) and orangutan (*Pongo pygmaeus*), indicating evolutionary relationships between these taxa (tips) (A). Each node represents the most recent common ancestor of the lineages to the right of that node. When node positions are conserved, the phylogenetic tree can be drawn in a variety of ways: trees A, B and C convey identical information. The collapse of the human/chimpanzee – gorilla node, to form a polytomy between the three species (D) however changes the information conveyed by the tree. In addition to information about evolutionary relationships, some trees provide information about divergence times between taxa. Tree E describes both the relationship and approximate divergence times between the great apes (in millions of years) as in Stauffer *et al.* (2001). Here, the common ancestor of humans and chimpanzees existed approximately 5.5 million years ago (MYA), and the common ancestor of humans, chimpanzees and gorillas approximately 6.5 MYA.

inexperienced user to remember that a single tree is simply a set of hypotheses whose strength needs to be evaluated using statistical tests, and should not be treated necessarily as demonstrating the true relationship between individuals.

Most phylogenetic methods were developed to estimate the relationships between distantly related organisms. Nonetheless, there has been a growing trend to apply these methods to population data, which in many cases may not be entirely appropriate. Coalescent theory provides a conceptual framework for analysing genetic diversity within a population (or within a species if the assumption of a single population can be tolerated), by considering the rate at which lineages coalesce, or share a common ancestor, starting from the present and going backward through time until only one lineage remains. The Kingman coalescent (Kingman, 1982) provides the statistical framework for assessing coalescent processes in populations. Software incorporating the coalescent can be used to estimate **genealogies** (phylogenies for a particular gene within a population) as well as to infer changes in population size through time (Drummond *et al.*, 2003; 2005), such as would be the consequence, for example, of a severe population bottleneck (Shapiro *et al.*, 2004).

Coalescent methods generally assume a constant mutation rate through time, often referred to as a **molecular clock** (Zuckerkandl and Pauling, 1962; Sarich and Wilson, 1967; Kimura, 1968). The assumption of a molecular clock makes it possible to turn genetic distances into chronological time, providing a calendar timescale along which the evolutionary history of that population can be explored. A molecular-clock model is not appropriate for all types of data; for divergent species in particular it is clear that different genes, different organisms, and even different nucleotides and amino acids evolve at different rates (Ho *et al.*, 2005; 2007).

Likelihood ratio tests, for example as implemented in the software package PAUP (Swofford, 1999), can be used to determine whether the molecular clock is appropriate for a particular data set. Statistical methods are also available that allow the 'relaxation' of the molecular clock (e.g. Drummond *et al.*, 2006), making it possible to provide a temporal scale on trees for which the strict assumption of the molecular clock is not appropriate. Despite the problems associated with the molecular clock, it makes it possible to estimate divergence times between DNA sequences, such as between species or between clades within populations, and has thus been important in human molecular evolution.

A major difficulty associated with the molecular clock is in choosing a rate of evolution. When sequences are all contemporaneous, rates are often estimated by choosing one or more calibration point based on fossil evidence of a common ancestor of two modern lineages. The palaeontological age of that common ancestor is then used to calibrate how much genetic divergence has occurred per unit time between the two modern lineages, and that rate is extrapolated to the rest of the tree. Because rates of mutation and substitution may change through time, however, it is important to remember that divergence estimates based on fossil calibration points may be inaccurate.

Incorporating ancient data

When genetic information is isolated from non-contemporaneous specimens (as is the case with aDNA), information about specimen age can be incorporated into the phylogenetic model. When aDNA data are available, it may be possible to estimate the evolutionary rate based on internal, rather than external, calibrations, for example from the radiocarbon dates of the DNA sequences themselves (Drummond *et al.*, 2003; Shapiro *et al.*, 2004). In this instance, the rate of evolution may be estimated simultaneously with the genealogy. If enough data are available, this method could lead to a much more accurate estimate of the rate of evolution along a tree.

Reconstructing the human phylogenetic history

DNA data and the primate phylogeny

A substantial body of work provides a molecular interpretation of the evolutionary history of the order Primates. While molecular and morphological studies are in general agreement as to the broad-scale evolutionary relationships in the order, relationships at the species and genus level are more heavily debated (e.g. gibbons; Hayashi *et al.*, 1995; Hall *et al.*, 1998; Roos and Geissmann, 2001). However, given the paucity of primate fossil information at this taxonomic level, any information seems useful, and a consensus is likely to arise as more data and better analysis techniques are deployed.

The majority of applications for a well-supported phylogeny of the primates lie outside, or at least tangential to, the interests of biological anthropologists. Such a phylogeny could be used, for example, to estimate the rate at which new clades are generated (Purvis *et al.*, 1995), or to better understand the relationships between evolution and geography. A well-characterised primate phylogeny is also crucial in underpinning studies of primate comparative genomics (Enard and Pääbo, 2004). However, a better understanding of the evolutionary history of those taxa most closely related to humans, the Hominidae, is important in providing a framework for human evolution. Fortunately, this family has been well studied by molecular phylogeneticists, and both the branching order and timing of divergence events are now well established.

The relationship of Neanderthals to humans

Fossil evidence has thus far been unable to indisputably define many of the relationships within genus *Homo*, and it seems unlikely that genetic evidence derived from ancient specimens will improve this situation. Most divergence events within this genus happened both a long time ago and in locations too hot for the successful recovery of aDNA with current technology. The earliest proposed dates for fossil *Homo sapiens* (modern humans) are around 190 000 years ago, potentially within the limits of DNA recovery, but the material comes from a subtropical environment: the Kibish formation of southern Ethiopia (McDougall *et al.*, 2005). Deposits thought to contain the fossils of *H. erectus* are only 50 000 years old (Swisher *et al.*, 1996), but are from a tropical site near the Solo River in Java. The same problem affects the more recent remains of *Homo floresiensis* (Brown *et al.*, 2004): 13 000 years BP, but a tropical environment.

However, for much of the late Pleistocene, *Homo neanderthalensis* (Neanderthals) survived in temperate Europe until around 30 000 BP. It is unsurprising, therefore, that Neanderthal DNA has been successfully recovered from fossil material several times. The interest in establishing a Neanderthal phylogeny lies in the debate about their relationship to modern Europeans. Neanderthals either (i) were the direct ancestors of modern Europeans (a notion derived from the multiregional hypothesis), or (ii) contributed some genes to modern Europeans (a kind of alternative multiregionalism), or (iii) were replaced by modern humans in Europe (the Out of Africa hypothesis). DNA sequences from a Neanderthal would therefore (i) include all human sequences (with some additional diversity), (ii) have some shared sequences with humans, with each species/subspecies also having some species/subspecies-specific sequences, or (iii) be substantially different to humans.

In July 1997, Matthias Krings and colleagues published the first Neanderthal sequence, from the mitochondrial control region (CR) of the Neanderthal type specimen, and demonstrated that it lay outside of all human CR variation. The result supports the Out of Africa model in suggesting a replacement of Neanderthals in Europe by modern humans, without significant gene flow between them. This paper constitutes a landmark in aDNA analysis, both because it was the first time that the technique had made a significant contribution to palaeobiology, and because it brought together a suite of aDNA methods to support the authenticity of the sequence. At the same time, the original interpretation of the data was strengthened by publication by other groups of additional Neanderthal sequences, all supporting a significant evolutionary difference between Neanderthals and humans (Krings *et al.*, 2000; Ovchinnikov *et al.*, 2000; Schmitz *et al.*, 2002; Orlando *et al.*, 2006). Figure 10.4 describes the current understanding of the relationship between Neanderthals and modern humans. Absolute confirmation that no Neanderthal genes survive in the human genome will depend on the recovery of sufficient nuclear DNA from Neanderthal remains, a process that has recently been commenced by Svante Pääbo and colleagues. Initial results indicate that this is the case (Noonan *et al.*, 2006).

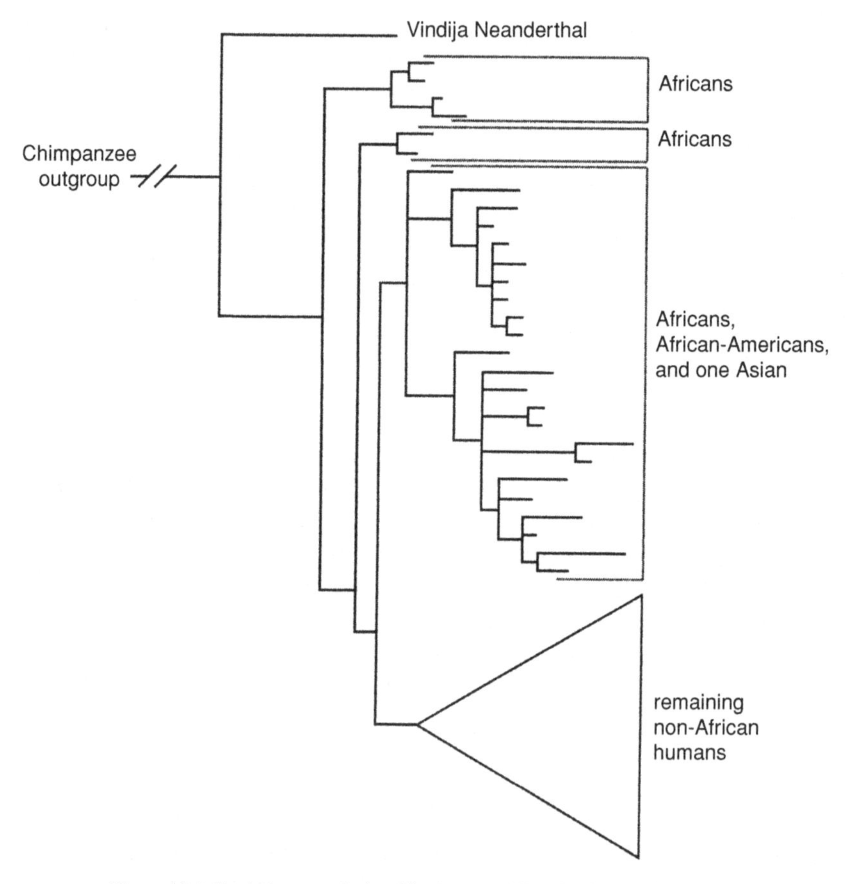

Figure 10.4 Evolutionary relationships between Neanderthals and modern humans. This phylogeny, by Hebsgaard *et al.* (2007), includes the Vindija Neanderthal, 7 chimpanzees, and 151 modern humans from around the globe. The Neanderthal sequence falls outside of the genetic diversity of all modern humans.

Human evolution 'Out of Africa'

Palaeoanthropological evidence indicates that roughly 100 000 years ago a morphologically diverse group of hominids inhabited the Old World: *Homo sapiens* in Africa and the Middle East, *Homo erectus* in Asia and *Homo neanderthalensis* in Europe. However, by 30 000 years ago the taxonomic diversity had vanished, and humans everywhere had evolved into the anatomically and behaviourally modern form.

How exactly this transformation occurred has given rise to two dominant schools of thought, one arguing for multiregional continuity between the forms (The Multiregional Continuity model), and the other for a single African origin of modern humans that ultimately resulted in the replacement of the other contemporary hominids (The Out of Africa model). Variations on these major themes have also been proposed, for example that several waves of migration out of Africa, rather than a single wave, occurred during human history (Templeton, 2005). Although there is a large amount of archaeological and anatomical evidence that can be drawn on in support of either hypothesis, the advent of genetic analyses provided a new tool with which to examine this question.

Evidence from mitochondrial DNA

Studies using mtDNA have generated considerable support for the Out of Africa hypothesis. In particular they have demonstrated that:

(1) Overall, the global diversity of mtDNA is surprisingly homogenous (Cann *et al.*, 1987; Ingman *et al.* 2000), consistent with a relatively recent origin of all modern populations.
(2) There is greater diversity among mtDNA in Africa than elsewhere in the world, and all non-African sequences form a subset of the African diversity. This phylogeny has been dated as having an origin 100 000 to 200 000 years ago (Cann *et al.*, 1987; Vigilant *et al.*, 1991; Ingman *et al.*, 2000). Molecular anthropologists have coined the term 'mitochondrial Eve' to describe the founder of this phylogeny, essentially the most recent matrilineal ancestor of all extant humans.
(3) The low amount of genetic variation observed in modern mtDNA also suggests that human origins outside of Africa stem from a small founding population, perhaps between 10 000 and 50 000 people who left Africa somewhere between 50 000 and 100 000 years ago (Rogers and Harpending, 1992).

Evidence from the Y-chromosome

Y-chromosome analyses provide additional support for the Out of Africa hypothesis. The term 'Y-Adam' has been coined to describe the male equivalent of mitochondrial Eve (that is, the male ancestor of the extant Y-chromosome phylogeny). A large number of studies have addressed the history of the Y-chromosome, and although the conclusions disagree on a number of issues, in general their results indicate that:

(1) No ancient (>200 000-year-old) Y lineages are found anywhere among extant populations, and the Y-chromosome phylogeny roots in Africa (Hammer *et al.*, 1998). Furthermore, the greatest diversity of Y-chromosomes is found among African populations, and all non-African lineages root within African clades.

(2) The age of the most recent common ancestor (MRCA), the restriction of the most divergent lineages to Africa, and evidence for an 'Out-of-Africa' range expansion (Hammer *et al.*, 1998) indicate that modern Y-chromosome diversity arose recently in Africa and replaced Y-chromosomes elsewhere in the world (derived from previous expansions).
(3) There is no evidence for any genetic contribution to the modern human Y-chromosome pool from other ancient *Homo* species (indeed there are few such signs observed using any other loci; Jobling and Tyler-Smith, 2003).

Summary and conclusions

Any study that uses genealogies to infer population history depends to a large extent on the assumptions and parameters of the evolutionary models under which they are constructed. For example, variation in population size (was the population a constant size, expanding, or even contracting?) will influence the shape of the genealogy and the rate at which lineages diverge. Other parameters that can influence results include random mating, generation time, and the rate of molecular evolution, which is often calibrated using a *Homo*–chimpanzee split of 5 Mya (a figure upon which a fair degree of uncertainty rests; Brunet *et al.*, 2002). Conclusions can also be affected by data choice.

For example, using 8 microsatellites from 445 Y-chromosomes Pritchard *et al.* (1999) estimate an origin of the human Y-chromosome lineage around 46 000 years ago (16 000–126 000 years ago) under a model of exponential population growth. Under a similar population model, but using variation in SNPs rather than microsatellite data, Thomson *et al.* (2000) find the MRCA of the Y-chromosome to be around 59 000 years ago (40 000–140 000 years ago).

An intuitive example of the importance of parameters is the choice of generation time applied to Y-chromosomal models (Jobling and Tyler-Smith, 2003). Often this is set at 20 years, with no uncertainty. Because men on average have offspring later than women, and continue to father children later in life, intergenerational time intervals are longer for male-only lines than for female-only lines. While choosing an appropriate generation time is naturally very difficult, this remains an important problem. If, for example, a value of 35 years is used instead of 20 years, one result would be an almost doubling of the estimated age for the MRCA of the Y-chromosome (Jobling and Tyler-Smith, 2003).

Using DNA data to reconstruct the history of recent populations

Over the last decade, perhaps the most significant contribution that genetics has made to the study of cultural change has been the identification of human movements. To demonstrate how genetic data can be used here, we will focus on a specific example: the Anglo-Saxon colonisation of England. We can use the extensive archaeological and historical literature to explore the types of genetic analyses that can be conducted. One additional advantage of focusing on this event is that, although long-debated by historians and archaeologists, it is covered by a relatively concise set of genetics literature. For more extensively studied migrations we direct the reader toward the literature on the peopling of the New World, the Neolithic colonisation of Europe, the Bantu expansion or the peopling of Polynesia (Jobling *et al.*, 2004).

The role of migration as an explanation for cultural changes observed in the archaeological record has been a major topic of controversy (Clark, 1966; Shennan, 2000). Culture-historical interpretations from the early part of the last century considered that evidence of wholesale cultural change automatically represented a substantial movement of people. However, the advent of the New Archaeology in the 1960s and 1970s led to a rejection of this view, favouring instead the movement of ideas through trade or the influx of a small ruling elite with minimal genetic input (e.g. Renfrew, 1987). More recently, this anti-migrationist view has been questioned, and migration models reconsidered (e.g. Burmeister, 2000). Despite the continuing increase in the quantity and quality of archaeological evidence, the role of migration continues to be unclear.

The Anglo-Saxon colonisation of England: background

The history of the British Isles is one of a series of cultural transitions, many of which stem from Continental Europe. In prehistory, the archaeological record shows a change to settled agricultural communities in the Neolithic (*c.* 4000 BC) and the spread of Celtic culture in the Late Bronze/Early Iron Age (*c.* 500 BC). This record is supplemented by documentary evidence for the Roman occupation (AD 43–*c.* 450), the spread of Anglo-Saxon language and culture (from *c.* AD 450) and the Viking invasions and settlements (*c.* AD 800–1000).

The role of migration in cultural, ethnic and demographic change in post-Roman Britain continues to be debated (Härke, 2002; Hills, 2003). It is generally accepted that between the fifth and seventh centuries AD, southern and eastern Britain were subject to immigration by various Germanic groups, collectively termed Anglo-Saxons, but the source and scale is questioned (Myres, 1986; Higham, 1992). The archaeological and historical evidence, which includes textual sources, the intrusive material culture of Continental type, the appearance of a new funerary rite (cremation), the introduction of a new language (Old English), and skeletal remains, has been interpreted as large-scale human movement (e.g. Cameron, 1975; Myres, 1986). In contrast, minimalist interpretations have been proposed, suggesting that immigrants were restricted predominantly to an elite that succeeded in influencing the natives to follow in their tastes and fashions, who thus became Anglo-Saxon (e.g. Arnold, 1984; Higham, 1992). A major problem with the latter interpretation is how such a small number of incomers succeeded in changing the language of a large native population (estimated to be up to 3.7 million by AD 300), and giving names to almost every place and field in England (Cameron, 1975). Similarly, the role of Viking immigrants on the current British population has also been questioned. Again, key to the debate is the number of settlers, with the conventional hypothesis of a large number of retired raiders and immigrants taking land (e.g. Stenton, 1971) being contested, e.g. Sawyer (1971).

Thus several potentially tractable questions exist for the genetic anthropologist:

(1) What was the scale of the migration? (Inferring demographic history)
(2) Where did the migrant population come from? (Phylogeography and the identification of a source population)
(3) To what extent did they interact with the existing autochthonous population? (Extent of admixture)
(4) Did both men and women invade, or was the migration a male-only affair? (Sex-biased migration)
(5) When did the migration happen? (Timing of events; use of the molecular clock.)

Early studies of patterns of variation in genetic systems within Britain established that variation existed, but could not formally test alternative explanations for these patterns. These include analyses of blood groups and Human Leukocyte Antigen (HLA) genes (Bodmer *et al.*, 1993), serum proteins and isoenzymes (Mastana *et al.*, 1993), other classical genetic markers (Falsetti and Sokal, 1993; Cavalli-Sforza *et al.*, 1994; Mastana and Sokol, 1998), and patterns of disease inheritance (Tyfield *et al.*, 1997).

More recently, both Y-chromosome and mtDNA data have been used to generate high resolution haplotypes that allow inferences to be made on the timing and scale of recent demographic events. Wilson *et al.* (2001) compared Y-chromosome, X-chromosome and mtDNA variation in eight population samples to investigate how genetic differences could be associated with cultural transitions in North Wales and Orkney, two areas at the fringes of the British Isles. The authors used the concept of signature haplotypes, or haplotypes occurring at high frequency in a particular population, as a means to identify past movements. In this study, mtDNA and X-chromosome markers failed to significantly differentiate between the sampled populations, likely due to female-mediated gene flow (i.e. female migration) during the later prehistoric period. A high degree of similarity between Irish and Welsh Y-chromosomes was identified, but these 'Celtic' male lineages and those from Norway, Friesland and Orkney were all significantly differentiated. The similarity between Wales and Ireland could not occur by chance, but suggests shared ancestry between the two populations. Interestingly, the Orkney sample seems to be intermediate between the Celtic and Norwegian samples, suggesting a combined (admixed) origin for this population, from both prehistoric and Viking sources.

Further comparisons used Basque data as a marker of Palaeolithic populations, and found high similarity to the male Celtic genetic component. This suggests that subsequent cultural transitions in North Wales were not associated with substantial incoming male gene flow. The study is important in identifying potential sources of migrants to the British Isles and in demonstrating an apparent continuity of Y-chromosome types in Wales and Ireland from early Prehistory to the present.

In order to address the effects of cultural transitions in England, the region of the Anglo-Saxon migration, Weale *et al.* (2002) compared the same set of Y-chromosome markers in samples along an East–West transect from East Anglia into North Wales. The authors limited the search to potential source populations, measuring the relative input from only two: Norway (as proxies for a Norse population) and Friesland (one of the most frequently proposed sources for the Anglo-Saxon migration). They found a striking similarity between English men and Frisian men, but a pronounced dissimilarity between English men and North Welsh men. This similarity cannot be due to 'background' migration (continuous low levels of migration) because the amount of background migration required to explain the data would be enormous: three times higher, in relative terms, than that occurring into and from the European Economic Area as a whole over the last 25 years.

Thus, in order to explain this pattern it was necessary to identify a migration event that (i) originated somewhere near to modern-day Friesland; (ii) was sufficiently recent that the two groups have not genetically 'drifted' away from each other; (iii) was of a scale large enough to result in a significant replacement of the native Y-chromosome type – the estimate given in the paper is of 50–100 % replacement; and (iv) led to population change that today affects populations throughout central England, but not in Wales. These criteria allowed the authors to reject the possibility that the data derive from a Neolithic or earlier migration, as this would be too long ago.

Equally, more recent historical events (such as the sixteenth century influx of Frisians into Norwich and surroundings) are too local and of insufficient scale. In conclusion, considering the known historical and archaeological migrations in England, these criteria are most convincingly satisfied by a large Anglo-Saxon mass migration event. A subsequent study by Capelli *et al.* (2003) compared Y-chromosome data from samples collected from Norway, Denmark, North Germany to those from 25 regularly spaced locations throughout the British

Isles. This study supports the notion of a significant demographic influence from incoming Anglo-Saxons, and possibly Vikings, whilst also identifying differences in the extent of population change in different parts of England.

To return to our set of questions, we find that genetic analysis has been useful in addressing hypotheses about source populations, timing and admixture. The data support the notion of a migration from the area around Friesland into England during the fifth to eighth centuries, leading to a substantial replacement of the indigenous population.

However, many questions remain unanswered. One obvious gap is the lack of a pan-British Isles mtDNA survey, which would allow comparison of both female and male migration history. A second weakness is the lack of samples from other Continental regions that may have had a role in British migration history. Although Friesland looks like the most viable source population, it may be that other areas, inland or around the coastline, are equally likely sources. Finally, although we know that the migration had a significant impact on the native population of England, this does not establish the scale of the migration. Disentangling the migration event from subsequent demographic changes is complex, and in this case perhaps best addressed by aDNA studies.

Using DNA sequences to identify individual samples

DNA can be a powerful tool for the identification of human remains, whether those remains are modern, historical or archaeological. Although the principles behind the tests are the same, the practicalities of collection, DNA extraction and analysis need to be considered in association with the preservation of the sample in question, with reference to potential problems associated with damage and contamination.

Historically, the standard DNA-based method for the identification of human samples has been DNA fingerprinting (Figure 10.5; Jeffreys *et al.*, 1985). The original technique involved the incubation of purified genomic DNA with restriction enzymes, which were chosen so as to ensure that, post digestion, every non-identical twin would exhibit a unique DNA profile or 'fingerprint'. DNA fingerprinting is often used as a means to determine the genetic relationship between individuals (for example parentage of children), and as a means to identify individuals from trace evidence (such as that found at crime scenes). In 1988, a suspect was convicted of the rape and murder of two schoolgirls in a small Leicestershire town. Where blood group and enzyme comparisons could only narrow down the list of suspects, comparison of DNA fingerprints taken from the crime scene with that from men in three local villages showed

(A)

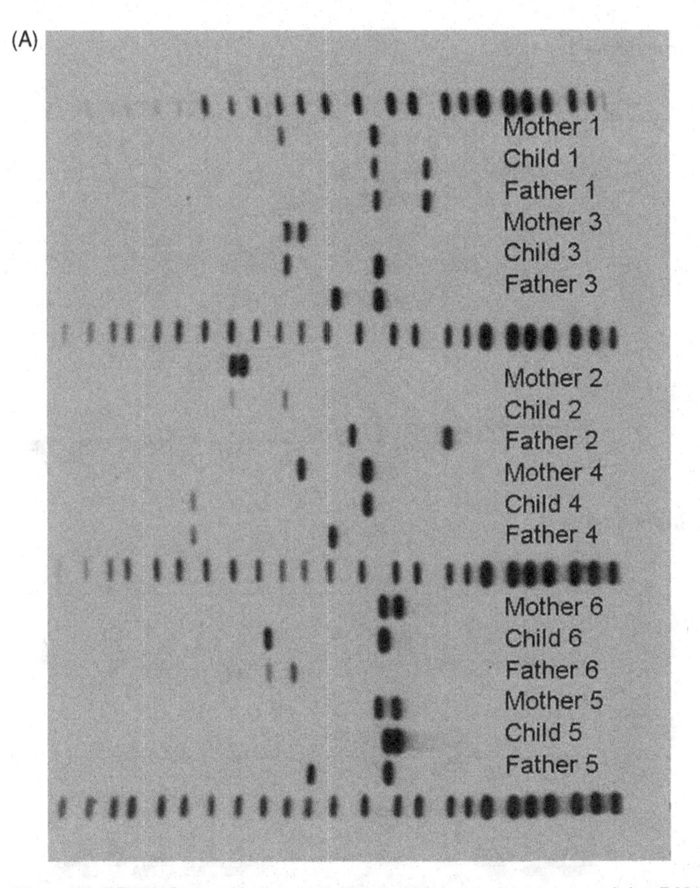

Figure 10.5 **DNA fingerprinting techniques.** (A) Autoradiogram containing DNA from six paternity cases, analysed using a gel-based DNA fingerprinting technique. The genomic DNA was first digested with the restriction enzyme *HinfI*, and fragments were separated on an agarose gel. The DNA was transferred to a membrane, then hybridised with a radiolabelled probe for the genomic marker D2S44 (YNH24). This marker is highly polymorphic (>97 % of people are heterozygotes), with alleles ranging from 50 to >5000 bp. This example demonstrates that one of each of the children's alleles is found in each mother, and the other in the putative genetic father (thus confirming paternity). However, putative genetic father 3 contains neither allele, excluding him as child 3's father. Autoradiogram provided courtesy of Gunilla Holmlund Ph.D., The National Board of Forensic Medicine, Department of Forensic Genetics and Forensic Toxicology, Linköping, Sweden. (B) DNA fingerprinting using simultaneous typing of multiple SNPs. Here, 4 SNPs typed in 2 individuals for 2 markers are shown. The markers are characteristically chosen so as to ensure that no two genetically non-identical individuals are the same, thus such methods have the power to rapidly compare and contrast purified DNA from individuals. Genescan trace provided courtesy of Juan J Sanchez Ph.D., Institute of Forensics, University of Copenhagen, Denmark.

(B)
Individual 1

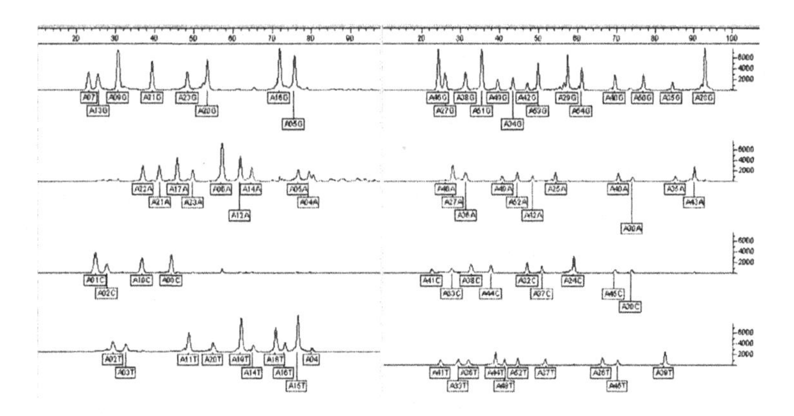

Individual 2

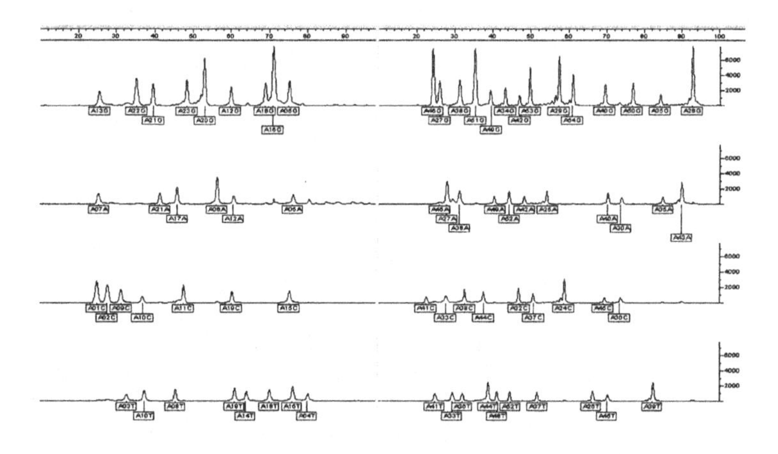

Figure 10.5 (*cont.*)

conclusively that Mr Pitchfork had committed the crime. This was the first use of DNA fingerprinting technology as a forensic tool (Bowyer *et al.*, 2004).

Recently, the methods underlying DNA fingerprinting analyses have changed. Instead of RFLP, studies routinely use alternative techniques such as microsatellites, direct sequence analysis of variable regions or, increasingly, sequencing of specific SNPs distributed across the genome, using inexpensive and powerful mass screening methods such as multiplex PCR with minisequencing (e.g. Sanchez *et al.*, 2003). MtDNA screening has been particularly useful in comparison of poorly preserved specimens with potential living maternal relatives. For example, mtDNA has been used to identify remains thought to belong to members of the Romanov royal family, who were executed as a result of the Russian Revolution of 1917 (Gill *et al.*, 1994). Screening methods have also been used to identify human remains from the Vietnam war (Holland *et al.*, 1993) and victims of the Bosnia genocide (Ladika, 2001) and the September 11th 2001 attacks on New York's World Trade Center (Budimlija *et al.*, 2003).

Nuclear DNA sequences can be used to identify the sex of a specimen (Sullivan *et al.*, 1993). The most commonly targeted gene to this effect is amelogenin, a gene located on both the X- and Y-chromosomes. The amelogenin gene on the Y-chromosome is 6 bp shorter than its counterpart on the X-chromosome. PCR amplification of this gene therefore results in one fragment from a female (XX) specimen, and two differently sized fragments in a male (XY) specimen. Problematically, this technique relies on an absence of the shorter Y-linked allele to confirm that a specimen is female. As such, contamination by modern male DNA can lead to inaccurate results, which is particularly important in studies involving ancient or poorly preserved specimens.

While DNA profiling is becoming increasingly common, it is not entirely foolproof, even when the specimens in question are modern. As DNA profiling has become increasingly popular in criminal cases, a large body of statistical work has arisen to try to determine how unique an individual DNA fingerprint actually is. The theoretical risk of a coincidental match (for which a common target is a 1 in 100 billion chance of an incorrect identification) is determined using information about relatedness between individuals as well as rates of laboratory error (Saks and Koehler, 2005). However, critics argue that the rate of laboratory error, for example arising due to sequencing errors (Brandstätter *et al.*, 2005), may be much higher than 1 in 100 billion, and that laboratory procedures often do not reflect the theory under which such probabilities were computed, and therefore reflect an artificially high level of confidence (see Saks and Koehler, 2005).

Using DNA to characterise natural selection in human populations

Most of the examples discussed above consider the accumulation of neutral mutations along DNA sequences, and how these mutations can be used to track changes occurring in diverging lineages. Mutations, however, can be neutral, deleterious or beneficial, and this will influence the rate at which that mutation goes to fixation in a population (becomes a substitution) or is lost from that population. It is believed that the majority of mutations are neutral (Kimura,

1983) or nearly neutral (Ohta, 1992), in which instance the probability that the mutation survives depends on a process called **genetic drift**, which is linked to population size. In small populations, neutral mutations will go to fixation in the population (or be lost from that population) more rapidly than in large populations. If the population size remains constant, neutral mutations are likely to arise and go to fixation at a constant rate (the clock rate), and this rate can be used to predict, for example, how long two populations have been isolated from each other.

Many non-neutral mutations occur in coding regions of the genome. In these regions, nucleotide triplets, or codons, each contain instructions for a specific amino acid. Coding fragments of DNA are instructions for strings of amino acids, which in turn become proteins. The human genetic code contains 20 amino acids, some more similar to each other than others, and three stop codons. Because there are 64 possible combinations of the four nucleotides in triplets, many combinations result in the same amino acid, an effect known as the 'degeneracy' of the genetic code. The majority of degenerate changes occur at third codon positions, and so may be neutral. First and second position changes, however, often result in an amino acid change, in which case a mutation occurring at that position may change the amino acid sequence, possibly in a deleterious or beneficial way. Other types of mutations may also be non-neutral in coding sequences, such as those that cause a shift in the reading frame or a premature stop codon.

Once a non-neutral mutation is introduced into a population, the rate at which it goes to fixation (or is lost) can be used to infer the strength of selection on that particular change. If a mutation has been introduced to a population but has yet to go to fixation, that population is said to be **polymorphic** for that mutation. Any one version of a gene which contains polymorphisms is referred to as an **allele**, and the relative abundance of different alleles in a population is called the **allele frequency**. Because different environments will produce different selective conditions, populations originating in different environments may differ in allele frequency for the same trait, and this information can be used to infer something about the history of each population and the environment in which they lived.

Lactase persistence and positive selection

One of the most frequently cited examples of positive selection in humans is the maintenance of the enzyme *lactase*, which is responsible for breaking down lactose, the major sugar of milk. In most humans, and in fact most mammals, production of lactase stops shortly after weaning, resulting in an inability to digest fresh milk. However, in some human populations, the ability to digest milk (due to the continued production of lactase) persists into adulthood. The genetic changes associated with lactase persistence have been well characterised using pedigree analysis, and are described in detail elsewhere (Enattah *et al.*, 2002). Lactase persistence is most commonly found in populations with a cultural history of drinking fresh milk, such as Europeans and African pastoralists, suggesting a link between the persistence of this mutation and a specific adaptive benefit to individuals carrying this mutation over those that do not. Specifically, the geographic distribution of lactase persistence suggests an evolutionary adaptation to changes in diet brought about by the development of dairying (Hollox and Swallow, 2002). As such, this is an example of how positive selection can act on a mutation to increase its frequency within a population.

From beneficial to deleterious: thrifty genes in modern society

Mutations that confer an adaptive advantage in one environment will not necessarily be beneficial in all environments, and this will lead to differences in the frequency of these mutations between populations. Data about allele frequency can be used therefore to discriminate between populations, and to learn about the environment in which different populations evolved. Equally, as the environment in which a population lives changes, so may the effect of accumulated mutations. For example, a recent global increase in non-insulin dependent diabetes mellitus (NIDDM, or type 2 diabetes), in particular among Amerindians, Australian Aborigines and Pacific Islanders, has been proposed to be the consequence of dietary changes brought about by modern society (Neel, 1962; 1982).This hypothesis suggests that while overproduction of insulin (the major cause of diabetes) was advantageous in societies where food sources were scarce or unpredictable, as it allowed for the build-up of fat deposits when food was abundant, this same mechanism is deleterious (leading to NIDDM, and even obesity) in the modern, resource-rich society. Other hypotheses suggest that poor nutrition of the mother during pregnancy and of the child in early infancy may increase the risk of NIDDM, by predisposing the child to being nutritionally 'thrifty' (Hales and Barker, 1992; Jobling *et al.*, 2004). In both instances, this is an example of how a gene may have switched from being beneficial to deleterious through a change in environment.

Fixation of a beneficial mutation: FOXP2

Some mutations that arise by chance may be so beneficial that they rapidly go to fixation in what is known as a **selective sweep**. A selective sweep is characterised by the elimination of alternative alleles at the gene in question, as well as a reduction in diversity in genes that are physically close to (linked to) that gene. Such beneficial mutations can be identified by comparing genes between humans and our close relatives, and have the potential to provide important insights into human evolution. An interesting example is the story surrounding the evolution of the FOXP2 transcription factor, a member of the Forkhead-box or FOX gene family. FOXP2 is highly conserved across vertebrates (Scharff and Haesler, 2005). For example, no amino acid differences are observed between chimpanzee, gorilla and rhesus monkey, and only one amino acid change is seen in the 75-million-year divergence between these three species and mouse (Enard *et al.*, 2002). In contrast, two amino acid changes have occurred along the human lineage in only six million years of divergence from the chimpanzee. Additionally, evidence from the region surrounding the FOXP2 gene supports the hypothesis that this region has undergone a recent selective sweep.

So what is so advantageous about these two mutations on the FOXP2 gene? Because of the complexities of interactions between genes, it is still unclear exactly what effect these mutations have. However, anecdotal evidence from individuals with defective copies of these genes suggests that it is linked to the development of speech and language (Lai *et al.*, 2001). People who lack two normal copies of the gene have considerable difficulty in language comprehension and speech production as well as limited motility of the mouth and face. Palaeoanthropological evidence also supports this hypothesis; the selective sweep observed in FOXP2 is predicted to have occurred within the last 200 000 years, which is consistent with the appearance of language proficiency (Enard *et al.*, 2002). In conclusion, FOXP2 appears to be important in some aspect of language evolution, and clearly represents an example of an adaptive human-specific trait. Importantly, FOXP2 is a transcription factor, which means that it plays an important role in regulating how and when other genes are expressed. As such, mutations in this single gene can directly affect multiple genes involved with a diversity of processes. While current research has shown an important link between FOXP2 and the evolution of language, more work will be required to fully understand this fundamental adaptation in human evolutionary history.

The future of DNA in biological anthropology

The future role of ancient DNA

Despite the severity of problems associated with working with ancient human specimens, data provided by aDNA analyses have contributed significantly to studies of human evolution over the last two decades. However, overcoming the limitations imposed by DNA damage and contamination will be key to taking full advantage of temporally sampled data. While methodological advances provide some relief from experimentally derived contaminants (Pääbo *et al.*, 2004; Salamon *et al.*, 2005), contaminants present within samples provide a much greater challenge (Gilbert *et al.*, 2005a), in particular as empirical evidence has shown that once a sample is contaminated, it is exceedingly difficult to decontaminate (Richards *et al.*, 1995; Kolman and Tuross, 2000; Hofreiter *et al.*, 2001b; Gilbert *et al.*, 2005b; 2006). A number of recent studies have begun to characterise the nature of these contaminants (e.g. Gilbert *et al.*, 2005b; Malmström *et al.*, 2005; Salamon *et al.*, 2005) with the goal of developing new techniques to avoid co-amplification.

A further limitation to aDNA is the amount and type of genetic information that can be extracted from degraded specimens. Technological developments will be required to expand both the size of DNA fragments that can be amplified, and sensitivity of the DNA extraction process (so as to make possible the amplification of nuclear DNA). Research into the character of damage undergone by DNA during the decay process (Pääbo *et al.*, 1989; Höss *et al.*, 1996; Poinar *et al.*, 1998; Hansen *et al.*, 2001; Hofreiter *et al.*, 2001a; Gilbert *et al.*, 2003a, b; Mitchell *et al.*, 2005) provides the first step toward the development of DNA repair mechanisms, with the potential to address both of these shortcomings. Techniques such as multiplex PCR (Ballabio *et al.*, 1990), which has recently been used to sequence the complete mitochondrial genome of the mammoth (Krause *et al.*, 2006), and advances in whole genome cloning that enable the recovery of large amounts of sequence from degraded sources (e.g. Noonan *et al.*, 2005) can significantly increase the amount of information that can be extracted from ancient samples, and therefore the range of hypotheses it is possible to address.

The future?

One of the biggest challenges to molecular anthropology is the development of appropriate analytical models for processing the huge amount of data being produced. This includes methods for analysing complete mitochondrial genomes (Ingman *et al.*, 2000; Ingman and Gyllensten, 2001; 2003), which, once a technological challenge, can now be recovered in a single day. New methods that target informative point mutations, such as multiplex PCR with minisequencing (Brandstätter *et al.*, 2003; Sanchez *et al.*, 2003) and hybridisation assays, also make possible the rapid production of large amounts of informative data (Pakendorf and Stoneking, 2005). Indeed the number of complete nuclear genomes currently being assembled attests to the recent increase in sequencing capability, which will need to be accompanied by corresponding increases in analytical power. Even without complete nuclear genomes, an ever-growing catalogue of identified SNPs is providing better resolution for forensic, medical and phylogenetic applications.

For example, the International HAPMAP project (www.hapmap.org) was launched in 2002, with the aim to provide a complete haplotype map of the human genome as a resource to assist in the identification of genes associated with human disease. In addition, the late 2005 launch of the Genographic Project further highlights the recent progress of human genetic research. A not-for-profit venture launched by National Geographic, IBM and the Waitt Family Foundation, and involving a global collaboration based in 10 regional centres, this project aims to sample 100 000 male volunteers of known ethnic ancestry, and type their Y-chromosomes and mtDNA genomes in order to refine existing phylogenies. As such, the project represents the largest-scale anthropological genotyping project ever, with the potential to revolutionise molecular anthropological research.

Finally, a note of caution: it is often important to consider the value of a sample when deciding whether DNA analysis is appropriate, particularly with ancient or rare specimens. Not only will the extraction have a financial and time cost, but the process itself is necessarily a destructive one. In many studies, the initial goal is simply to recover the DNA sequence of the sample, which while in some situations can provide scientifically useful information, in others is done simply because it can be. This raises an important issue: in many precious samples, should such data be recovered if it serves no scientific purpose?

Note

1. Author to whom correspondence should be addressed.

References

Ansorge, W., Sproat, B. S., Stegemann, J. and Schwager, C. (1986). A non-radioactive automated method for DNA sequence determination. *Journal of Biochemical and Biophysical Methods*, **13**, 315–23.

Arnold, C. J. (1984). *Roman Britain to Saxon England*. London: Croom Helm.

Ballabio, A., Ranier, J. E., Chamberlain, J. S., Zollo, M. and Caskey, C. T. (1990). Screening for steroid sulfatase (STS) gene deletions by multiplex DNA amplification. *Human Genetics*, **84**, 571–3.

Bodmer, J. G., Marsh, S. G. E., Albert, E. D. *et al.* (1993). Nomenclature for factors of the HLA system, 1991. *Immunobiology*, **187**, 51–69.

Bowyer, V. L., Graham, E. A. and Rutty, G. N. (2004). 9649 forensic web watch – DNA in forensic science. *Journal of Clinical Forensic Medicine*, **11**, 271–3.

Brandstätter, A., Parsons, T. J. and Parson, W. (2003). Rapid screening of mtDNA coding region SNPs for the identification of west European Caucasian haplogroups. *International Journal of Legal Medicine*, **117**, 291–8.

Brandstätter, A., Sanger, T., Lutz-Bonengel, S. *et al.* (2005). Phantom mutation hotspots in human mitochondrial DNA. *Electrophoresis*, **26**, 3414–29.

Brown, P., Sutikna, T., Morwood, M. J. *et al.* (2004). A new small-bodied hominin from the Late Pleistocene of Flores, Indonesia. *Nature*, **431**, 1055–61.

Brown, W. M. (1980). Polymorphisms in mitochondrial DNA of humans as revealed by restriction endonuclease analysis. *Proceedings of the National Academy of Science*, **77**, 3605–9.

Brunet, M., Guy, F., Pilbeam, D. *et al.* (2002). A new hominid from the Upper Miocene of Chad, central Africa. *Nature*, **418**, 145–51.

Budimlija, Z. M., Prinz, M. K., Zelson-Mundorff, A. *et al.* (2003). World trade center human identification project: experiences with individual body identification cases. *Croatian Medical Journal*, **44**, 259–63.

Burmeister, S. (2000). Archaeology and migration: approaches to an archaeological proof of migration. *Current Anthropology*, **41**, 539–67.

Cameron, K. (1975). *Place-Name Evidence for the Anglo-Saxon Invasion and Scandinavian Settlements*. Nottingham: English Place-Name Society.

Cann, R. L., Stoneking, M. and Wilson, A. C. (1987). Mitochondrial DNA and human evolution. *Nature*, **325**, 31–6.

Capelli, C., Redhead, N., Abernethy, J. K. *et al.* (2003). A Y chromosome census of the British Isles. *Current Biology*, **13**, 979–84.

Cattaneo, C., Craig, O. E., James, N. T. and Sokol, R. J. (1997). Comparison of three DNA extraction methods on bone and blood stains up to 43 years old and amplification of three different gene sequences. *Journal of Forensic Sciences*, **42**, 1126–35.

Cavalli-Sforza, L. L., Menozzi, P. and Piazza, A. (1994). *The History and Geography of Human Genes*. Princeton: Princeton University Press.

Clark, G. (1966). The invasion hypothesis in British archaeology. *Antiquity*, **40**, 172–89.

Cooper A. and Poinar, H. (2000). Ancient DNA: do it right or not at all. *Science*, **289**, 1139.

Deamer, D. W. and Akeson, M. (2000). Nanopores and nucleic acids: prospects for ultrarapid sequencing. *Trends in Biotechnology*, **18**, 147–51.

Drummond, A. J., Pybus, O. G., Rambaut, A., Forsberg, R. and Rodrigo, A. G. (2003). Measurably evolving populations. *Trends in Ecology and Evolution*, **18**, 481–8.

Drummond, A. J., Rambaut, A., Shapiro, B. and Pybus, O. G. (2005). Bayesian coalescent inference of past population dynamics from molecular sequences. *Molecular Biology and Evolution*, **22**, 1185–92.

Drummond, A. J., Ho, S. Y. W., Philips, M. J. and Rambaut, A. (2006). Relaxed phylogenetics and dating with confidence. *PLoS Biology*, **4**, e88.

Enard, W. and Pääbo, S. (2004). Comparative primate genomics. *Annual Review of Genomics and Human Genetics*, **5**, 351–78.

Enard, W., Przeworski, M., Fisher, S. E. *et al.* (2002). Molecular evolution of FOXP2, a gene involved in speech and language. *Nature*, **418**, 869–72.

Enattah, N. S., Sahi, T., Savilahti, E. *et al.* (2002). Identification of a variant associated with adult-type hypolactasia. *Nature Genetics*, **30**, 233–7.

Endicott, P., Gilbert, M. T. P., Stringer, C. *et al.* (2003). The genetic origins of the Andaman Islanders. *American Journal of Human Genetics*, **72**, 178–84.

Falsetti, A. B. and Sokal, R. R. (1993). Genetic structure of human populations in the British Isles. *Annals of Human Biology*, **20**, 215–29.

Felsenstein, J. (1989). PHYLIP (Phylogeny Inference Package). *Cladistics*, **5**, 164–6.

Felsenstein, J. (2003). *Inferring Phylogenies*. Sunderland: Sinauer Associates.

Gilbert, M. T. P., Hansen, A. J., Willerslev, E. *et al.* (2003a). Characterization of genetic miscoding lesions caused by postmortem damage. *American Journal of Human Genetics*, **72**, 48–61.

Gilbert, M. T. P., Willerslev, E., Hansen, A. J. *et al.* (2003b). Distribution patterns of postmortem damage in human mitochondrial DNA. *American Journal of Human Genetics*, **72**, 32–47.

Gilbert, M. T. P., Wilson, A. S., Bunce, M. *et al.* (2004). Ancient mitochondrial DNA from hair. *Current Biology*, **14**, R463–R464.

Gilbert, M. T. P., Bandelt, H. J., Hofreiter, M. and Barnes, I. (2005a). Assessing ancient DNA studies. *Trends in Ecology and Evolution*, **20**, 541–4.

Gilbert, M. T. P., Rudbeck, L., Willerslev, E. *et al.* (2005b). Biochemical and physical correlates of DNA contamination in archaeological human bones and teeth excavated at Matera, Italy. *Journal of Archaeological Science*, **32**, 785–93.

Gilbert, M. T. P., Menez, L., Janaway, R. C. *et al.* (2006). Resistance of degraded hair shafts to contaminant DNA. *Forensic Science International*, **156**, 201–7.

Gill, P., Ivanov, P. L., Kimpton, C. *et al.* (1994). Identification of the remains of the Romanov family by DNA analysis. *Nature Genetics*, **6**, 130–5.

Haak, W., Forster, P., Bramanti, B. *et al.* (2005). Ancient DNA from the first European farmers in 7500-year-old Neolithic sites. *Science*, **310**, 1016–18.

Hales, C. N. and Barker, D. J. P. (1992). Type-2 (non-insulin dependent) diabetes mellitus: the thrifty phenotype hypothesis. *Diabetologia*, **35**, 595–601.

Hall, L. M., Jones, D. S. and Wood, B. A. (1998). Evolution of the gibbon subgenera inferred from cytochrome b DNA sequence data. *Molecular Phylogenetics and Evolution*, **10**, 281–6.

Hammer, M. F., Karafet, T., Rasanayagam, A. *et al.* (1998). Out of Africa and back again: nested cladistic analysis of human Y chromosome variation. *Molecular Biology and Evolution*, **15**, 427–41.

Hansen, A. J., Willerslev, E., Wiuf, C., Mourier, T. and Arctander, P. (2001). Statistical evidence for miscoding lesions in ancient DNA templates. *Molecular Biology and Evolution*, **18**, 262–5.

Härke, H. (2002). Kings and warriors: population and landscape from post Roman to Norman Britain. In *The Peopling of Britain: the Shaping of a Human Landscape*, ed. P. S. A. R. Ward. Oxford: Oxford University Press, pp. 145–75.

Hayashi, S., Hayasaka, K., Takenaka, O. and Horai, S. (1995). Molecular phylogeny of gibbons inferred from mitochondrial DNA sequences: Preliminary report. *Journal of Molecular Evolution*, **41**, 359–65.

Hebsgaard, M. B., Wiuf, C., Gilbert, M. T. P., Glenner, H. and Willerslev, E. (2007). Evaluating Neanderthal genetics and phylogeny. *Journal of Molecular Evolution*, **64**, 50–60.

Higham, N. (1992). *Rome, Britain and the Anglo-Saxons*. London: Seaby.

Higuchi, R., Bowman, B., Freiberger, M., Ryder, O. A. and Wilson, A. C. (1984). DNA sequences from the Quagga, an extinct member of the horse family. *Nature*, **312**, 282–4.

Hills, C. (2003). *Origins of the English*. London: Duckworth.
Hirzfeld, L. and Hirzfeld, H. (1919). Essai d'application des methodes au probléme des races. *Anthropologie*, **29**, 505–37.
Ho, S. Y. W., Phillips, M. J. Cooper, A. and Drummond, A. J. (2005). Time dependency of molecular rate estimates and systematic overestimation of recent divergence times. *Molecular Biology and Evolution*, **22**, 1561–8.
Ho, S. Y. W., Shapiro, B. Phillips, M. J., Cooper, A. and Drummond, A. J. (2007). Evidence for time dependency of molecular rate estimates. *Systematic Biology*, **56**, 515–22.
Hofreiter, M., Jaenicke, V., Serre, D., von Haeseler A. and Pääbo, S. (2001a). DNA sequences from multiple amplifications reveal artefacts induced by cytosine deamination in ancient DNA. *Nucleic Acids Research*, **29**, 4793–9.
Hofreiter, M., Serre, D., Poinar, H. N., Kuch, M. and Pääbo, S. (2001b). Ancient DNA. *Nature Reviews Genetics*, **2**, 353–9.
Holland, M. M., Fisher, D. L., Mitchell, L. G. *et al.* (1993). Mitochondrial DNA sequence analysis of human skeletal remains – identification of remains from the Vietnam war. *Journal of Forensic Sciences*, **38**, 542–53.
Hollox, E. J. and Swallow, D. M. (2002). Lactase deficiency – biological and medical aspects of the adult human lactase polymorphism. In *Genetic Basis of Common Diseases*, eds. R. A. King, J. I. Rotter and A. G. Motulsky. Oxford: Oxford University Press, pp. 250–65.
Höss, M., Jaruga, P., Zastawny, T. H., Dizdaroglu, M. and Pääbo, S. (1996). DNA damage and DNA sequence retrieval from ancient tissues. *Nucleic Acids Research*, **24**, 1304–7.
Huelsenbeck, J. P. and Ronquist, F. (2001). MrBayes: Bayesian inference of phylogeny. *Bioinformatics*, **17**, 754–5.
IHGSC (2004). Finishing the euchromatic sequence of the human genome. *Nature*, **431**, 931–45.
IMGSC (2001). Initial sequencing and analysis of the human genome. *Nature*, **409**, 860–921.
Ingman, M. and Gyllensten, U. (2001). Analysis of the complete human mtDNA genome: methodology and inferences for human evolution. *Journal of Heredity*, **92**, 454–61.
Ingman, M. and Gyllensten, U. (2003). Mitochondrial genome variation and evolutionary history of Australian and New Guinean aborigines. *Genome Research*, **13**, 1600–6.
Ingman, M., Kaessmann, H., Pääbo, S. and Gyllensten, U. (2000). Mitochondrial genome variation and the origin of modern humans. *Nature*, **408**, 708–13.
Jeffreys, A. J., Wilson, V. and Thein, S. L. (1985). Individual-specific 'fingerprints' of human DNA. *Nature*, **316**, 76–9.
Jobling, M. A. and Tyler-Smith, C. (2003). The human Y chromosome: an evolutionary marker comes of age. *Nature Reviews Genetics*, **4**, 598–612.
Jobling, M. A., Hurles, M. E. and Tyler-Smith, C. (2004). *Human Evolutionary Genetics: Origins, Peoples and Disease*. New York: Garland Science.
Johnson, M. J., Wallace, D. C., Ferris, S. D., Rattazzi, M. C. and Cavalli-Sforza, L. L. (1983). Radiation of human mitochondrial DNA types analyzed by restriction endonuclease cleavage patterns. *Journal of Molecular Evolution*, **19**, 255–71.
Jukes, T. H. and Cantor, C. R. (1969). Evolution of protein molecules. In *Mammalian Protein Metabolism*, ed. N. H. Munro. New York: Academic Press, pp. 21–123.

Khaitovich, P., Hellmann I., Enard, W. *et al.* (2005). Parallel patterns of evolution in the genomes and transcriptomes of humans and chimpanzees. *Science*, **309**, 1850–4.

Kimura, M. (1968). Evolutionary rate at the molecular level. *Nature*, **217**, 624–6.

Kimura, M. (1983). *Neutral Theory of Molecular Evolution*. Cambridge: Cambridge University Press.

Kingman, J. F. C. (1982). On the genealogy of large populations. *Journal of Applied Probability*, **19A**, 27–43.

Kolman, C. J. and Tuross, N. (2000). Ancient DNA analysis of human populations. *American Journal of Physical Anthropology*, **111**, 5–23.

Krause J., Dear, P. H. Pollack, J. L. *et al.* (2006). Multiplex amplification of the mammoth mitochondrial genome and the evolution of the Elephantidae. *Nature*, **439**, 724–7.

Krings, M., Capelli, C., Tschentscher, F. *et al.* (2000). A view of Neandertal genetic diversity. *Nature Genetics*, **26**, 144–6.

Kumar, S., Tamura, K. and Nei, M. (2004). MEGA3: Integrated software for molecular evolutionary and genetic analysis and sequence alignment. *Briefings Bioinformatics*, **5**, 150–63.

Ladika, S. (2001). DNA forensics – laying ghosts to rest in Bosnia. *Science*, **293**, 1422–3.

Lai, C. S. L., Fisher, S. E., Hurst, J. A., Vargha-Khadem, F., and Monaco, A. P. (2001). A forkhead domain gene is mutated in a severe speech and language disorder. *Nature*, **413**, 519–23.

Lalueza-Fox, C., Calderon, F. L., Calafell, F., Morera, B. and Bertranpetit, J. (2001). MtDNA from extinct Tainos and the peopling of the Caribbean. *Annals of Human Genetics*, **65**, 137–51.

Lalueza-Fox, C., Sampietro, M. L., Gilbert, M. T. P. *et al.* (2004). Unravelling migrations in the steppe: mitochondrial DNA sequences from ancient Central Asians. *Proceedings of The Royal Society of London Series B Biological Sciences*, **271**, 941–7.

Landsteiner, K. (1900). Zur Kenntnis der antifermentativen lytisichen und agglutinierenden Wirkungen des Blutserums end der Lymphe. *Zenralblatt für Bakteriologie*, **27**, 357–62.

Lió, P. and Goldman, N. (1998). Models of molecular evolution and phylogeny. *Genome Research*, **8**, 1233–44.

Malmström, H., Stora, J., Dalen, L., Holmlund, G. and Gotherstrom, A. (2005). Extensive human DNA contamination in extracts from ancient dog bones and teeth. *Molecular Biology and Evolution*, **22**, 2040–7.

Margulies, M., Egholm, M., Altman, W. E. *et al.* (2005). Genome sequencing in microfabricated high-density picolitre reactors. *Nature*, **437**, 376–80.

Mastana, S. S. and Sokol, R. J. (1998). Genetic variation in the East Midlands. *Annals of Human Biology*, **25**, 43–68.

Mastana, S. S., Fisher, P., Sokol, R. J. and Boam, S. (1993). Genetic diversity in the human-populations of the East Midlands, United Kingdom. *American Journal of Human Genetics*, **53**, 829.

McDougall, I., Brown, F. H. and Fleagle, J. G. (2005). Stratigraphic placement and age of modern humans from Kibish, Ethiopia. *Nature*, **433**, 733–6.

Mitchell, D., Willerslev, E. and Hansen, A. (2005). Damage and repair of ancient DNA. *Mutation Research–Fundamental and Molecular Mechanisms of Mutagenesis*, **571**, 265–76.

Morin, P. A., Chambers, K. E., Boesch, C. and Vigilant, L. (2001). Quantitative polymerase chain reaction analysis of DNA from noninvasive samples for accurate microsatellite genotyping of wild chimpanzees (Pan troglodytes verus). *Molecular Ecology*, **10**, 1835–44.

Mullis, K. B. and Faloona, F. (1987). Specific synthesis of DNA in vitro via a polymerase-catalysed chain reaction. *Methods in Enzymology*, **155**, 335–50.

Myres, J. (1986). *The English Settlements*. Oxford: Clarendon.

Neel, J. V. (1962). Diabetes mellitus: a thrifty genotype rendered detrimental by "progress"? *American Journal of Human Genetics*, **14**, 353–62.

Neel, J. V. (1982). The thrifty genotype revisited. In *The Genetics of Diabetes Mellitus*, eds. J. Kobberling and R. Tattersall. London: Academic Press, pp. 283–93.

Noonan, J. P., Hofreiter, M., Smith, D. *et al.* (2005). Genomic sequencing of Pleistocene cave bears. *Science*, **309**, 597–600.

Noonan, J. P., Coop, G., Kudaravalli, S. *et al.* (2006). Sequencing and analysis of Neanderthal genomic DNA. *Science*, **314**, 1113–18.

O'Rourke, D. H., Hayes, M. G. and Carlyle, S. W. (2000). Ancient DNA studies in physical anthropology. *Annual Review of Anthropology*, **29**, 217–42.

Ohlin, M., Kristensson, K., Carlsson, R. and Borrebaeck, C. A. K. (1992). Epstein-Barr virus induced transformation of human B lymphocytes: the effect of L-leucyl-L-leucine methyl ester on inhibitory T cell populations. *Immunology Letters*, **34**, 221–8.

Ohta, T. (1992). The nearly neutral theory of molecular evolution. *Annual Review of Ecology and Systematics*, **23**, 263–86.

Orlando, L., Darlu, P., Toussaint, M. *et al.* (2006). Revisiting Neanderthal diversity with a 100,000 year old mtDNA sequence. *Current Biology*, **16**, R400–R402.

Ovchinnikov, I. V., Gotherstrom, A., Romanova, G. P. *et al.* (2000). Molecular analysis of Neanderthal DNA from the northern Caucasus. *Nature*, **404**, 490–3.

Pääbo, S. (1985a). Preservation of DNA in ancient Egyptian mummies. *Journal of Archaeological Science*, **12**, 411–17.

Pääbo, S. (1985b). Molecular cloning of ancient Egyptian mummy DNA. *Nature*, **314**, 644–5.

Pääbo, S. (1986). Molecular genetic investigations of ancient human remains. *Cold Spring Harbor Symposium on Quantitative Biology*, **51**, 441–6.

Pääbo, S., Higuchi, R. G. and Wilson, A. C. (1989). Ancient DNA and the polymerase chain reaction – the emerging field of molecular archaeology. *Journal of Biological Chemistry*, **264**, 9709–12.

Pääbo, S., Poinar, H., Serre, D. *et al.* (2004). Genetic analyses from ancient DNA. *Annual Review of Genetics*, **38**, 645–79.

Pakendorf, B. and Stoneking, M. (2005). Mitochondrial DNA and human evolution. *Annual Review of Genomics and Human Genetics*, **6**, 165–83.

Poinar, H. N., Hofreiter, M., Spaulding, W. G. *et al.* (1998). Molecular coproscopy: Dung and diet of the extinct ground sloth Nothrotheriops shastensis. *Science*, **281**, 402–6.

Poinar, H. N., Schwarz, C., Qi, J. *et al.* (2006). Metagenomics to paleogenomics: large-scale sequencing of mammoth DNA. *Science*, **311**, 392–4.

Pritchard, J. K., Seielstad, M. T., Perez-Lezaun, A. and Feldman, M. W. (1999). Population growth of human Y chromosomes: a study of Y chromosome microsatellites. *Molecular Biology and Evolution*, **16**, 1791–8.

Prober, J. M., Trainor, G. L., Dam, R. J. *et al.* (1987). A system for rapid DNA sequencing with fluorescent chain-terminating dideoxynucleotides. *Science*, **238**, 336–41.

Purvis, A., Nee, S. and Harvey, P. H. (1995). Macroevolutionary inferences from primate phylogeny. *Proceedings of the Royal Society of London Series B Biological Sciences*, **260**, 329–33.

Renfrew, C. (1987). *Archaeology and Language: the Puzzle of Indo-European Origins*. London: Jonathan Cape.

Richards, M. B., Sykes, B. C. and Hedges, R. E. M. (1995). Authenticating DNA extracted from ancient skeletal remains. *Journal of Archaeological Science*, **22**, 291–9.

Rogers, A. R. and Harpending, H. (1992). Population growth makes waves in the distribution of pairwise genetic differences. *Molecular Biology and Evolution*, **9**, 552–69.

Ronaghi, M., Karamohamed, S., Pettersson, B., Uhlen, M. and Nyren P. (1996). Real-time DNA sequencing using detection of pyrophosphate release. *Analytical Biochemistry*, **242**, 84–9.

Ronaghi, M., Uhlen, M. and Nyren, P. (1998). A sequencing method based on real-time pyrophosphate. *Science*, **281**, 363–5.

Ronquist, F. and Huelsenbeck, J. P. (2003). MrBayes 3: Bayesian phylogenetic inference under mixed models. *Bioinformatics*, **19**, 1572–4.

Roos, C. and Geissmann, T. (2001). Molecular phylogeny of the major hylobatid divisions. *Molecular Phylogenetics and Evolution*, **19**, 486–94.

Rudbeck, L., Gilbert, M. T. P., Willerslev, E. *et al.* (2005). mtDNA analysis of human remains from an early Danish Christian cemetery. *American Journal of Physical Anthropology*, **128**, 424–9.

Saiki, R. K., Gelfand, D. H., Stoffel, S. *et al.* (1988). Primer-directed enzymatic amplification of DNA with thermostable DNA polymerase. *Science*, **239**, 487–91.

Saks, M. J. and Koehler, J. J. (2005). The coming paradigm shift in forensic identification science. *Science*, **309**, 892–5.

Salamon, M., Tuross, N., Arensburg, B. and Weiner, S. (2005). Relatively well preserved DNA is present in the crystal aggregates of fossil bones. *Proceedings of the National Academy of Science*, **102**, 13 783–8.

Sanchez, J. J., Borsting, C., Hallenberg, C. *et al.* (2003). Multiplex PCR and mini-sequencing of SNPs – a model with 35 Y chromosome SNPs. *Forensic Science International*, **137**, 74–84.

Sanger, F. and Coulson, A. R. (1975). Rapid method for determining sequences in DNA by primed synthesis with DNA-polymerase. *Journal of Molecular Biology*, **94**, 441–8.

Sarich, V. M. and Wilson, A. C. (1967). Immunological time scale for homonid evolution. *Science*, **158**, 1200–3.

Sawyer, P. (1971). *The Age of the Vikings*. London: Edward Arnold.

Scharff, C. and Haesler, S. (2005). An evolutionary perspective on FoxP2: strictly for the birds? *Current Opinion in Neurobiology*, **15**, 694–703.

Schmitz, R. W., Serre, D., Bonani, G. *et al.* (2002). The Neandertal type site revisited: interdisciplinary investigations of skeletal remains from the Neander Valley, Germany. *Proceedings of the National Academy of Science*, **99**, 13 342–7.

Shapiro, B., Drummond, A. J., Rambaut, A. *et al.* (2004). Rise and fall of the Beringian steppe bison. *Science*, **306**, 1561–5.

Shennan, S. (2000). Population, culture, history, and the dynamics of culture change. *Current Anthropology*, **41**, 811–35

Smith, L. M., Sanders, J. Z., Kaiser, R. J. *et al.* (1986). Fluorescence detection in automated DNA sequence analysis. *Nature*, **321**, 674–9.

Stauffer, R. L., Walker, A., Ryder, O. A., Lyons-Weiler, M. and Hedges, S. B. (2001). Human and ape molecular clocks and constraints on paleontological hypotheses. *Journal of Heredity*, **92**, 469–74.

Stenton, F. (1971). *Anglo-Saxon England*. Oxford: Oxford University Press.

Sullivan, K. M., Mannucci, A., Kimpton, C. P. and Gill, P. (1993). A rapid and quantitative DNA sex test: fluorescence-based PCR analysis of X-Y homologous gene amelogenin. *Biotechniques*, **15**, 637–41.

Swisher, C. C., Rink, W. J., Anton, S. C. *et al.* (1996). Latest Homo erectus of Java: potential contemporaneity with Homo sapiens in southeast Asia. *Science*, **274**, 1870–4.

Swofford, D. (1999). *PAUP* Phylogenetic Analysis Using Parsimony (*and Other Methods)*. Sunderland: Sinaur.

Swofford, D. L., Olsen, G. J., Waddell, P. J. and Hillis, D. M. (1996). Phylogenetic inference. In *Molecular Systematics*, eds. D. M. Hillis, C. Moritz and B. K. Mable. Sunderland: Sinauer Associates, pp. 407–514.

Syvanen, A. C. (2005). Toward genome-wide SNP genotyping. *Nature Genetics*, **37**, S5–S10.

TCSaAC (2005). Initial sequence of the chimpanzee genome and comparison with the human genome. *Nature*, **437**, 69–87.

Templeton, A. R. (2005). Haplotype trees and modern human origins. *Yearbook of Physical Anthropology*, **128**, 33–59.

Thomson, R., Pritchard, J. K., Shen, P. D., Oefner, P. J. and Feldman, M. W. (2000). Recent common ancestry of human Y chromosomes: evidence from DNA sequence data. *Proceedings of the National Academy of Science*, **97**, 7360–5.

Töpf, A. L., Gilbert, M. T. P., Dumbacher, J. P. and Hoelzel, A. R. (2006). Tracing the phylogeography of human populations in Britain based on 4th–11th century mtDNA genotypes. *Molecular Biology and Evolution*, **23**, 152–61.

Tyfield, L., Stephenson, A. and Holton, J. B. (1997). Mutation analysis and psychometric assessments in the galactosaemia population of the British Isles. *Journal of Medical Genetics*, **34**, SP17.

Venter, J. C., Adams, M. D., Myers, E. W. *et al.* (2001). The sequence of the human genome. *Science*, **291**, 1304–51.

Vigilant, L., Stoneking, M., Harpending, H., Hawkes, K. and Wilson, A. C. (1991). African populations and the evolution of human mitochondrial DNA. *Science*, **253**, 1503–7.

Weale, M. E., Weiss, D. A., Jager, R. F., Bradman, N. and Thomas, M. G. (2002). Y chromosome evidence for Anglo-Saxon mass migration. *Molecular Biology and Evolution*, **19**, 1008–21.

Wilson, J. F., Weiss, D. A., Richards, M. *et al.* (2001). Genetic evidence for different male and female roles during cultural transitions in the British Isles. *Proceedings of the National Academy of Science*, **98**, 5078–83.

Zuckerkandl, E. and Pauling, L. (1962). Molecular disease evolution, and genetic heterogeneity. In *Horizons in Biochemistry*, eds. M. Kasha and B. Pullman. New York: Academic Press, pp. 189–225.

11 *Isotopes and human migration: case studies in biogeochemistry*

T. DOUGLAS PRICE

Isotopic tracers have an enormous potential to tell us about past human behaviour, particularly in terms of diet, interaction and migration. Focus in the discussion here is on human movement in the past. In principle, any isotope that varies geographically and is deposited and retained in the human skeleton may be of interest. The method is straightforward. Tooth enamel is formed in early childhood and its chemical composition does not change during life or substantially during burial. If the chemistry of tooth enamel is different from the place of death, the individual must have moved or been moved to the area. To date, the isotopes of strontium, oxygen and lead have been used in a variety of archaeological contexts for investigations of residence change. Two general kinds of studies have been undertaken, focused either on groups or individuals. For purposes of this discussion, examples from a late prehistoric pueblo in the American Southwest and the ancient city of Teotihuacán in Mexico are used to document studies of groups. Investigations of the Iceman of Neolithic Italy and the first ruler of the Maya city of Copán illustrate the application of the method to the life history of individuals. The isotopic investigation of residential change is becoming a commonplace procedure in bioarchaeological studies.

Instrumental methods have long been a part of the repertoire of bioarchaeological enquiry. In the last 25 years, however, the chemistry of human bone, both at the molecular and atomic level, has become an essential means for studying past human behaviour and activity. Studies of ancient DNA, for example, concern the molecular contents of bone and are described elsewhere in this volume. The elemental and isotopic chemistry of bone is important for the study of diet, interaction and human movement in the past. The use of light isotopes of carbon and nitrogen for information on palaeodiet is described as well in another section of the book. This chapter will focus on the use of strontium and oxygen isotopes for the investigation of human migration in the past.

Isotopic tracers have of course been known for some time. Their use in the human body is well established in the field of medicine. Isotopic tracers

Between Biology and Culture, ed. Holger Schutkowski. Published by Cambridge University Press.

have also been employed in the environmental sciences and ecological studies to map the geographic movement of various materials and species (e.g. Gosz *et al.*, 1983; Rundel *et al.*, 1989; Koch *et al.*, 1994; Lajtha and Michener, 1994; Åberg, 1995; Chamberlain *et al.*, 1997; Gannes *et al.*, 1998; Blum *et al.*, 2000). In the last 30 years, isotope provenancing of archaeological materials has become almost commonplace. Applications have included a range of prehistoric artefacts and involved questions of sources and the provenance of stone such as marble (e.g. Matthews *et al.*, 1995), sea shells (e.g. Shackleton and Elderfield, 1990), ceramics (e.g. Carter *et al.*, 2003) and other materials.

The isotopic provenancing of human remains was introduced in archaeology more than two decades ago for the investigation of residential change among prehistoric humans (Ericson, 1981; 1985). Stable isotope ratios of several different elements have since been investigated as possible indicators of movement by prehistoric peoples (Katzenberg and Harrison, 1996). Oxygen (Stuart-Williams *et al.*, 1995; White *et al.*, 1998; 2000), lead (Carlson, 1996; Gulson *et al.*, 1997; Budd *et al.*, 1998) and strontium (e.g. Ericson, 1989; Sealy, 1989; van der Merwe *et al.*, 1990; Sealy *et al.*, 1991; Price *et al.*, 1994a, b; 1998; 2000; 2001; Jones *et al.*, 2003; Bentley, 2006) have been considered to date. Strontium and oxygen isotopes have also been used in the study of archaeological fauna such as sheep and pigs (Balasse *et al.*, 2002; Schweissing and Grupe, 2003; Bentley and Knipper, 2005). Some preliminary work with sulphur isotopes in human provenancing has begun (Richards *et al.*, 2003).

Studies have been conducted on human remains from a wide range of past times and places, and demonstrate the utility of strontium isotope analysis. These studies come from archaeological sites belonging to the Anasazi period in Arizona (Ezzo, 1993; Price *et al.*, 1994a; Ezzo *et al.*, 1997) the Later Stone Age and the historic period in South Africa (Sealy *et al.*, 1991; 1995), the Neolithic Linearbandkeramik and Bell Beaker periods in southern Germany (Price *et al.*, 1994b; 1998; 2001; Grupe *et al.*, 1997), the Classic period of the central highlands of Mexico (Price *et al.*, 2000), the prehispanic period in the Central Andes of South America (Knudson *et al.*, 2004; 2005), the Neolithic in southern England (Montgomery *et al.*, 2000), Norse settlement of the northern isles of Britain (Montgomery *et al.*, 2003b) and Iceland (Price and Gestsdóttir, 2005) and the colonial period in Campeche, Mexico (Price *et al.*, 2005), among a number of others. These studies have generally focused on one of two questions: the identity of a specific individual or the question of migration in a group or population from a particular site or area.

In the paragraphs below, some basic principles, details on strontium and oxygen isotope analysis, a few problem areas, and other considerations of the isotopic provenancing of humans are discussed. This introduction is followed by a series of case studies from various times and places that document the application and utility of these methods. A few thoughts on such studies conclude this paper.

Isotopic provenancing and human mobility

Direct evidence of human mobility in the past has been difficult to obtain. Archaeological approaches to questions regarding prehistoric conquest, colonisation, diffusion and the like have depended largely on materials such as distinctive stone tools, pottery or other items as evidence that people moved. Artefact shapes, styles of decoration or architecture have been employed indirectly as a proxy for the movement of people. The problem, of course, is that artefacts and ideas can move independently of the people that created them, through borrowing, exchange or theft.

Today, however, it is possible to track the movement of specific individuals in the chemistry of prehistoric human skeletons using geographically constrained isotope ratios. The basic principles are straightforward. The enamel in teeth forms during infancy and early childhood (Hillson, 2005) and undergoes relatively little subsequent change (Budd *et al.*, 2001; Schoeninger *et al.*, 2003). Thus, differences in isotopic ratios between tooth enamel (place of birth) and the local (place of burial) isotopic value document a change in residence for the individual.

Isotopic ratios are measured using a mass spectrometer. The analysis is destructive but only a very small amount of material is required. The permanent first molar is preferred both for consistency and the fact that the enamel of this tooth forms during gestation and the first few years of childhood. The isotopic signal of the place of burial can be measured in several ways: in human bone from the individuals whose teeth are analysed, from other human bones at the site, from archaeological fauna at the site, or from modern fauna in the vicinity.

There is a variety of information available in the isotopic composition of human bone and enamel. In addition to the question of residential change, various aspects of life history and diet may also be present. In principle, it should be possible to look at changes during the life of an individual. While dental enamel undergoes virtually no remodelling, bone tissues themselves remodel at different rates depending upon the ratio of active osteoblasts (responsible for calcium precipitation) and osteoclasts (responsible for calcium dissolution) (Cox and Sealy, 1997; Teitelbaum, 2000). Hence trabecular tissue, with more active osteoclasts, remodels quite rapidly; dense, cortical tissue much more slowly. Bones, such as the femur and tibia, with a high ratio of cortical to trabecular tissue, remodel over a period of decades, while bones with abundant trabecular tissue, such as ribs and the iliac crest, can remodel on a scale of several years. Dental enamel from different teeth likewise develops at different ages, as do different teeth (Smith, 1994; Hillson, 2005). Thus the human skeleton preserves not only a childhood/adulthood contrast in isotopes, but essentially records the isotopic life history of the individual. Thus it becomes theoretically possible to assess whether an immigrant moved within the last few years of their life (fast-remodelling bones resemble enamel more than local isotopes), moved as a child (bones have local signature in contrast to enamel) or at some intermediate interval, by assessing how closely various skeletal tissues have remodelled toward local isotopic ratios.

These principles have been employed in several studies to examine individual life histories. Müller *et al.* (2003), for example, measured isotope ratios in the enamel, bone and gut of the Iceman to obtain information on various points in his life. Tooth enamel as a signal of the place of birth was compared with both cortical and trabecular bone (as signals from the later years of the Iceman's life) and with his intestinal contents which should reflect the last hours of his life. Current research is focused on the isolation of osteoblasts as indicators of specific moments in an individual's life (e.g. Bell *et al.*, 2001).

Another important consideration in isotopic provenancing is the combined use of two or more isotopic systems for distinguishing place of origin. Müller *et al.* (2003) utilised strontium, lead and oxygen isotopes in their examination of the Iceman, described below. The use of both strontium and oxygen isotope ratios at Teotihuacán and the Maya site of Copán is described in later sections of this paper. Budd *et al.* (1999; 2001) have used oxygen, strontium and lead in the investigation of residential mobility among Pacific Islands and in the Neolithic of Britain. The application of multiple isotopic ratios for provenancing in the coming years will greatly enhance our understanding of patterns of migration and provide higher resolution for discerning places of origin.

Strontium isotopes

Strontium isotopes vary with geology and enter the skeleton through the food chain. Strontium concentrations, both elemental and isotopic, differ among types of rock. Strontium in bedrock moves into soil and ground water and through the food chain (e.g. Comar *et al.*, 1957; Elias *et al.*, 1982). Strontium is incorporated into bone as a substitute for calcium in the mineral hydroxy-apatite (e.g. Schroeder *et al.*, 1972; Rosenthal, 1981). Although local levels of elemental strontium in plant and animal tissue vary due to many factors, the isotopic composition of strontium is not changed (or fractionated) by biological processes because of the very small relative mass differences among strontium isotopes. Virtually all strontium in vertebrate organisms is found in the skeleton.

The strontium isotope composition of human bones and teeth, therefore, matches the diet of the individual, which in turn reflects the strontium isotope composition of the local geology.

The stable isotopes of strontium include ^{84}Sr (*c.* 0.56 %), ^{86}Sr (*c.* 9.87 %) and ^{88}Sr (*c.* 82.53 %). ^{87}Sr is formed over time by the radioactive decay of rubidium (^{87}Rb) and comprises approximately 7.04 % of total strontium (Faure and Powell, 1972). Variations in strontium isotope compositions in natural materials are conventionally expressed as ^{87}Sr/^{86}Sr ratios, as the abundance of ^{86}Sr is similar to that of ^{87}Sr. Strontium isotope ratios vary with the age and type of rock as a function of the original ^{87}Rb/^{86}Sr ratio of a source and its age (Faure and Powell, 1972; Faure, 1986). Geologic units that are very old ($>$100 MYA) and had very high original Rb/Sr ratios will have very high ^{87}Sr/^{86}Sr ratios today as well as in the recent past ($<$1 MYA). In contrast, rocks that are geologically young ($<$1–10 MYA) and that have low Rb/Sr ratios, such as late Cenozoic volcanic zones, generally have ^{87}Sr/^{86}Sr ratios less than 0.706 (e.g. Rogers and Hawkesworth, 1989). Differences in the third decimal place are usually significant in terms of human movement. These variations may seem small, but they are exceptionally large from a geological standpoint and far in excess of analytical error using a Thermal Ionisation Mass Spectrometer, or TIMS ($\pm$0.00001 for ^{87}Sr/^{86}Sr). Analytical procedures are described in detail in Ezzo *et al.* (1997).

A small portion of the variability in strontium isotopes is accounted for by differences in diet rather than geographic mobility. Sealy *et al.* (1991) have discussed the utility of ^{87}Sr/^{86}Sr as a dietary indicator in modern and archaeological bone. A pair of examples provide some illustration. In a study of the first settlers of Iceland (Price and Gestsdóttir, 2005), there were two surprises. Iceland is some of the youngest land on earth and the human isotope ratios there were expected to be the same as the very low ^{87}Sr/^{86}Sr value of the new rock, *c.* 0.704. Measurement of almost 90 human samples, however, revealed that the lowest ratios were around 0.706 (Figure 11.1). To verify this discrepancy, we measured modern sheep teeth from various parts of the island and found a mean of 0.7063, almost identical to the lowest human values. We suspect that there is a seaspray effect on Iceland raising the ratios of the diet of the humans and the sheep closer to values of seawater (0.7092).

The second surprise was the wide range in local human values from 0.7059 to approximately 0.7096, shown in the plot of rank ordered values in Figure 11.1. Migrant individuals to Iceland were obvious above the higher value as seen by the break in the trend of the line. However, why was there a continuous increase in values between 0.7059 and 0.7096? We believe that this range reflects differences in diet among the early Icelanders, with a range from almost completely terrestrial to almost completely marine-based diets. This difference

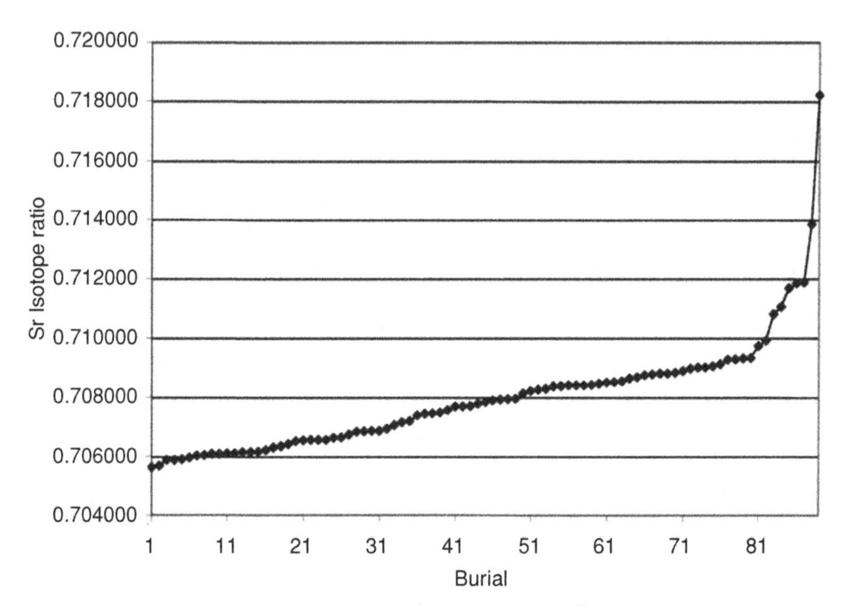

Figure 11.1 Strontium isotope ratios in early human burials from Iceland. Values range from 0.704 to 0.718. Local values of Iceland natives range from approximately 0.704 to 0.7092. The remainder of individuals in the sample are likely immigrants.

is explained in part by location, individuals buried closer to the coast of Iceland tended to have more marine values. There may also be some chronological variation reflected if diets changed over time from terrestrial to marine as has been noted for the Norse on Greenland (Fricke *et al.*, 1995; Arneborg *et al.*, 1999).

A second example of how strontium isotopes can reflect diet comes from the site of Tikal in the Maya region (Wright, 2005). Many of the Maya region sites are located on a massive limestone shelf in northern Guatemala and the Yucatan of Mexico. This limestone shows a regular trend in strontium isotope ratios from 0.7075 to 0.709, from south to north. Tikal, in the northern part of Guatemala, has a $^{87}Sr/^{86}Sr$ value of 0.7081 measured in modern and archaeological fauna. A variety of samples of rock, soil and plants within a 50 km range of the site show a value between 0.7078 and 0.7081. It was, therefore, surprising to learn that almost half of the human samples from Tikal have a $^{87}Sr/^{86}Sr$ value greater than the local bioavailable ratio. Wright (2005) has proposed that the only feasible explanation for these higher values is the consumption of significant amounts of sea salt (with an isotopic ratio of 0.7092) in the diet. Ethnographic records in the area note high salt consumption with a predominantly maize diet, and the salt trade was an important aspect of Maya commerce.

Oxygen isotopes

Oxygen isotopes in human teeth and bones come primarily from drinking water (Longinelli, 1984; Luz *et al.*, 1984; 1990; Luz and Kolodny, 1985). Rainfall is the major source of drinking water; water from food and atmospheric oxygen are minor, secondary sources (Schwarcz and Schoeninger, 1991). The oxygen isotope content of rainfall depends on climatic and environmental variables such as temperature, humidity, altitude and distance from the sea where clouds form (Yurtsaver and Gat, 1981; Ayliffe and Chivas 1990). Polar and inland rain is isotopically lighter than tropical and coastal rain.

Oxygen isotopes are measured as a ratio between ^{18}O and ^{16}O; the ratio ($\delta^{18}O$) is expressed in units per mil, ‰. Clouds that form over tropical seas gradually lose heavier $H_2{}^{18}O$ through rainfall as they move over land masses and toward higher latitudes and elevations, retaining disproportionately the lighter $H_2{}^{16}O$, to fall in later showers. Thus, the $\delta^{18}O$ of drinking water declines with increasing latitude, elevation and distance from the coast (or source of evaporated water) (Longinelli, 1984; Luz and Kolodny, 1985; Schwarcz *et al.*, 1991; Bowen and Wilkinson, 2002). The relative balance of evaporation and recharge of lakes and other surface water reservoirs also contributes to the $\delta^{18}O$ of available drinking water. Oxygen isotope ratios in prehistoric human enamel provide information on mobility in much the same way as strontium isotopes (Schwarcz *et al.*, 1991). However, because breast milk is enriched in ^{18}O when compared to the water a mother consumes, selection of teeth to sample must take account of the fact that cultural and individual variation in nursing behaviour may affect the $\delta^{18}O$ of teeth formed in infancy (Wright and Schwarcz, 1998).

The hydroxyapatite in enamel and bone contains oxygen, both in phosphate groups (PO_4) and in carbonates (CO_3). Both phosphate and carbonate oxygen have been used in studies of human provenance (Sponheimer and Lee-Thorp, 1999). Ratios are designated with a subscript ‘p’ for phosphate ($\delta^{18}O_p$) and subscript ‘c’ for carbonate ($\delta^{18}O_c$). Phosphate oxygen ratios are reported relative to the VSMOW (Vienna Standard Mean Ocean Water) standard; carbonate oxygen ratios are reported relative to PDB (Pee Dee Belemnite), a Cretaceous marine carbonate. Values in carbonate are generally negative; values in phosphate are generally positive and range between 10 and 30 $‰_{VSMOW}$. Carbonate and phosphate values measured on tooth enamel from Kaminaljuyu are highly correlated; carbonate ratios reported in PDB can be converted to VSMOW by adding 21.40‰ (Wright and colleagues, personal communication). Phosphate and carbonate produce similar results, but less sample is needed for carbonate, preparation is less demanding, and results between laboratories are more comparable (Bryant *et al.*, 1996; Lee-Thorp, 2002; Lee-Thorp and Sponheimer, 2003). Enamel shows less diagenetic change than bone apatite, in which both phosphate and carbonate $\delta^{18}O$ may be significantly modified (Stuart-Williams *et al.*, 1995; Iacumin *et al.*, 1996; Wright and Schwarcz, 1998). Analytical procedures for phosphate are described in detail in Schwarcz *et al.* (1991) and White *et al.* (1998); analysis of carbonate oxygen is outlined in Fricke and O'Neil (1996), Fricke *et al.* (1995), Kohn (1996), and Wright and Schwarcz (1998).

Potential problems

There are several areas of problem or uncertainty in human isotopic provenancing. These include establishing isotopic levels in the local area, diagenesis and distinguishing foreign individuals. These issues are discussed in the following paragraphs.

Local isotopic levels

In order to determine if 'foreign' isotopic values are present in an area it is essential to know what the local values are. This simple and obvious requirement is more easily stated than met. Expected values for oxygen, strontium and lead isotope ratios can often be predicted based on the sources and location of rainfall or the known geological units present. However, these theoretical expectations are often incorrect.

The direct use of strontium isotope ratios from bedrock geology is confounded by several factors. Isotopic ratios in the local environment are composed of a mixture of strontium derived from both atmospheric sources and mineral weathering (Miller *et al.*, 1993). Biologically available strontium isotope ratios can differ substantially between bedrock and other environmental values. As Sillen *et al.* (1998, p. 2466) noted, 'The large difference in strontium isotope composition between plant and available Sr on the one hand, and whole soil Sr, on the other, suggest that potential applications of $^{87}Sr/^{86}Sr$ relationships should use biologically available strontium as a starting point, rather than substrate geology *per se*.' It is essential to measure the baseline bioavailable levels of these isotopes in a study area (Price *et al.*, 2000). Determination of baseline bioavailable strontium isotope ratios is best done through measurement of local fauna, either modern or archaeological. At the same time, samples are difficult to come by and measurements are expensive. As a consequence, baseline studies have rarely been done at an adequate level of intensity.

Similar problems with baseline information are inherent in oxygen isotope studies. Local values are often unknown. An example of this situation can be seen in the study of the so-called Amesbury Archer from Wessex, England (Fitzpatrick, 2003). Measurement of oxygen isotope ratios in the tooth enamel of this individual revealed a value that did not correspond to known $\delta^{18}O$ levels for the British Isles. The investigator, in a widely publicised statement, argued for a birthplace in a colder climate than in Britain today for the 'archer', and suggested somewhere in central Europe. Such an assertion, based on a single isotopic ratio, unmeasured local values and the assumption that modern and ancient values are identical, is unfounded.

Climatic changes can have a major impact on $\delta^{18}O$. Oxygen isotope ratios have been used for many years as a proxy for temperature change in a variety of palaeoclimate studies (e.g. Craig, 1965; Williams, 1984; Dansgaard, 1994). In addition to a lack of baseline information and complications due to climatic change, there are additional sources of variation in oxygen isotope ratios in human skeletal tissue beyond local bioavailability. Variation within a single human population is not well known. Oxygen, a light isotope, is subject to various fractionation processes and varies in humans with breast feeding, metabolic rates and relative humidity, among other factors (Koch *et al.*, 1994; Kohn, 1999).

Diagenesis

Diagenesis is a continuing undercurrent in bone chemistry studies in archaeology. Some initial studies (e.g. Nelson *et al.*, 1986) were particularly pessimistic about the potential for reliable biogenic isotopic signals in buried material. Continuing investigations of diagenesis, however, have clarified this picture. Bone is subject to substantial diagenesis and postmortem change. Tooth enamel, on the other hand, a denser, harder and more inert substance than bone, is much less susceptible to diagenesis (e.g. Driessens and Verbeeck, 1990; Wang and Cerling, 1994; Kolodny *et al.*, 1996; Hillson, 1997; Sharp *et al.*, 2000; Hoppe *et al.*, 2003). The higher organic content and reduced crystal size in dentine increase its susceptibility to contamination (Kohn *et al.*, 1999). There is some suggestion of in vivo uptake of strontium on the surface of enamel (Grupe *et al.*, 1999; Hörn and Müller-Söhnius, 1999), but this has not been identified analytically (Brudevold and Söremark, 1967; Montgomery *et al.*, 1999).

Enamel thus has been shown to be generally resistant to contamination and a reliable container of biogenic levels of isotopes (e.g. Kohn, 1996; Koch *et al.*, 1997; Åberg *et al.*, 1998; Kohn *et al.*, 1999; Budd *et al.*, 2000b; Montgomery *et al.*, 2003a; Schoeninger *et al.*, 2003). Diagenesis in bone and enamel appears to follow a similar pattern in terms of oxygen isotopes. Trueman *et al.* (2003) document extensive diagenesis in terrestrial dinosaur bones buried in marine deposits. A number of studies have documented the more resistant nature of enamel to postmortem chemical and physical changes. Sharp *et al.* (2000) note that enamel provides a more reliable biogenic signal for oxygen phosphate values than bone or bulk tooth data from Triassic fossil animals, due to diagenesis. Boecherens *et al.* (1994) in another study of fossil dinosaur teeth have demonstrated that enamel was more resistant than dentine to trace element contamination, even in material more than 100 000 000 years old.

Kohn *et al.* (1999) have undertaken perhaps the most detailed study of diagenesis in fossil teeth to date, using the latest instrumentation to examine elemental abundances in 4 MYA fossil and modern animal tooth enamel from East Africa, at a μm to sub-μm scale. Their study documented physical contamination of enamel by a number of elements (Fe, Mn, Si, Al, Ba and possibly Cu) and chemical alteration of U, REE, F and 'possibly' Sr. Isotopic ratios were not measured. Concentrations of secondary minerals ranged from 0.3 % in enamel to *c.* 5 % in dentine, documenting the more resistant nature of enamel and very low levels of contamination. Evidence for strontium contamination in enamel was equivocal; no significant differences could be observed between modern and fossil specimens. This study unfortunately compared samples from separate areas in East Africa where natural abundance and availability of strontium may have differed. Variation in biologically available strontium, barium and other elements incorporated into the skeleton would mean that comparison of different areas was futile without knowledge of baseline local levels. Nevertheless, the fact that strontium cannot clearly be shown to have contaminated fossil enamel supports an argument for the use of this material.

Our own studies have demonstrated that modern fauna closely mimic values seen in local prehistoric humans from the same place. For example, at Grasshopper Pueblo (see below) the mean value for $^{87}Sr/^{86}Sr$ in modern mouse tooth enamel (0.71004) was virtually identical to that for the prehistoric local humans (0.71009). As a final consideration in studies of human migration, it is important to note that contamination of bone or tooth would only be with local strontium isotope ratios; migration would not be inferred in cases of diagenesis. Isotope ratios from exotic geological contexts in bone or tooth indicate that contamination is not masking such information.

Distinction of foreign individuals

Another problem that often arises in isotope studies of past human migration involves the distinction between local and migrant individuals. In an ideal situation, the enamel values of the migrants might be completely different from those of indigenous persons. In reality, however, this is not usually the case. Extreme values are not a problem, but there is no objective criterion for distinguishing among individuals with values close to the local range. A range of isotope ratio values for tooth enamel is often observed in such studies, such as the strontium isotope data from Grasshopper Pueblo, shown in Figure 11.2. Because of

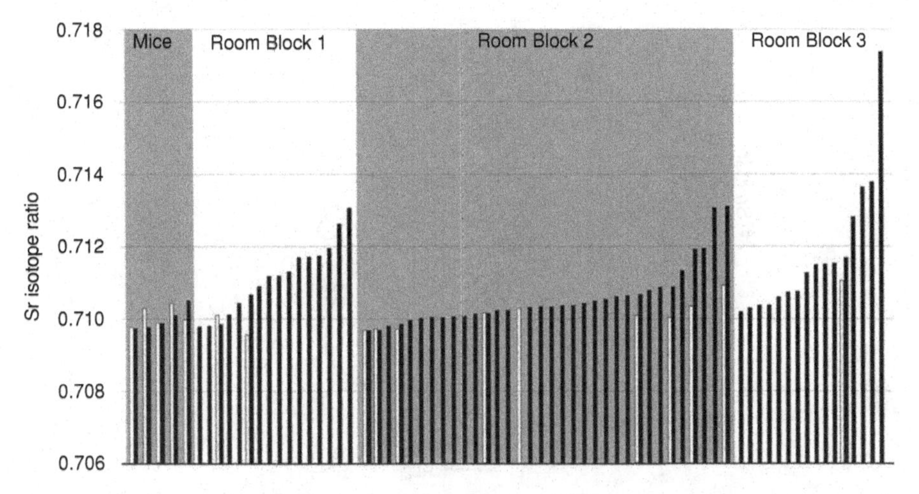

Figure 11.2 Strontium isotope ratios in humans and modern fauna from Grasshopper Pueblo, Arizona. Black bars are for tooth enamel, open bars for bone.

the range in values present in the samples at Grasshopper, two different criteria were employed in the original study to distinguish between migrants and locals: (1) a cut-off value based on human bone values which are assumed to largely reflect the local ratio, and (2) a cut-off value derived from the $^{87}Sr/^{86}Sr$ ratios in local fauna (Price *et al.*, 1994a; Ezzo *et al.*, 1997). In both cases, a range of local variation was defined using one or two standard deviations from the mean. Individuals with enamel values outside those limits were declared as foreign. These solutions assisted in making decisions about exotic values, but will not work well in cases of small samples or situations where the range of values is continuous. In many cases, the distinction of uncertain individuals in isotopic provenancing is going to remain a judicious combination of common sense, archaeological correlates and caution.

In spite of these difficulties and unknown sources of variation, most isotopic provenancing studies appear to be surprisingly robust and reliable. Differences are recognised and often corroborated by other archaeological information. Success stories far outnumber failed investigations. Four such examples are described in the following pages, including Grasshopper Pueblo in Arizona, the ancient city of Teotihuacán in Mexico, the Iceman from the Italian Alps and the first king of the Maya city of Copán in Honduras.

Grasshopper Pueblo, Arizona

This case study involves the site of Grasshopper Pueblo in north-central Arizona where ceramic and architectural evidence suggested that several human groups had settled in the latter part of the thirteenth century AD, and departed

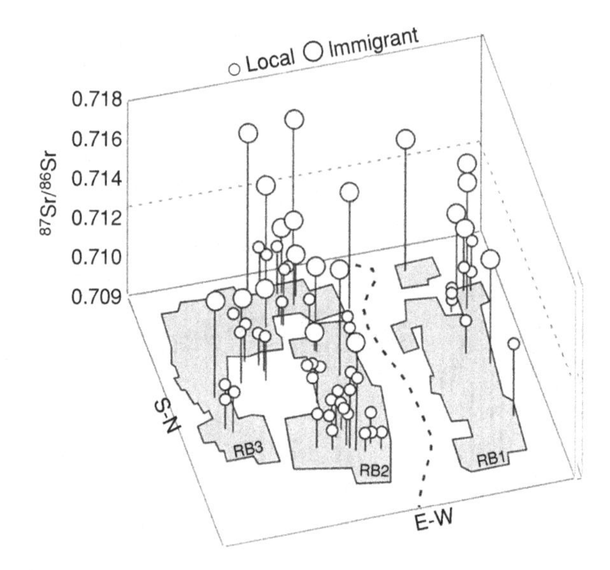

Figure 11.3 Strontium isotope ratios in humans from Grasshopper Pueblo, Arizona, plotted by room block (RB). The length of the line to the circle represents the strontium isotope ratio. Large circles show migrants; small circles show local individuals.

approximately 125 years later. Grasshopper was a complex 500 room pueblo, composed of three major room blocks (1, 2 and 3) and a number of additional units. Maximum population at the site was somewhere between 600 and 700 residents (Reid and Whittlesey, 1999). Almost 700 burials have been recovered at Grasshopper beneath room floors, in the plazas and away from residential areas. Approximately one-third of these are individuals of reproductive age. For the study of Grasshopper Pueblo, a total of 69 enamel and 16 bone samples were analysed, normally from the first molar and the femur. Strontium isotope ratios were also measured in samples of soil, rock and modern field mice from the site area for comparative information. This area is geologically diverse in terms of lithological units and Sr isotope ratios. Rocks types range from Quaternary and Tertiary volcanics to Precambrian formations as much as 1.7 billion years old.

The results of this study provided strong evidence for the presence of both migrants and locals at Grasshopper Pueblo (Figure 11.2). Two different criteria were used to designate migrants to the site. Depending on the criteria, between one-third (using a more cautious range of values in bone at Grasshopper) and more than one-half (using a value based on local fauna) of the residents at Grasshopper had migrated to the site. Both estimates are likely conservative. Archaeological evidence supports the higher estimate.

Information on the spatial distribution of these individuals is also of interest (Figure 11.3). The majority of local individuals are associated with Room Block

2 and the Great Kiva. Individuals in Room Blocks 1 and 3 tend to be migrants. Immigrants came largely from two geological areas: those underlain by Precambrian rocks to the west and south of Grasshopper, and those underlain by Phanerozoic sedimentary rocks to the north and immediate east of the pueblo. There are a number of regions of volcanic rock in the area around Grasshopper from which the migrants did not come.

Chronological evidence indicates that migration continued at about the same rate (50 %) through the occupation of the pueblo, during both the early and late period of the settlement. No differences were noted between sexes. All individuals analysed were adults; there are indications that the proportion of immigrants increases with age of the individual, peaking during the 30s and 40s.

The results from 10 individuals with distinctive amounts and kinds of burial goods also provided some interesting information. These individuals of higher status were a mix of migrants and locals. The artefacts were markers of sodalities, such as warrior groups or dance societies, that cross-cut ethnic and social boundaries and also indicate that social relationships in the pueblo overlapped areas of origin, and integrated locals and migrants.

Teotihuacán, Mexico

The ancient city of Teotihuacán is located in the Central Highlands of Mexico, approximately 50 km northeast of modern Mexico City. During the Classic horizon (AD 1–650), a major population centre developed here with numbers estimated at more than 100 000 people (Cowgill, 1992; Storey, 1992). The city grew very quickly and this rapid increase must have involved immigration into the city (Millon, 1973; 1981; Sanders *et al.*, 1979). There are a number of distinctive areas in the city, including ceremonial precincts, administrative areas, 'palaces', market areas, avenues, tunnels and residential compounds.

Studies of strontium isotopes at Teotihuacán compared burials from several different areas in the ancient city (Figure 11.4), including the Barrio de los Comerciantes, the Oaxaca Barrio, Oztoyahualco, and Cuevas de las Varillas and del Pirul (Price *et al.*, 2000). The Barrio de los Comerciantes, or the Merchant's Compound, is an area of the city with architecture and artefacts characteristic of the Gulf Coast region. Individuals buried in the Oaxaca compound are thought to represent migrants from the Monte Albán area in Oaxaca (Millon, 1973; Sempowski and Spence, 1994). Oztoyahualco is an early residential compound in the city thought to contain local inhabitants. The Cuevas de las Varillas and del Pirul are located beneath the ceremonial centre of the city. Individuals buried in these caves are later inhabitants of the city from the period after collapse. These individuals may have come to the city from outside areas or they may be local inhabitants who remained after the collapse of Teotihuacán.

Results of strontium isotope analyses (Figure 11.4) show substantial variation in enamel $^{87}Sr/^{86}Sr$ ratios. The local ratio at Teotihuacán is estimated from the

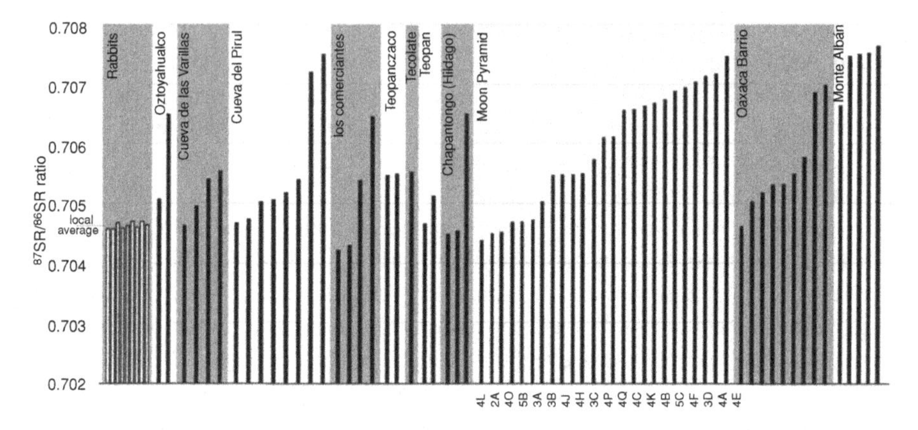

Figure 11.4 Bone vs. Tooth $^{87}Sr/^{86}Sr$ ratios in the individuals from Teotihuacán. Black bars are tooth enamel; open bars are bone. Paired bone and tooth bars are from the same individual. Values grouped by location of burials. Local average is based on measurements of nine rabbit bones from Teotihuacán. The named places in the graph are different parts of the ancient city with the exception of Monte Albán, a contemporary city in the state of Oaxaca. Most of the named places are described in the text.

nine rabbit bones shown at the left in the diagram. Interpretation of these values suggests that many of the individuals in our initial samples were migrants. Comparison of several individuals in the Oaxaca barrio at Teotihuacán shows a strong similarity of some individuals with samples from Monte Albán itself.

There are a number of published reports on the application of oxygen isotopes to examine place of origin in ancient Mesoamerica (Stuart-Williams *et al.*, 1995; Wright and Schwarcz, 1998; Valdés and Wright, 2003; Spence *et al.*, 2004). White and her colleagues (White *et al.*, 1998; 2000; 2001; 2002; 2004) have published a series of studies documenting migration and foreign origin at Teotihuacán using oxygen isotopes in phosphate. A combination of strontium and oxygen isotopes has been used in the study of sacrificial burials found inside the Pyramid of the Moon at Teotihuacán. This massive structure was built in a series of stages, with each new building covering the previous. A series of tunnels, excavated by the Pyramid of the Moon project under the direction of S. Sugiyama and R. Cabrera, have exposed seven superimposed structures in the interior of the pyramid and six groups of distinctive, sacrificial interments of almost 40 individuals buried with other dedicatory items. These offerings date between AD 150 and AD 350 and mark various phases of construction. Burial groups 2–5 have been analysed and reported (White *et al.*, 2007).

The strontium isotope data for the Moon Pyramid sacrifices are graphed in Figure 11.4 along with other enamel values from Teotihuacán. These values

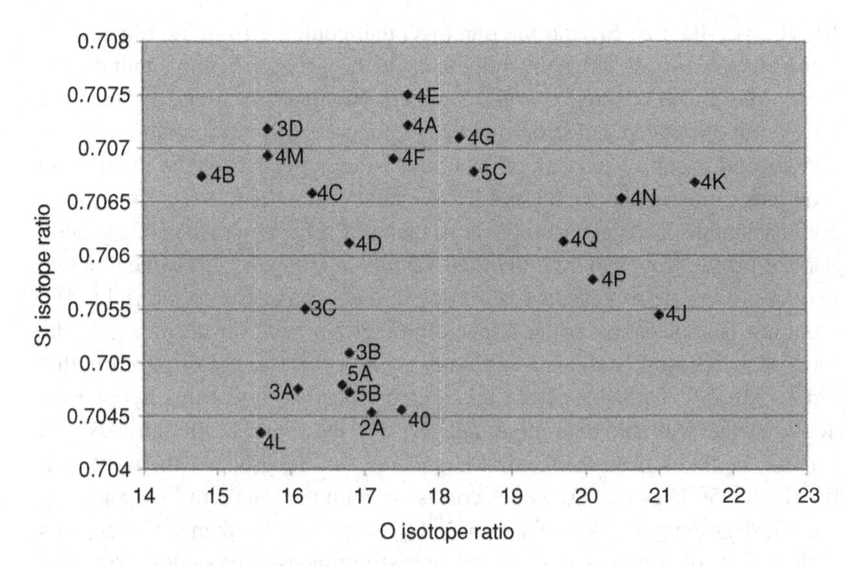

Figure 11.5 Strontium and oxygen isotope ratios in the human burials from the Moon Pyramid, Teotihuacán, Mexico.

are arranged in rank order by location within the site and individual burial designations are indicated for the individuals from the Moon Pyramid. These data provide information on the place of origin of the sacrificial victims in the Moon Pyramid. The expected local $^{87}Sr/^{86}Sr$ value is *c.* 0.7045 based on the measured value in archaeological fauna (rabbits) from the site. It is clear that most of the individuals from the Moon Pyramid were not born at Teotihuacán.

Only two of the victims (5B, 3A) have strontium isotope ratios directly comparable to the local signature. There are three values lower than the local Teotihuacán signature (4L, 2A and 4O). These values fall in the range of the Central Highlands and are comparable to values from the site of Chapantongo in Hildago. The remainder of the values are higher than the Teotihuacán signature. The distribution of the remaining values suggests that several different areas provided these sacrificial victims. Individual 3B has a ratio of around 0.7050, still within the range reported for the Central Highlands. Three values around 0.7055 (4J, 4H and 3C) suggest these three individuals came from the same location. This impression is strengthened by the fact that the burials were found directly adjacent to one another in burial 4.

A scatterplot of the combined oxygen and strontium isotope values for these samples is shown in Figure 11.5. These results give a slightly different perspective on place of origin. Three major clusters of points are discernible in this plot. There is a group of 8 points in the lower left of the graph including 2A,

3B, 3C, 4L, 4O, 5A, 5B and the one (3A) that could be from Teotihuacán. It seems entirely probable that the individuals in this group are from either nearby areas of the Basin of Mexico or the Southern Highlands of Mexico. The close correlation of oxygen and strontium values supports this interpretation.

A second group of ten individuals lies in the upper left of the graph with strontium ratios above 0.706 and $\delta^{18}O_p$ values less than 19‰. This group exhibits a range of oxygen isotope ratios that may reflect more than one general place of origin. If we ignore individual 4D in this group, it is possible that two distinct groups are represented here, one to the left and one to the right. The strontium isotope ratios in the larger, single group however all fit within the range of the Central Highlands and probably reflect a place of origin from that region. More $\delta^{18}O_p$ values from this area are needed to test this hypothesis. However, the available data might suggest that the group to the left (4B, 4C, 4M, 3D) could be from the Central Highlands, but the group to the right (4A, 4E, 4F, 4G, 5C) have $\delta^{18}O_p$ values consistent with the Southern Lowlands.

A third group of individuals with $\delta^{18}O_p$ values greater than 19‰ appears distinct. The combination of recent volcanic strontium isotope values and higher oxygen isotope ratios suggests an area of volcanic geology in more coastal lowland regions and/or greater heat, humidity and rainfall. We suspect the Gulf Coast of Mexico is a likely homeland for these individuals.

The combination of strontium and oxygen isotopes in the study of the Pyramid of the Moon provided some insight on place of origin but also created some confusion. Certain individuals, those identified by higher oxygen isotope values, were not distinguished by strontium isotopes in most cases. More study is necessary to establish baseline information so that we can identify geographically the places that exhibit this combination of isotopic values.

The Iceman, Italian Alps

One of the most extraordinary archaeological finds of the last century was made in 1991. Two hikers noticed a human body, half-frozen, face down, in the snow and ice at 3200 m in the high Alps on the border between Italy and Austria. The Iceman was approximately 46 years old at the time of his death. Analysis of pollen in the stomach contents indicates he died between March and June. His death probably was caused by an arrow wound in his back. Radiocarbon dates from the body and the equipment the Iceman carried indicate an age of around 4300 BC, clearly in the Neolithic period.

More than 150 specialists have been examining all aspects of the Iceman. One of the more difficult questions concerns from where he came. The high Alps are uninhabitable in winter; the Iceman's home must have been elsewhere at lower altitude. Two research groups have used different methods to arrive at contrasting conclusions regarding the homeland of the Iceman. Müller *et al.*

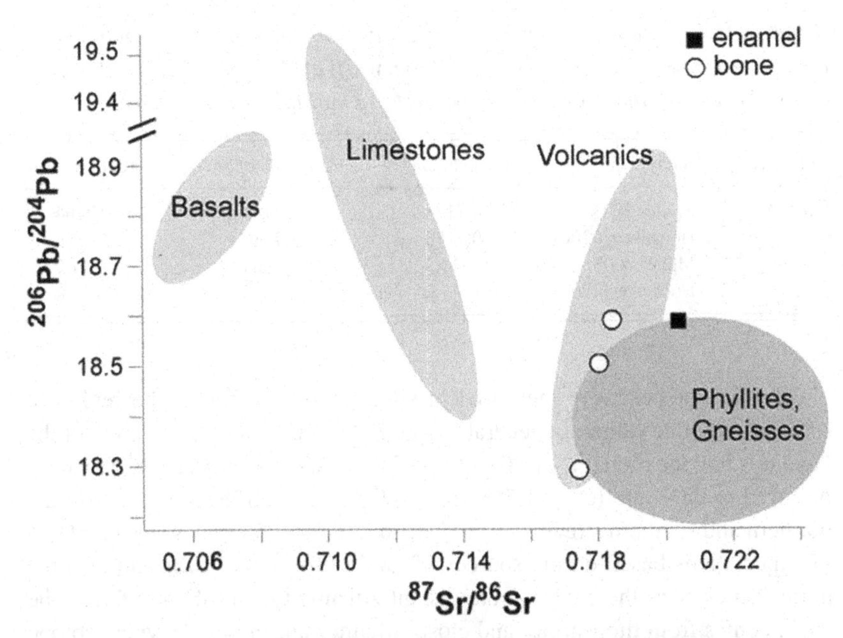

Figure 11.6 Lead 206/204 and strontium 87/86 isotopes in soil leachates from lithological units in the larger region around the find location of the Iceman. Four distinct units can be identified with some overlap between the volcanics and phyllites/gneisses. Four samples from the enamel and bone of the Iceman are shown on the graph, indicating that the Iceman was born and lived on the volcanics and phyllites/gneisses which are found south of the find location in Italy. From Müller *et al.* (2003), p. 864, Figure 2B.

(2003) undertook an isotopic study of the Iceman's tooth enamel, bone and stomach contents to determine his birthplace, habitat and range. Their study focused on strontium, lead and oxygen isotopes to determine the place of origin. Hoogewerff *et al.* (2001) used strontium and oxygen isotopes and trace elements to compare the bones of the Iceman to isotopically mapped geographic regions. This study will be discussed following a consideration of Müller *et al.*

Müller *et al.* (2003) identified four major lithological units in the potential homelands of the Iceman on the basis of strontium and lead isotope ratios (Figure 11.6). These include Eocene basalts (*c.* 0.705–0.708), Mesozoic limestones (*c.* 0.710–0.714) and Permian volcanics (*c.* 0.717–0.719), and a mixed unit of phyllites and gneisses (*c.* 0.720–0.724). The last two units overlap slightly in their ratios. Strontium isotope ratios in the enamel of the Iceman averaged 0.721 and values in bone were approximately 0.718. These values are most congruent with the volcanics and phyllites/gneisses which are found largely to the south of the find location in Italy and suggested to the researchers that the Iceman spent most of his life in this area no more than 60 km from where he was found.

Table 11.1 *Strontium and oxygen isotope ratios in bone and tooth enamel of the Iceman, reported in two studies (Hoogewerff* et al., *2001; Müller* et al., *2003). Several of these values are averages of multiple measurements*

Isotopes	Study	Cortical bone	Trabecular bone	Tooth enamel
$^{87}Sr/^{86}Sr$	Müller 2003	0.7178	0.7184	0.7204
	Hoogewerff 2001	0.7178	0.7186	na
$\delta^{18}O$	Müller 2003	−11.5	na	−11.0
	Hoogewerff 2001	−7.0	−7.2	na

Oxygen isotopes were measured in streams and springs in the region to establish baseline values. In general, lower $\delta^{18}O$ values were found north of the Iceman's find location (*c.* −13.0‰ to −15.0‰) and less negative $\delta^{18}O$ values occurred to the south (*c.* −11.5‰ to −14.0‰). The differences between the northern and southern streams were on the order of 1–2‰. These observations fit expectations based on the sources of rainfall for this area. Rainfall to the north comes from the cooler, more distant Atlantic Ocean; the rainfall to the south comes from the warmer and closer Mediterranean Sea. Oxygen isotope ratios in the skeletal tissue of the Iceman ($\delta^{18}O = 11$‰ for enamel carbonate and 11.5‰ for bone carbonate) better match the water values to the south of the find locality and support the inference based on strontium and lead.

In the second study, Hoogewerff *et al.* (2001) measured strontium, oxygen and carbon isotopes along with a number of trace elements on bone samples from the Iceman. Their conclusions were somewhat different from those of Müller *et al.* although their results are surprisingly similar. A comparison of the two studies (Table 11.1) is valuable here to learn more about the methods, the results and their interpretation.

It is important to remember that the Hoogewerff study only analysed bone from the Iceman, not tooth enamel. The $^{87}Sr/^{86}Sr$ results on cortical and trabecular bone of the two studies are almost identical and document the replicability of analyses. There is very little difference between the two types of bone; results for trabecular bone are slightly higher. Oxygen isotope measurements on bone are reported differently in the two studies and difficult to compare. Oxygen also appears to be inconclusive with regard to place of origin.

Müller *et al.* argue that their strontium isotope ratios indicate that the Iceman's place of origin was to the south in Italy. Hoogewerff *et al.* argue on the basis of the somewhat lower values in the bone, and comparison with a number of historical burials from both areas, that the Iceman spent the later years of his life to the north in Austria. Both studies may be correct. Müller *et al.* focus on place of origin, while Hoogewerff *et al.* emphasise the residence of the last 20 years of the life of the Iceman. It is possible that the Iceman was born to the south but spent his later years north of his place of death. It is clear from the different values in enamel and bone tissue that he moved.

Copán, Honduras

The ancient Maya city of Copán is located in western Honduras not far from the Guatemalan border. The ruins of the site cover approximately 15 hectares, centred on the great acropolis which is composed of five plazas, major pyramids, temples and other structures. The major occupation of the site took place during the Classic period from AD 300 to AD 900 (e.g. Fash and Fash, 2000; 2001). Recent tunnel excavations into the great acropolis have revealed a series of temples and major tombs that appear to represent the burial place of Copán's earliest rulers (Sharer *et al.*, 1999; Bell *et al.*, 2003).

These tombs vary from the elaborate crypts of rulers to the simple interment of a sentinel at the entrance to a passageway leading into one of the structures. One elaborate tomb was found to the west of the major group. This grave (designated as SubJaguar) is possibly associated with Ruler 7 at Copán who died in AD 544. Another tomb, designated as Motmot, included a primary interment of a young adult female and several peripheral skulls, possibly sacrifices. A single male burial (called Bubba), 18–25 years of age, was found to the north of the royal tombs at the base of an Acropolis façade, dating to *c.* AD 480. The so-called Tlaloc Warrior tomb, dating to *c.* AD 450, contained an individual buried with cut shell 'goggles', and other artefacts that explicitly link him to central Mexico. The Tlaloc remains are those of a robust male, apparently over 40 years of age. Interred at approximately the same time, the Northern Guardian was a large, robust male at least 30–40 years of age, found near the outer entrance to the Margarita tomb, perhaps also a guardian. The Margarita tomb was an elaborate structure containing the single skeleton of a female more than 50 years of age. The nearby Hunal tomb contained the partially disarticulated remains of an older adult male. The tomb's architectural and artefactual associations suggest that he was Yax K'uk Mo (Sharer *et al.*, 1999). Epigraphic information states that Yax K'uk Mo was a warrior who came from the north, and the founder of the Copán dynasty. The interment in the Margarita tomb is thought to have been his wife. The identification of Yax K'uk Mo, however, is inferential. There are several lines of evidence, including isotopic provenancing, that support this contention.

Strontium isotopes have been measured in the individuals buried under the great acropolis (Sharer *et al.*, 1999; Buikstra *et al.*, 2003) as well as from other graves at Copán and in local fauna. The strontium isotope analysis of these burials and of comparative human remains from Tikal and Kaminaljuyu, and fauna from the Northern Yucatan and the Central Peten, shed some light on the place of origin of Copán individuals (Figure 11.7). Strontium isotope ratios in

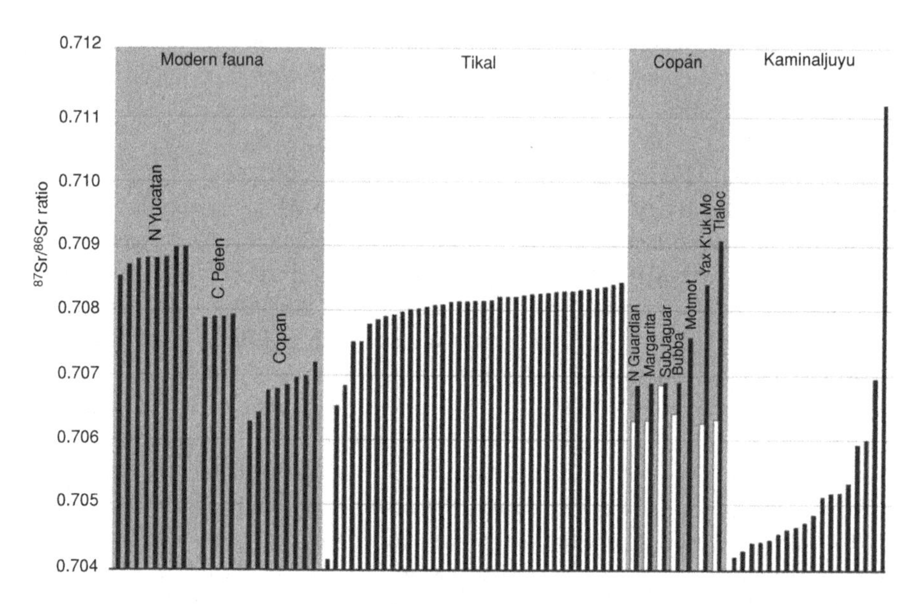

Figure 11.7 Bone and tooth $^{87}Sr/^{86}Sr$ ratios in fauna and humans from the Yucatan, Tikal, Copán, and Kaminaljuyu. Black bars are tooth enamel; open bars are bone. Paired bone and tooth bars are from the same individual. Values grouped by location of burials.

modern fauna from Copán are variable, ranging between 0.7062 and 0.7072. Modern fauna from the Central Peten average almost 0.7080; fauna from the northern Yucatan have ratios between 0.7085 and 0.7090 (Hodell *et al.*, 2004; Price *et al.*, 2005). Human samples from Tikal exhibit an average $^{87}Sr/^{86}Sr$ value of 0.7080, while individuals from the highland Maya site of Kaminaljuyu average 0.7050.

In this context, the seven burials from the underground tombs at Copán are particularly interesting. Four of the individuals exhibit bone and enamel values within the range of local fauna at the site. These individuals (Northern Guardian, Margarita, SubJaguar and Bubba) are likely indigenous inhabitants of the site. Three individuals appear to have come to Copán from elsewhere. The origins of these individuals, as reflected in the strontium isotope ratio for enamel, are geographically separate and impossible to determine precisely as yet. The ratio for the enamel from Motmot suggests an area in the lowlands of Guatemala. The Tlaloc individual may have come from a highland area with older lithological units, but at the same time we cannot rule out a coastal region. The strontium isotope ratio for Yax K'uk Mo is consistent with an origin in the Central Peten, a place that corresponds closely with epigraphic records for this individual and helps to identify him as the first ruler of Copán (Buikstra *et al.*, 2003).

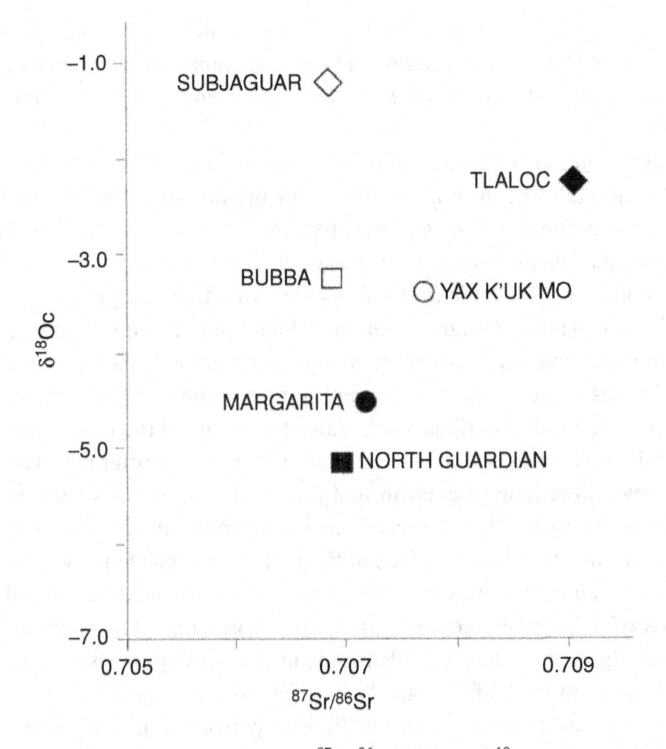

Figure 11.8 M1 enamel values for $^{87}Sr/^{86}Sr$ ratios and $\delta^{18}O$ (carbonate) for six burials from the Copán acropolis (after Buikstra *et al.*, 2003).

The measurement of carbonate oxygen isotopes on the burials from Copán did not improve the geographic discrimination among these individuals (Figure 11.8). Three of the four local individuals, based on strontium isotopes, fall fairly close together in terms of oxygen isotope ratios. The fourth, SubJaguar, has a much less negative oxygen signature. The two 'foreign' individuals for whom oxygen isotopes were measured, Yax K'uk Mo and Tlaloc, also have somewhat distinctive values. The significance of these differences in oxygen isotopes is unclear.

Thoughts and conclusions

When I first began to work with isotopic provenancing, I was a serious skeptic about the extent of human movement in the past and I imagined that much of the discussion about long-distance movement, residential change and migration in the past was exaggerated or mistaken. Over the course of the last 15 years, however, as a result of a series of isotopic provenance studies, my skepticism has been replaced by an appreciation of how much human movement there must have been in the past, how mobile our ancestors were in many different times and places.

That appreciation is coupled with the knowledge that isotopic provenancing will continue to be an important tool for the investigation of past human behaviour. Strontium and oxygen isotopes are of proven utility in addressing questions of past residential change. The case studies presented here document the power of these tools for learning about the past. Studies of groups or populations tell us about rates of migration among other things. At Grasshopper Pueblo, the high number and spatial distribution of migrant individuals was of particular interest as was some of the supplementary information on gender and social differences. Studies at Teotihuaćan revealed that many of the inhabitants of the site, both in local areas and at 'foreign' compounds were migrants. The strontium isotopes were able to confirm that individuals in the Oaxaca compound likely came from that region. The combination of strontium and oxygen isotopes, demonstrated that the majority of the individuals sacrificed at the Pyramid of the Moon were not local and may have been brought from some distance to the site.

Studies of individuals are also extremely informative. The isotopic provenancing of the Iceman has provided a means of determining where he spent his childhood and later life, probably in different places, but all within a few tens of kilometres of the place of his discovery. Isotopic investigations of the burials inside the great acropolis at Copán have helped to identify the first ruler of the city, Yax K'uk Mo, by pinpointing his place of origin some distance to the north. Other individuals buried beneath the acropolis likely came from an even greater distance.

Strontium isotopes are being tested, along with lead, in some contemporary forensic situations (Gulson *et al.*, 1997; Katzenberg and Krause, 1999; Beard and Johnson, 2000). Lead isotopes have been used to examine questions of mobility (e.g. Carlson, 1996; Budd *et al.*, 1999; 2000a), but the utility of lead may be limited to areas where the element is abundant in nature. Sulphur isotopes are currently being explored as another system of interest for isotopic provenancing (Richards *et al.*, 2003). Isotopes of copper and zinc have some potential as geographically distinct signals of residence (Gale *et al.*, 1999; Zhu *et al.*, 2000) and need further investigation. I look forward to the exploration and application of new methods in this exciting field of bioarchaeology.

References

Åberg, G. (1995). The use of natural strontium isotopes as tracers in environmental studies. *Water, Air, and Soil Pollution*, **79**, 309–22.

Åberg, G., Fosse, G. and Stray, H. (1998). Man, nutrition and mobility: a comparison of teeth and bone from the Medieval era and the present from Pb and Sr isotopes. *The Science of the Total Environment*, **224**, 109–19.

Arneborg, J., Heinemeier, J., Lynnerup, N. *et al.* (1999). Change of diet of the Greenland Vikings determined from stable carbon isotope analysis and ^{14}C dating of their bones. *Radiocarbon*, **41**, 157–68.

Ayliffe, L. K. and Chivas, A. R. (1990). Oxygen isotope composition of bone phosphate of Australian kangaroos: potential as a palaeoenvironmental recorder. *Geochimica et Cosmochimica Acta*, **54**, 2603–9.

Balasse, M., Ambrose, S. H., Smith, A. B. and Price, T. D. (2002). The seasonal mobility model for prehistoric herders in the south-western Cape of South Africa assessed by an isotopic study of sheep tooth remains. *Journal of Archaeological Science*, **29**, 917–32.

Beard, B. L. and Johnson, C. M. (2000). Strontium isotope composition of skeletal material can determine the birth place and geographic mobility of humans and animals. *Journal of Forensic Sciences*, **45**, 1049–61.

Bell, E. E., Canuto, M. A. and Sharer, R. J. (eds.) (2003). *Understanding Early Classic Copan*. Philadelphia: University of Pennsylvania Museum.

Bell, L. S., Cox, G. and Sealy, J. (2001). Determining isotopic life history trajectories using bone density fractionation and stable isotope measurements: A new approach. *American Journal of Physical Anthropology*, **116**, 66–79.

Bentley, R. A. (2006). Strontium isotopes from the earth to the archaeological skeleton: a review. *Journal of Archaeological Method and Theory*, **13**, 135–87.

Bentley, R. A. and Knipper, C. (2005). Geographic patterns in biologically available strontium, carbon and oxygen isotopes signatures in prehistoric SW Germany. *Archaeometry*, **47**(3), 629–44.

Blum, J. D., Taliaferro, E. H. Weisse, M. T. and Holmes, R. T. (2000). Changes in Sr/Ca, Ba/Ca and $^{87}Sr/^{86}Sr$ ratios between two forest ecosystems in the northeastern U. S. A. *Biogeochemistry*, **49**, 87–101.

Boecherens, H., Brinkman, D. B., Dauphin, Y. and Mariotti, A. (1994). Microstructural and geochemical investigations on Late Cretaceous archaeosaur teeth from Alberta, Canada. *Canadian Journal of Earth Science*, **31**, 783–92.

Bowen, G. J. and Wilkinson, B. (2002). Spatial distribution of ^{18}O in meteoric precipitation. *Geology*, **30**, 315–8.

Brudevold, F. and Söremark, R. (1967). Chemistry of the mineral phase of enamel. In *Structural and Chemical Organisation of Teeth*. Vol. 2, ed. A. E. W. Miles. New York: Academic Press, pp. 247–77.

Bryant, J. D. and Froelich, P. N. (1996). Oxygen isotope composition of human tooth enamel from medieval Greenland: linking climate and society: comment and reply. *Geology*, **24**, 477–9.

Budd, P., Montgomery, J., Cox, A. *et al.* (1998). The distribution of lead within ancient and modern human teeth: implications for long-term and historical exposure monitoring. *The Science of the Total Environment*, **220**, 121–36.

Budd, P., Montgomery, J., Rainbird, P., Thomas, R. G. and Young, S. M. M. (1999). Pb- and Sr-isotope composition of human dental enamel: an indicator of Pacific Islander population dynamics. In *The Pacific from 5000 to 2000 BP: Colonisations and Transformations*, eds. J. C. Galipaud and I. Lilley. Paris: Institut de Recherche pour le Développement, pp. 301–11.

Budd, P., Montgomery, J., Evans, J. and Barreiro, B. (2000a). Human tooth enamel as a record of the comparative lead exposure of prehistoric and modern people. *The Science of the Total Environment*, **263**, 1–10.

Budd, P., Montgomery, J., Barriero, B. and Thomas, R. G. (2000b). Differential diagenesis of strontium in archaeological human dental tissues. *Applied Geochemistry*, **15**, 687–94.

Budd, P., Montgomery, J., Evans, J. and Chenery, C. (2001). Combined Pb-, Sr- and O-isotope analysis of human dental tissue for the reconstruction of archaeological residential mobility. In *Plasma Source Mass Spectrometry: The New Millennium.* Special publication 267, eds. J. G. Holland and S. D. Tanner. Cambridge: Royal Society of Chemistry, pp. 311–326.

Buikstra, J. E., Price, T. D., Wright, L. E. and Burton, J. H. (2003). Tombs from the Copán acropolis: a life history approach. In *Understanding Early Classic Copan*, eds. E. E. Bell, M. A. Canuto and R. J. Sharer. Philadelphia: University of Pennsylvania Museum, pp. 185–205.

Carlson, A. K. (1996). Lead isotope analysis of human bone for addressing cultural affinity: a case study from Rocky Mountain House, Alberta. *Journal of Archaeological Science*, **23**, 557–68.

Carter, S. W., Wiegand, B., Mahood, G. A. *et al.* (2003). Strontium isotopic analysis as a tool for ceramic provenance research. *Geological Society of America Abstracts with Programs*, **35**(6), 398.

Chamberlain, C. P., Blum, J. D., Holmes, R. T. *et al.* (1997). The use of isotope tracers for identifying populations of migratory birds. *Oecologia*, **109**, 132–41.

Comar, C., Russell, R. S. and Wasserman, R. H. (1957). Strontium-calcium movement from soil to man. *Science*, **126**, 485–96.

Cowgill, G. L. (1992). Toward a political history of Teotihuacan. In *Ideology and Pre-Columbian Civilizations*, eds. A. A. Demarest and G. W. Conrad. Santa Fe: School of American Research Press, pp. 87–114.

Cox, G. and Sealy, J. (1997). Investigating identity and life histories: isotopic analysis and historical documentation of slave skeletons found on the Cape Town foreshore, South Africa. *International Journal of Historical Archaeology*, **1**, 207–24.

Craig, H. (1965). The measurement of oxygen isotope paleotemperatures. In *Stable Isotopes in Oceanographic Studies and Paleotemperatures*, ed. E. Tongiorgi. Spoleto: Consiglio Nazionale delle Ricerche, Laboratorio di Geologica Nucleare, Pisa, pp. 161–82.

Dansgaard, W. (1994). Iskerner og isotoper. *Geologisk Nyt*, **3/94**, 19–23.

Driessens, F. C. M. and Verbeeck, R. M. (1990). *Biominerals*. Boca Raton: CRC Press.

Elias, R. W., Hirao, Y. and Patterson, C. C. (1982). The circumvention of the natural biopurification of calcium along nutrient pathways by atmospheric inputs of industrial lead. *Geochimica et Cosmochimica Acta*, **46**, 2561–80.

Ericson, J. E. (1981). Residence patterns by isotopic characterization. Paper presented at the Society for American Archaeology, San Diego, CA.

Ericson, J. E. (1985). Strontium isotope characterization in the study of prehistoric human ecology. *Journal of Human Evolution*, **14**, 503–14.

Ericson, J. E. (1989). Some problems and potentials of strontium isotope analysis for human and animal ecology. In *Stable Isotopes in Ecological Research*, eds. P. W. Rundel, J. R. Ehleringer, and K. A. Nagy. Berlin: Springer-Verlag, pp. 252–9.

Ezzo, J. A. (1993). *Human Adaptation at Grasshopper Pueblo, Arizona: Social and Ecological Perspectives*. Ann Arbor: International Monographs in Prehistory.

Ezzo, J. A., Johnson, C. M. and Price, T. D. (1997). Analytical perspectives on prehistoric migration: a case study from east-central Arizona. *Journal of Archaeological Science*, **24**, 447–66.

Fash, W. L. and Fash, B. (2000). Teotihuacan and the Maya: A Classic Heritage. In *Mesoamerica's Classic Heritage: From Teotihuacan to the Aztecs*, eds. D. Carrasco, J. Jones and S. Sessions. Boulder: University of Colorado Press, pp. 433–63.

Fash, W. L. and Fash, B. (2001). *Scribes, Warriors and Kings: The City of Copán and the Ancient Maya*. New York: Thames and Hudson.

Faure, G. (1986). *Principles of Isotope Geology*. New York: John Wiley.

Faure, G. and Powell, T. (1972). *Strontium Isotope Geology*. New York:Springer-Verlag.

Fitzpatrick, A. P. (2003). The Amesbury Archer. *Current Archaeology*, **184**, 146–52.

Fricke, H. C. and O'Neil, J. R. (1996). Inter- and intra-tooth variation in the oxygen isotope composition of mammalian tooth enamel phosphate: implications for paleoclimatological and paleobiological research. *Palaeogeography, Palaeoclimatology, Palaeoecology*, **126**, 91–9.

Fricke, H. C., O'Neil, J. R and Lynnerup, N. (1995). Oxygen isotope composition of human tooth enamel from medieval Greenland: Linking climate and society. *Geology*, **23**, 869–72.

Gale, N. H., Woodhead, A. P., Stos-Gale, Z. A., Walder, A. and Bowen, I. (1999). Natural variations detected in the isotopic composition of copper: possible applications to archaeology and geochemistry. *International Journal of Mass Spectrometry*, **184**(1), 1–9.

Gannes, L. Z., Martínez del Rio, C. and Koch, P. (1998). Natural abundance variations in stable isotopes and their potential uses in animal physiological ecology. *Comparative Biochemistry and Physiology*, **119A**, 725–37.

Gosz, J. R., Brookins, D. G. and Moore, D. I. (1983). Using strontium isotope ratios to estimate inputs into ecosystems. *Bioscience*, **33**, 23–30.

Grupe, G., Price, T. D., Schörter, P. *et al.* (1997). Mobility of Bell Beaker people revealed by stable strontium isotope ratios of teeth and bones. A study of southern Bavarian skeletal remains. *Applied Geochemistry*, **12**, 517–25.

Grupe, G., Price, T. D. and Söllner, F. (1999). Mobility of Bell Beaker people revealed by strontium isotope ratios of tooth and bone: a study of southern Bavarian skeletal remains. A reply to the comment by Peter Horn and Dieter Müller-Sohnius. *Applied Geochemistry*, **14**, 271–5.

Gulson, B. L., Jameson, C. W. and Gillings, B. R. (1997). Stable lead isotopes in teeth as indicators of past domicile – a potential new tool in forensic science? *Journal of Forensic Science*, **42**, 787–91.

Hillson, S. (1997). *Dental Anthropology*. Cambridge: Cambridge University Press.

Hillson, S. (2005). *Teeth*. Cambridge: Cambridge University Press.

Hodell, D. A., Quinn, R. L., Brenner, M. and Kamenov, G. (2004). Spatial variation of strontium isotopes (^{87}Sr/ ^{86}Sr) in the Maya region: a tool for tracking ancient human migration. *Journal of Archaeological Science*, **31**, 585–601.

Hoogewerff, J., Papesch, W., Kralik, M. *et al.* (2001). The last domicile of the Iceman from Hauslabjoch: A geochemical approach using Sr, C and O isotopes and trace element signatures. *Journal of Archaeological Science*, **28**, 983–9.

Hoppe, K. A., Koch, P. L. and Furutani, T. T. (2003). Assessing the preservation of biogenic strontium in fossil bones and tooth enamel. *International Journal of Osteoarchaeology*, **13**, 20–8.

Hörn, P. and Müller-Söhnius D. (1999). Comment on 'Mobility of Bell Beaker people revealed by Sr isotope ratios of tooth and bone: a study of southern Bell Beaker skeletal remains' by Gisela Grupe, T. Douglas Price, Peter Schrörter, Frank Söllner, Clark M. Johnson, and Brian L. Beard. *Applied Geochemistry*, **14**, 263–9.

Iacumin, P., Bocherens, H., Mariotti, A. and Longinelli, A. (1996). Oxygen isotope analyses of co-existing carbonate and phosphate in biogenic apatite: a way to monitor diagenetic alteration of bone phosphate? *Earth and Planetary Science Letters*, **142**, 1–6.

Jones, J. L., Goodman, A. H., Reid, J. *et al.* (2003). Permanent molars and shifting landscapes: elemental signature analysis of natality at the New York African Burial Ground. *American Journal of Physical Anthropology*, **120**(S36), 124.

Katzenberg, M. A. and Harrison, R. G. (1996). What's in a bone? Recent advances in archaeological bone chemistry. *Journal of Archaeological Research*, **5**, 265–93.

Katzenberg, M. A. and Krause, H. R. (1999). Application of stable isotope variation in human tissues to problems in identification. *Canadian Society of Forensic Science Journal*, **22**, 7–19.

Knudson, K. J., Price, T. D., Buikstra, J. E. and Blom, D. E. (2004). The use of strontium isotope analysis to investigate Tiwanaku migration and mortuary ritual in Bolivia and Peru. *Archaeometry*, **46**, 5–18.

Knudson, K. J., Nystrom, K. C., Tung, T. A. *et al.* (2005). The origin of the Juch'uypampa cave mummies: strontium isotope analysis of archaeological human remains from Bolivia. *Journal of Archaeological Science*, **32**, 903–13.

Koch, P. L., Fogel, M. L. and Tuross, N. (1994). Tracing the diets of fossil animals using stable isotopes. In *Stable Isotopes in Ecology and Environmental Science*, eds. R. Michener and K. Lajtha. Oxford: Blackwell Scientific, pp. 63–92.

Koch, P. L., Tuross, N. and Fogel, M. L. (1997). The effects of sample treatment and diagenesis on the isotopic integrity of carbonate in biogenic hydroxylapatite. *Journal of Archaeological Science*, **24**, 417–29.

Kohn, M. J. (1996). Predicting animal δ^{18}O: Accounting for diet and physiological adaptation. *Geochimica et Cosmochimica Acta*, **60**, 4811–29.

Kohn, M. J. (1999). You are what you eat. *Science* **283**: 335–6.

Kohn, M. J., Schoninger, M. J. and Barker, W. W. (1999). Altered states: effects of diagenesis on fossil tooth chemistry. *Geochimica et Cosmochimica Acta*, **63**, 2737–47.

Kolodny, Y., Luz, B., Sander, M. and Clemens, W. A. (1996). Dinosaur bones: fossils or pseudomorphs? The pitfalls of physiology reconstruction from apatitic fossils. *Palaeogeography, Palaeoclimatology, Palaeoecology*, **126**, 161–71.

Lajtha, K. and Michener, R. H. (eds.) (1994). *Stable Isotopes in Ecology and Environmental Science*. Oxford: Blackwell Scientific Publications.

Lee-Thorp, J. A. (2002). Two decades of progress towards understanding fossilisation processes and isotopic signals in calcified tissue minerals. *Archaeometry*, **44**, 435–46.

Lee-Thorp, J. A. and Sponheimer, M. (2003). Three case studies used to reassess the reliability of fossil bone and enamel isotope signals for paleodietary studies. *Journal of Anthropological Archaeology*, **22**, 208–16.

Longinelli, A. (1984). Oxygen isotopes in mammal bone phosphate: A new tool for palaeohydrological and palaeoclimatological research? *Geochimica et Cosmochimica Acta*, **48**, 385–90.

Luz, B. and Kolodny, Y. (1985). Oxygen isotope variations in phosphate of biogenic apatites. IV. Mammal teeth and bones. *Earth and Planetary Science Letters*, **72**, 29–36.

Luz, B., Kolodny, Y. and Horowitz, M. (1984). Fractionation of oxygen isotopes between mammalian bone-phosphate and environmental drinking water. *Geochimica et Cosmochimica Acta*, **48**, 1689–93.

Luz, B. Cormie, A. B. and Schwarcz, H. P. (1990). Oxygen isotope variations in phosphate of deer bones. *Geochimica et Cosmochimica Acta*, **54**, 1723–8.

Matthews, K. J., Leese, M. N., Hughes, M. J., Herz, N. and Bowman, S. G. E. (1995). Establishing the provenance of marble using statistical combinations of stable isotope and neutron activation analysis data. In *The Study of Marble and Other Stones Used in Antiquity*, eds. Y. Maniatis, N. Herz and Y. Basiakos. London: Archetype Publications, pp. 171–80.

Miller, E. K., Blum, J. A. and Friedland, A. J. (1993). Determination of soil exchangable-cation loss and weathering rates using Sr isotopes. *Nature*, **362**, 438–41.

Millon, R. (1973). *Urbanization at Teotihuacan, Mexico. Vol. 1, The Teotihuacan Map. Part One: Text*. Austin: University of Texas Press.

Millon, R. (1981). Teotihuacan: city, state, and civilization. In *Supplement to the Handbook of Middle American Indians. Vol. 1, Archaeology*, eds. V. Bricker and J. Sabloff. Austin: University of Texas Press, pp. 198–243.

Montgomery, J., Budd, P., Cox, A., Krause, P. and Thomas, R. G. (1999). LA-ICP-MS evidence for the distribution of Pb and Sr in Romano-British medieval and modern human teeth: implications for life history and exposure reconstruction. In *Metals in Antiquity: Proceedings of the International Symposium*. eds. S. M. M. Young, A. M. Pollard, P. Budd and R. A. Ixer. Oxford: Archaeopress.

Montgomery, J., Budd, P. and Evans, J. (2000). Reconstructing the lifetime movements of ancient people: a Neolithic case study from southern England. *European Journal of Archaeology*,, **3**, 407–22.

Montgomery, J., Evans, J. and Roberts, C. A. (2003a). Mineralization, preservation and sampling of teeth: Strategies to optimise comparative study and minimise age-related change for lead and strontium analysis. *American Journal of Physical Anthropology*, **S36**, 153–4.

Montgomery, J., Evans, J. A. and Neighbour, T. (2003b). Sr isotope evidence for population movement within the Hebridean Norse community of NW Scotland. *Journal of the Geological Society*, **160**, 649–53.

Müller, W., Fricke, H. Halliday, A. N., McCulloch, M. T. and Wartho, J.-A. (2003). Origin and migration of the Alpine Iceman. *Science*, **302**, 862–6.

Nelson, B. K., DeNiro, M. J., Schoeninger, M. J., DePaolo, D. J. and Hare, P. E. (1986). Effects of diagenesis on strontium, carbon, nitrogen, and oxygen concentration and isotopic composition of bone. *Geochimica et Cosmochimica Acta*, **50**, 1941–9.

Price, T. D. and Gestsdóttir, H. (2005). The first settlers of Iceland: an isotopic approach to colonization. *Antiquity*, **80**, 130–44.

Price, T. D., Johnson, C. M., Ezzo, J. A., Burton, J. H. and Ericson, J. A. (1994a). Residential mobility in the Prehistoric Southwest United States. A preliminary study using strontium isotope analysis. *Journal of Archaeological Science*, **24**, 315–30.

Price, T. D., Grupe, G. and Schröter, P. (1994b). Reconstruction of migration patterns in the Bell Beaker Period by stable strontium isotope analysis. *Applied Geochemistry*, **9**, 413–17.

Price, T. D., Grupe, G. and Schröter, P. (1998). Migration and mobility in the Bell Beaker period in Central Europe. *Antiquity*, **72**, 405–11.

Price, T. D., Manzanilla, L. and Middleton. W. H. (2000). Residential mobility at Teotihuacan: a preliminary study using strontium isotopes. *Journal of Archaeological Science*, **27**, 903–13.

Price, T. D., Bentley, R. A., Lüning, J., Groenborn, D. and Wahl, J. (2001). Migration in the Linearbandkeramik of Central Europe. *Antiquity*, **75**, 593–603.

Price, T. D., Tiesler, V. and Burton, J. H. (2005). Conquerors, conquered, and slaves: bioarchaeological evidence of the Early Colonial inhabitants of Campeche, Mexico. In *Los Investigadores de la Cultura Maya II*. Campeche: University of Campeche Press, pp. 357–64.

Reid, R. J. and Whittlesey, S. M. (1999). *Grasshopper Pueblo: a Story of Archaeology and Ancient Life*. Tucson: University of Arizona Press.

Richards, M. P., Fuller, B., Sponheimer, M., Robinson, T. and Ayliffe L. (2003). Sulphur isotopes in palaeodietary studies: review and results from a feeding experiment. *International Journal of Osteoarchaeology*, **13**, 37–45.

Rogers, G. and Hawkesworth, C. J. (1989). A geochemical traverse across the North Chilean Andes: evidence for crust generation from the mantle wedge. *Earth and Planetary Science Letters*, **91**, 271–85.

Rosenthal, H. L. (1981). Content of stable strontium in man and animal biota. In *Handbook of Stable Strontium*, ed. S. C. Skoryna. New York: Plenum Press, pp. 503–14.

Rundel, P. W., Ehleringer, J. R. and Nagy, K. A. (eds.) (1989). *Stable Isotopes in Ecological Research*. New York: Springer-Verlag.

Sanders, W. T., Parsons, J. and Santley, R. (1979). *The Basin of Mexico*. New York: Academic Press.

Schoeninger, M. J., Hallin, K., Reeser, H., Valley, J. W. and Fournelle, J. (2003). Isotopic alteration of mammalian tooth enamel. *International Journal of Osteoarchaeology*, **13**, 11–19.

Schroeder, H. A., Nason, A. P. and Tipton, I. H. (1972). Essential metals in man: strontium and barium. *Journal of Chronic Diseases*, **25**, 491–517.

Schwarcz, H. P. and Schoeninger, M. J. (1991). Stable isotope analyses in human nutritional ecology. *Yearbook of Physical Anthropology*, **34**, 283–321.

Schwarcz, H. P., Gibbs, L. and Knyf, M. (1991). Oxygen isotope analysis as an indicator of place of origin. In *Snake Hill: An Investigation of a Military Cemetery from the War of 1812*, eds. S. Pfeiffer and R. Williamson. Toronto: Dundurn Press, pp. 263–8.

Schweissing, M. M. and Grupe, G. (2003). Tracing migration events in man and cattle by stable strontium isotope analysis of appositionally grown mineralized tissue. *International Journal of Osteoarchaeology*, **13**, 96–103.

Sealy, J. C. (1989). Reconstruction of later Stone Age diets in the South-Western Cape, South Africa: evaluation and application of five isotopic and trace element techniques. Ph.D. thesis, University of Capetown.

Sealy, J. C., van der Merwe, N. J., Sillen, S., Kruger, F. J. and Krueger, H. W. (1991). $^{87}Sr/^{86}Sr$ as a dietary indicator in modern and archaeological bone. *Journal of Archaeological Science*, **18**, 399–416.

Sealy, J., Armstrong, R. and Schrire, C. (1995). Beyond lifetime averages: tracing life histories through isotopic analysis of different calcified tissues from archaeological human skeletons. *Antiquity*, **69**, 290–301.

Sempowski, M. L. and Spence, M. W. (eds). (1994). *Mortuary Practices and Skeletal Remains at Teotihuacan*. Salt Lake City: University of Utah Press.

Shackleton, J. and Elderfield, H. (1990). Strontium isotope dating of the source of Neolithic European Spondylus shell artefacts. *Antiquity*, **64**, 312–15.

Sharer, R. J., Traxler, L. P., Sedat, D. W. *et al.* (1999). Early classic architecture beneath the Copán acropolis. *Ancient Mesoamerica*, **10**, 3–23.

Sharp, Z. D., Atudorei, V. and Furrer, H. (2000). The effects of diagenesis on oxygen isotope ratios of biogenic phosphates. *Science*, **300**, 222–37.

Sillen, A., Hall, G. and Armstrong, R. (1998). $^{87}Sr/^{86}Sr$ ratios in modern and fossil food-webs of the Sterkfontein Valley: Implications for early hominid habitat preference. *Geochimica et Cosmochimica Acta*, **62**, 2463–78.

Smith, B. H. (1994). Sequence of emergence of the permanent teeth in *Macaca, Pan, Homo*, and *Australopithecus*: Its evolutionary significance. *American Journal of Human Biology*, **6**, 61–76.

Spence, M. White, C. D. and Longstaffe, F. J. (2004) Victims of the victims: human trophies worn by the sacrificed soldiers from the Feathered Serpent Pyramid. *Ancient Mesoamerica*, **15**, 1–15.

Sponheimer, M. and Lee-Thorp, J. (1999). Oxygen isotopes in enamel carbonate and their ecological significance. *Journal of Archaeological Science*, **26**, 723–8.

Storey, R. (1992). Patterns of susceptibility to dental defects in the deciduous dentition of a Precolumbian skeletal population. In *Recent Contributions to the Study of Enamel Developmental Defects*, eds. A. L. Goodman and L. L. Capasso. Chieti, Italy: Edigrafital Teramo.

Stuart-Williams, H. L. Q., Schwarcz, H. P., White, C. D. and Spence, M. W. (1995). The isotopic composition and diagenesis of human bone from Teotihuacan and Oaxaca, Mexico. *Paleogeography, Paleoclimatology, Paleoecology*, **126**, 1–11.

Teitelbaum, S. L. (2000). Bone resorption by osteoclasts. *Science*, **289**, 1504.

Trueman, C., Chenery, C., Eberth, D. A. and Spiro, B. (2003). Diagenetic effects on the oxygen isotope composition of bones of dinosaurs and other vertebrates recovered from terrestrial and marine sediments. *Journal of the Geological Society*, **160**, 895–901.

Valdés, J. A. and Wright, L. E. (2003). The Early Classic and its antecedents at Kaminaljuyu: A complex society, with complex problems. In *Understanding Early Classic Copan*, eds. E. E. Bell, M. A. Canuto and R. J. Sharer. Philadelphia: University of Pennsylvania Museum, pp. 327–46.

van der Merwe, N. J., Lee-Thorp, J. A., Thackeray, J. F. *et al.* (1990). Source-area determination of elephant ivory by isotopic analysis. *Nature*, **346**, 744–6.

Wang, Y. and Cerling, T. E. (1994). A model of fossil tooth and bone diagenesis: Implications for paleodiet reconstruction from stable isotopes. *Palaeogeography, Palaeoclimatology, Palaeoecology*, **107**, 281–9.

White, C. D., Spence, M. W. Stuart-Williams, H. LeQ. and Schwarcz, H. P. (1998). Oxygen isotopes and the identification of geographical origins: the Valley of Oaxaca versus the Valley of Mexico. *Journal of Archaeological Science*, **25**, 643–55.

White, C. D., Spence, M. W. and Law, K. R. (2000). Testing the nature of Teotihuacan imperialism at Kaminaljuyu using phosphate oxygen-isotope ratios. *Journal of Anthropological Research*, **56**, 535–58.

White, C. D., Longstaffe, F. J. and Law, K. R. (2001). Revisiting the Teotihuacan connection at Altun Ha: Oxygen-isotope analysis of Tomb f-8/1. *Ancient Mesoamerica*, **12**, 65–72.

White, C. D., Spence, M. and Longstaffe, F. (2002) Geographic identities of the sacrificial victims at the Temple of Quetzalcoatl: Implications for the nature of state power. *Latin American Antiquity*, **13**, 217–36.

White, C. D., Storey, R., Spence, M. and Longstaffe, F. J. (2004). Immigration, assimilation and status in the ancient city of Teotihuacan: Isotopic evidence from Tlajinga 33. *Latin American Antiquity*, **15**, 176–98.

White, C. D., Price, T. D. and Longstaffe, F. (2007). Residential histories of the human sacrifices at the Moon Pyramid: evidence from oxygen and strontium isotopes. *Ancient Mesoamerica*, **18**(1), 159–72.

Williams, D. F. (1984). Correlation of Pleistocene marine sediments of the Gulf of Mexico and other basins using oxygen isotope stratigraphy. In *Principles of Pleistocene Stratigraphy Applied to the Gulf of Mexico*, ed. N. Healy-Williams. Boston: International Human Resources Development Corp, p. 65–118.

Wright, L. E. (2005). Identifying immigrants to Tikal, Guatemala: Defining local variability in strontium isotope ratios of human tooth enamel. *Journal of Archaeological Science*, **32**(4), 555–66.

Wright, L. E. and Schwarcz, H. P. (1998). Stable carbon and oxygen isotopes in human tooth enamel: Identifying breastfeeding and weaning in prehistory. *American Journal of Physical Anthropology*, **106**(1), 1–18.

Yurtsaver, Y. and Gat, J. R. (1981). Atmospheric waters. In *Stable Isotope Hydrology: Deuterium and Oxygen-18 in the Water Cycle. Technical Report Series 210*, eds. J. R. Gat and R. Gonfiantini. Vienna: International Atomic Energy Agency, pp. 103–42.

Zhu, X. K., O'Nions, R. K. Guo, Y., Belshaw, N. S. and Rickard, D. (2000). Determination of natural Cu-isotope variation by plasma-source mass spectrometry: implications for use as geochemical tracers. *Chemical Geology*, **163**, 139–49.

12 *From bodies to bones and back: theory and human bioarchaeology*

KIRSI LORENTZ

> [T]he body as simultaneously a physical and symbolic artefact, as both naturally and culturally produced, and as securely anchored in a particular historical moment
>
> *(Scheper-Hughes and Lock, 1987, p. 7).*

> Examining . . . bodies in this view becomes a matter of tracing through the means, the varied array of materials and practices involved in their construction and maintenance – and in some circumstances their unravelling and disintegration
>
> *(Prout, 2000, p. 15).*

The body forms a current focus of intense research both in the humanities and sciences, and is one of the main preoccupations of contemporary popular culture and social discourse (Sweeney and Hodder, 2002a). While the physiology of the body has a long history of research, the focus on cultural construction of bodies is of more recent origin (Bourdieu, 1977; Mauss, 1979 [1934]).

This chapter aims to address a range of themes reflecting the close relationship between human culture and biology, body and society. The focus is on the relationships between the physical body and socio-cultural practices, aspects of which can be identified in the skeletal record. The chapter will explore new research agendas relating to the body, understood both as biological and cultural entity, and the role of biological anthropology and bioarchaeology in the investigation of the socio-cultural body and the body politic. For example, bodily manipulations and modifications have been used cross-culturally, to denote various forms of socio-cultural difference, leading to specific forms of biocultural identity. Other phenomena bringing to focus the complex relationships between human culture and biology include various socio-cultural practices affecting, and affected by, the body (such as eating, or body techniques – that is the culturally contingent way humans move their bodies), interpersonal violence, and gender. Theories of the body in the social sciences focus mainly on the living body, taking a constructivist view, while osteoarchaeology is often perceived as science-based, focusing on the dead body. The perceived theory gap between

Between Biology and Culture, ed. Holger Schutkowski. Published by Cambridge University Press.

the way osteoarchaeology has viewed the body through the skeleton, and the cultural and biological reality of the living person and the body that the skeleton once formed a part of, needs to be addressed. The analytical repertoire for the investigation of the body, understood both as cultural and physical entity, living and dead, lies at the interface of science and humanities.

Human remains constitute the most direct means of accessing conditions of life in the past (Larsen, 2002, p. 145). Further, human remains constitute a rich data resource for recognising attitudes towards the body in archaeological contexts, whether through physical anthropological methods relating to aspects of the skeleton affected by socio-cultural practices and dispositions, or analysis of disposal of the dead. In non-literate contexts, human remains are often the only means of accessing such dispositions, particularly where anthropomorphic depictions are lacking.

There is a vast literature on the body ranging through various disciplines (see e.g. Douglas, 1966; Bourdieu, 1977; 1990; Foucault, 1977; Turner, 1984; Merleau-Ponty, 1989; Laqueur, 1990; Featherstone *et al.*, 1991; Butler, 1993; Shilling, 1993; Csordas, 1994; Grosz, 1994; Lambek and Strathern, 1998; Strathern, 1999; Weiss and Haber, 1999; Featherstone, 2000; Sweeney and Hodder, 2002a), a recently established journal *Body and Society*, and a growing amount of archaeological literature on the body (e.g. Sørensen, 1991; 1997; 2000; Kus, 1992; Marcus, 1993; Yates, 1993; Treherne, 1995; Lucas, 1996; Meskell, 1996; Joyce, 1998; 2002; 2003; Montserrat, 1998; Hamilakis, 1999; Rautman, 2000; Tarlow, 2000; Hamilakis *et al.*, 2002; Meskell and Joyce, 2003; Sofaer, 2006). It is not the aim of this chapter to review this literature, but to focus on material and theoretical directions relevant to understanding the position of archaeological bodies between biology and culture.

Various related strands of scientific enquiry focus on the physiology and biology of the human body, while avenues of enquiry into the cultural construction of human bodies have remained largely separate from these (Sweeney and Hodder, 2002b, p. 4). In no other discipline is this divide more visible (and untenable) than in archaeology, with osteoarchaeological approaches aligned with medicine, biology and other science-based disciplines, and interpretive approaches (see Hodder, 2005, pp. 209–10) to archaeological bodies tapping into theoretical approaches to the human body within philosophy, sociology and other humanities (Sofaer, 2006, pp. 1–11). These divisions in the investigation of the material remains of past human bodies frequently lead to tensions. The potential of archaeological bodies to elucidate fundamental questions about the human body, in a long-term perspective, are currently not realised. Once researchers integrate theoretical discourse, archaeological practice, analytic methodologies and concrete case studies, bioarchaeology of the body has the potential to contribute in a major way to the wider discussion on human bodies, beyond its own disciplinary boundaries. Such bioarchaeological approaches to the body afford a uniquely long temporal vision, and access to societies different from the present, taking full account of the materiality of the body – all aspects sorely lacking in contemporary social theory (Dobres and Robb, 2000, p. 14).

First, the question of what constitutes the body needs to be addressed. Second, an attempt to move from the dry bones that form the bulk of archaeological bodies back to the bodies of the living, culturally situated individuals is necessary. The section that follows discusses various strands of contemporary body theory that have relevance to the understanding of archaeological bodies. Care should be taken not to import concepts and theoretical dispositions wholesale, but to formulate concepts and theories that take into account the unique time-depth, cultural plurality, and materiality of archaeological remains.

A group of themes illustrating particularly poignantly the location of human bodies between biology and culture are discussed. These include a variety of cultural practices and dispositions, the bodily effects of which the past acting agents would not necessarily have been aware of (gendered work patterns, body techniques), as well as those which were conducted with the conscious aim of shaping or transforming the body or its parts (body modifications, interpersonal violence). The range of themes chosen bring to the fore questions of intentionality and consciousness of the bodily changes in the past, as well as illustrating the potential for bioarchaeological applications of the theoretical considerations on the body that follow.

Body: mindful, non-static, active

What exactly is the body? Commonsense Western notions see the body as the physical entity 'containing' the mind, a popular version of the Cartesian mind-body dualism. However, cross-cultural studies show a wide variety of ways in which the body and its boundaries are viewed by human societies (Moore, 1994), and various strands of academic enquiry use the concept in multiple ways (see, e.g. contributions in Sweeney and Hodder, 2002a). Many researchers are striving to move away from the Cartesian mind-body dualism by eroding the culture/biology split (for a brief history see Sweeney and Hodder, 2002b, pp. 4–8; DeMarrais *et al.*, 2005).

Can the skeleton be equated with the body? What about the fleshed corpse? Do depicted bodies constitute human bodies? What constitutes a 'body part'? Defining a certain entity as a 'body part' is a cultural process (M. Strathern, personal communication, 2002). The difficulties in defining the boundaries of the human body are thrown into relief by research on contemporary intensive care wards, where human tissues are supplemented and integrated with various technological devices (Prout, 2000). While Western societies see the outer skin as the boundary of body and self, in some societies the boundary of self may extend to include objects (Moore, 1994; Hodder, 2000), or body parts may be taken to represent the whole body (Busby, 1997). While the problems of defining the boundaries and character of the human body in past cultural contexts differ from contemporary medical practice, both hinge on the deconstruction of simplistic Cartesian dualisms (Bendelow and Williams, 1995). Within bioarchaeology, the crux of the matter is how to link remains of dead bodies (bones) with the living body (Sofaer, 2006, p. 42).

Sofaer explores how the 'dead body has come to be identified as a biological object and the living body with the cultural subject' (Sofaer, 2006, p. 40; Ingold, 1998, p. 35), critiques the binary conventions of archaeological practice allocating the dead, biological body to osteoarchaeology, and the living, cultural body to interpretive archaeology (Sofaer, 2006, pp. 31–61), and proposes a framework that links osteoarchaeological and interpretive approaches to the body, through the concepts of materiality, material bodies, and bodies as material culture (Sofaer, 2006, pp. 62–88; the concept of materiality is central to the material engagement theory, where materialisation is used to denote the way human societies use aspects of the material world to give expression to symbolic concepts [Renfrew, 2005, pp. 159–63]; cf. DeMarrais *et al.*, 1996; 2005; Jones, 2002; 2004). This chapter concentrates on new agendas in reconstructing the living body through skeletal remains, rather than exploring what can be said about attitudes towards the dead body through contextual analysis of mortuary remains. There is a wealth of interesting work on the manipulation of the dead body and its parts (Strouhal, 1973; Arensburg and Hershkovitz, 1989; Bienert, 1991; Cauwe, 2001), but this lies outside the scope of this paper. The following discussion focuses on the aspects of the living body, rather than the dead body – the latter perhaps better theorised as the corpse, as a transformation of the living, mindful, non-static, and active body.

Mindful

Strathern (1999) discusses mental illness as a case study eroding the Cartesian mind-body dualism. Emotion is also both 'bodily' and 'mental', simultaneously (Lyon, 1997). Neither mental illness nor emotion can be fully understood within the Cartesian framework. Pain is yet another phenomenon that is never the sole creation of anatomy and physiology (Bendelow and Williams, 1995). Thus the body is not divorced from the conscious, thinking and intentional mind (James, 2000, p. 27). Bioarchaeology needs to take off 'from an assumption of the body as simultaneously a physical and symbolic artefact, as both naturally and culturally produced, and as securely anchored in a particular historical moment' (Scheper-Hughes and Lock, 1987, p. 7). Scheper-Hughes and Lock (1987) suggest that it is this 'mindful body' that should be explored, and not just the embodied mind. Bioarchaeology is uniquely placed as regards to such an undertaking, due to its data source affording a long temporal vision, investigation of cultural plurality, as well as the materiality of the body. Strathern (1999) refers to three levels of the mindful body: the individual, experiencing body, the social body (as in Douglas, 1966) and the body politic, i.e. the regulated body, hastening to add that this list can be expanded. These different levels of the mindful body may be accessed in the past, to an extent, through skeletal remains, burials and depictions. While emotion and pain remain largely inaccessible in non-literary archaeological contexts despite some recent efforts (Meskell, 1994; Tarlow, 2000), there are other phenomena that enable us to locate the body between biology and culture, and that can be explored archaeologically. These include gender, body techniques, intentional cultural modifications of the body and interpersonal violence.

Non-static

The body is not static, but ever changing. It grows, ages, is trained, acquires body techniques and is culturally modified. The focus on the bodies of the young in comparison to those of adults of various ages enables one to investigate the human body as a profoundly changing entity as opposed to a static one, shaped by biological (Scheuer and Black, 2000) as well as cultural factors (Valsiner, 2000). It allows for the investigation of the processes, which go into the making of adult bodies, making it impossible to take the traditional archaeological view of the human body simply as a static biological base on which cultural elaborations may be expressed (Knapp and Meskell, 1997). In the following, some pertinent concepts deriving from body theory are discussed, including Mauss' 'body techniques', Strathern's 'becoming body', and Shilling's 'physical capital'.

The very way people walk is a product of cultural processes (Ingold, 1998, p. 26; Valsiner, 2000). While the capability to move the body is physical, sociocultural processes lead to the internalisation of particular, desired or required characteristics of movement. Posture, and the way individuals move their bodies are thus a combination of physiological and cultural factors. The cultural and natural processes that shape, reshape and formalise the body start prior to birth and continue in infancy and childhood (Valsiner, 2000). Mauss first formulated the concept 'techniques of the body' in the 1930s (Mauss, 1979), to denote the culturally contingent way humans move their bodies. Even a function as fundamental to the survival of the body as respiration can be influenced by emotion or by will, and can thus be theorised as a body technique (Lyon, 1997). Some techniques of the body can reshape the form and structure of the body and the skeleton or their parts, thus being accessible through bioarchaeological analysis. Examples include squatting facets (Huard and Montagne, 1950), and other musculoskeletal stress markers (Knüsel, 2000; Eshed *et al.*, 2004).

By using the expression 'the becoming body' Strathern focuses on 'both the sense of the body involved in change over the life cycle and the sense of the body as good, desirable, suitable' (1999, p. 63). He goes on to state that '[w]hat we find is that the good condition of the body is never aimed at in isolation from other actions and values but always in conjunction with these' (Strathen, 1999, pp. 63–4). These considerations are pertinent to the discussion of bodily remains in archaeological contexts. Bodily changes over the life cycle, and bodily appearance that is seen as desirable or suitable during particular periods, in particular geographical regions, on particular sites, can be accessed, to an extent, through the analysis of human remains, burial contexts and anthropomorphic depictions (Sørensen, 1991; 1997; Robb, 2002; Lorentz, 2003; 2004; 2005; 2006; 2008), firmly rooted in appreciation of their socio-political and economic contexts.

Physical capital and its display is a key element in complex societies. The term is used by Shilling to denote acquired bodily manifestations with socio-political and/or economic advantage (1993, p. 127). Specific bodily practices and modifications can be used by social agents to accumulate physical capital. In Prendergast's words: '[T]he body offers us the ground of fabulous potentiality, drawn over by the mappings of a cultural biology on which we act as gendered individuals "endlessly becoming".' (Prendergast, 2000, p. 124). Physical capital can thus be accumulated over time, throughout the life course, while its forms and meanings may change in the *longue durée*. Some bodily practices leading to accumulation of physical capital mark the body permanently (e.g. tattoos). Some of these even affect the human skeleton (e.g. headshaping, Chinese foot-binding). Thus this phenomenon lends itself to archaeological analysis, through human remains. Human remains are the only direct form of evidence available on the past human bodies and individuals. Human remains provide unmediated access to the form and physical make-up of individual bodies, unlike visual representations that are subject to cultural conventions of depiction, or texts with various socio-political motivations. Further, focus on physical capital and its accumulation allows us to investigate the processes of change in the construction of difference, in that the body is viewed here as 'endlessly becoming', rather than as the static biological constant that many archaeological approaches have previously employed, implicitly or explicitly. Both long-term diachronic change, as well as short-term changes over the life course of an individual, can be investigated through focusing on the accumulation of physical capital.

Active

Agency as a term has multiple meanings (Dobres and Robb, 2000, pp. 8–9). What is clear however is that '[b]odily discipline is intrinsic to the competent social agent' (Giddens, 1991, p. 57). In archaeology, the agent is often conceptualised as an individual, a single body (Barrett, 2000). Researchers cross-fertilising their work on agency with aspects deriving from both processual and post-processual archaeology hold that '(1) humans reproduce their being and their social relations through everyday practices; (2) practices take place in material conditions and through material culture', including the body; (3) practices are situated in (aiding and constraining) historical settings with inherited cultural beliefs, attitudes and habits; and (4) 'in action, humans do not simply reproduce their material conditions, inherited structures of meaning, and historical consciousness, but change, reinterpret and redefine them' (Robb, 2005, p. 5).

Thus people not only create the conditions and structures in which they live (Giddens, 1979; 1984; Dobres and Robb, 2000), but shape their very bodies while doing so (Bourdieu, 1977; Lorentz, 2003). Therefore,

(1) The body is not a given, or external to the mind. It is understood here as matter and mind, as a mindful body. Critical works on emotions and pain highlight this issue.
(2) The body is non-static, ever changing. It grows, ages, is trained, acquires body techniques and is modified.
(3) The body is not passive. It is not just a tool or 'man's technical means', but active. Critical works on agency highlight the active nature of bodies.

From bodies to bones and back: studies in theoretical bioarchaeology

The following themes take us from general theoretical considerations on the body to its remains (bones), and back, illustrating current attempts to reconstruct aspects of the living bodies through theoretically informed bioarchaeological analyses of human skeletal remains. First, I turn to sex and gender, an area where the dual split between biology and culture is beginning to erode.

Sex, gender and materiality

Past approaches to gender and sex have viewed sex as natural, based on physiology, and gender as cultural elaboration of physiological difference. More recent archaeological approaches, particularly within interpretive archaeology, view both sex and gender as culturally constructed (after Butler, 1990; 1993). Many such approaches, however, fail to demonstrate how these concepts can be fully explored materially, through actual archaeological remains. Further, the ways in which the physical body is linked to sex and gender, if at all, have been debated extensively (Gibbs, 1987; Nordbladh and Yates, 1990; Meskell, 1996; 1998; 2001; 2002a, b; Conkey and Gero, 1997; Knapp and Meskell, 1997; Gilchrist, 1999; Joyce, 2000a, b; 2002; Sørensen, 2000; Sofaer, 2006, p. 89).

Two largely opposing positions towards sex and gender are discernible in current archaeological practice. Interpretive archaeological approaches have embraced the view that both sex and gender are cultural constructions, and in their extreme these approaches advocate conflating sex and gender, or detaching sex and gender from the physical body (Sofaer, 2006, pp. 89–105). On the contrary, within osteoarchaeology sex is seen as an unambiguously bodily feature, and estimation of sex from the remains of individual bodies, using the sexual dimorphism of the human skeleton (Molleson and Cox, 1993; Schwartz, 1995; Mays and Cox, 2000) is one of the main foci in traditional osteological reports. Osteoarchaeological approaches to gender are few, ranging from the very traditional to some more theoretically aware approaches (Pearson, 1996; Armelagos, 1998; Grauer and Stuart-Macadam, 1998; Walker and Cook, 1998; Mays and Cox, 2000; Sofaer Derevenski, 2000a). Thus, osteoarchaeological approaches are based on a view of sex as a physical feature, while interpretive archaeological approaches advocate seeing not just gender, but also sex as a cultural construct (Sofaer, 2006, p. 98). In contradiction to their own theoretical position, interpretive archaeological approaches often employ osteoarchaeological sex estimates (implying that sex indeed is a bodily feature, rather than a cultural construct) as a basis for analyses of gender, associating artefacts in burial contexts with osteologically sexed bodies, thus superimposing culturally constructed gender onto biological sex, and therefore conflating sex and gender (Sofaer Derevenski, 1998; Sofaer, 2006, pp. 89–90). This results in tensions between theoretical position (both sex and gender seen as cultural constructs) and methodological practice (sex seen as a bodily feature, gender as cultural construct) within interpretive archaeology (Sofaer Derevenski, 1998; Sofaer, 2006, pp. 89–90), leading to conceptual confusion and methodological error. How can these opposing positions be resolved?

While some attempts have been made to offer a theoretical framework for exploring sex and gender within the bioarchaeological realm (Geller, 2005), it is the theoretical position formulated by Sofaer (2006) that seems most apt for resolving tensions between differing approaches to sex and gender employed within interpretive archaeology and osteoarchaeology. Together with Sørensen (2000), Sofaer argues that 'sex and gender need not be equivalent', and that 'maintaining the possibility of both sex and gender allows for a range of possibilities in analysis that collapsing sex with gender as a single construction seems to preclude' (2006, p. 98). Sofaer (2006) sees sex as an analytical category estimated through osteological techniques (skeletal morphology and metrics) and gender as a cultural construction, however with a firm material and bodily basis. Rather than investigating gender by using osteologically derived sex estimates in connection to artefact distribution in mortuary contexts, Sofaer advocates a more complex approach to gender by exposing the biological body as a non-static entity (Sofaer, 2006, pp. 105–16). Processes such as growth, senescence and plasticity constantly modify the form, structure and composition of the human body. The plasticity of the human body (Lasker, 1969; Hulse, 1981; Lerner, 1984; Garruto, 1995; Mascie-Taylor and Bogin, 1995; Roberts, 1995; Schell, 1995) allows environmental processes and cultural practices (including gender-differential practices) to cause changes in the human body and its form and functions. Thus the body is affected through its plasticity by various cultural practices, such as gender-differential work practices (Merbs, 1983; Bridges, 1989; Molleson, 1989; 1994; Lovell and Dublenko, 1999), gendered ways of carrying loads (Sofaer Derevenski, 2000a), and nutrition (White *et al.*, 1993; Schulting and Richards, 2001; Privat *et al.*, 2002; Ambrose *et al.*, 2003). Therefore the human body itself contains material evidence for gender, in the form of changes in its very shape, structure and composition. Sofaer Derevenski provides an example of this in her study of activity-induced osseous change in the skeletal remains from sixteenth to nineteenth century Ensay in the Outer Hebrides, where gender-differential work practices resulted in a material expression of gender on the skeletal bodies (2000a). Gender-differential headshaping, known from various archaeological and ethnographic contexts ranging from Africa to the Americas and the Near East, forms another example of the materialisation of gender (see further discussion below; Özbek, 1974; Lorentz, 2003; 2008). Recognising the human body as a 'project', endlessly becoming, in a very material sense, allows us to move away from the static concept of the biological body present in previous archaeological analyses of gender or other forms of socio-cultural difference. This position is attractive in that it allows for a truly materially grounded investigation of gender, its construction, and the processes of change through the life course.

Body techniques such as ways of moving and eating are learned in infancy and childhood. While there is an emerging archaeology of childhood (Moore and Scott, 1997; Scott, 1999; Sofaer Derevenski, 2000b), approaches taking into account the whole human age range (Sofaer Derevenski, 2000c; Lorentz, 2004; 2005), and its bodily correlates, are needed to provide more fully comprehensive analyses of how bodily changes, and their intentional manipulation (see, e.g.

discussion on body modification below), articulate with cultural perceptions of age in particular socio-cultural contexts.

Social relations, cultural practices and the body

While studies of sex and gender are beginning to bridge the split between osteoarchaeology and interpretive approaches within archaeology, there are many areas where the full potential of integrated biocultural approaches has yet to be realised.

Movement

The dual division of labour between interpretive archaeology and osteoarchaeology discussed above is visible in the analyses of human movement in the past. On the one hand, there are detailed analyses of gesture and posture through anthropomorphic depictions (Karageorghis, J., 1977; 1999; Karageorghis, V., 1991; 1993a, b; 1995; 1998; Morris and Peatfield, 2002; Bachand *et al.*, 2003; Meskell and Joyce, 2003), while on the other there are physical anthropological analyses of musculoskeletal stress markers (Huard and Montagne, 1950; Larsen, 1995; 1997; Robb, 1998b; Mays, 1999; Knüsel, 2000; Eshed *et al.*, 2004) and handedness (Steele, 2000).

The concept of body techniques becomes useful here, aiding us to reconcile the bodies forming the focus of interpretive studies of posture and gesture with the bodies at the centre of osteoarchaeological studies of musculoskeletal activity markers. While it may not be possible to investigate all kinds of human movements through both sets of analytical methods, and while a significant amount of future research is needed to understand the relationship between osseous change and activity more fully (Knüsel, 2000; Pearson and Lieberman, 2004), a common theoretical framework is necessary. Theorising the movement of the body through the concept of techniques of the body, focusing on their physical effects and cultural correlates, allows for more nuanced approaches and interpretations, taking into account the effects of culturally informed ways of movement. Such an approach has the potential to exploit fully the various kinds of archaeological data available, without reproducing the artificial split between the physical and the cultural in the body.

Soafer combined the osteoarchaeological study of activity-induced osseous change in the skeletal remains from sixteenth to nineteenth century Ensay in the Outer Hebrides with investigation of ethnographic and historical sources to arrive at a material investigation of gendered division of labour (Sofaer Derevenski, 2000a; Sofaer, 2006). This study, and others focusing on the physical effects of culturally sanctioned gender-differential practices (Merbs, 1983; Bridges, 1989; Molleson, 1989; 1994; Lovell and Dublenko, 1999), clearly indicates the non-static, culturally contingent nature of the physical body and its movement. The cultural becomes physical in the body.

Food

The dual division of labour between interpretive archaeology and osteo-archaeology is also visible in the analyses of human food consumption. On the one hand, there are the innovative analyses of food consumption and feasting (Sherratt, 1991; 1995; Dietler, 1996; Hamilakis, 1999; 2002), and on the other the increasing corpus of scientific analyses of nutrition (Chapter 7; White *et al.*, 1993; Schulting and Richards, 2001; Privat *et al.*, 2002; Ambrose *et al.*, 2003; Honch *et al.*, 2006; Lösch *et al,* 2006).

Theories of the consuming body formulated within the contemporary social sciences (Featherstone, 1982; Falk, 1997) may not be the most suitable source of theoretical inspiration, as they are based on an individualistic view of the social actor, and thus are inherently Western and contemporary in origin. However, insights into the cross-culturally variable social functions of food, 'a highly condensed social fact' (Appadurai, 1981, p. 494), are useful when attempting to understand food consumption in the past. When making inferences on inequality through differences in food consumption patterns, derived through stable isotope analyses, bioarchaeologists should be wary of simplistic, culturally contingent interpretations, as many cross-cultural studies show that it is not only the types and relative quantities of foodstuffs consumed, but where they originate, the way food is prepared and served, and how it is ingested (body techniques) that lends prestige to the consuming, embodied agent (de Boeck, 1994; Eves, 1996; Wiessner and Schiefenhövel, 1996; Ervynck *et al.*, 2003; Van der Veen, 2003). This points to the need for multifaceted, theoretically informed studies including data from human bioarchaeological analysis of dietary patterns and related phenomena (stable isotopes, dental microwear, antemortem tooth loss and caries) by sex, age and socio-cultural group, as well as data on vessel forms (cooking ware, serving bowl shapes and elaboration), flora and fauna, any residues in containers and cooking ware, and textual and pictorial accounts where available. Archaeobotany and zooarchaeology have taken a head start over human bioarchaeology in integrating theoretical insights on food and its social consumption with scientific analyses of food remains, crystallised in a recent volume of *World Archaeology* (see e.g. Van der Veen, 2003; Ervynck *et al.*, 2003). The interesting results gained within these fields may encourage parallel undertakings in human bioarchaeology.

Body modification

This section focuses on intentional cultural body modification, performed during the lifetime of an individual. Such modifications form examples of the materialisation of symbolic concepts and social relationships (such as gender, status and socio-cultural group membership) through the manipulation of the human body (Lorentz, 2008; in press). Body modifications can be performed in various ways, involving a range of soft tissues as well as skeletal elements. Body modifications may be permanent or reversible, painful or painless (Featherstone, 2000). They may need constant attention (for example Chinese foot-binding) or none at all once instigated (headshaping). They can be results of long-term processes (headshaping), or one-off activities (severing of fingers, dental evulsion). Pitts (1999) has analysed the representation of body modification in current Western media, and criticised the recurrent connection made between body modification as pathological mutilation (see Favazza, 1996) and deformation. It is interesting that much of the academic research on headshaping and dental modifications employs terminology that equates these modifications with deformity, and even mutilation (Dingwall, 1931; Moss, 1958; Ossenberg, 1970; Schendel *et al.*, 1980; Milner and Larsen, 1991; Konigsberg *et al.*, 1993; Özbek, 2001). This is so even where attempts to explore socio-cultural significance are made (Meiklejohn *et al.*, 1992; Watson, 2000). It is argued here that these concepts are unhelpful, and form a hindrance for the understanding of headshaping as well as other forms of intentional cultural modification of the body.

While the literature on modern forms of body modification rapidly expands (Pitts, 2003), most archaeological theorising, including a recent bioarchaeological treatise (Sofaer, 2006) has scarcely addressed intentional cultural body modification and its theoretical implications. Theorising on modern forms of body modification is not directly applicable to the past as it deals with body modification in cultural contexts where the subject makes the decisions on the kinds of modifications performed on his or her body. The range of motivations for body modifications in the contemporary West falls mostly within the enhancement of personal, individualised identity (Favazza, 1996; Featherstone, 2000; Pitts, 2003). The key is individualistic choice, by an agent situated in a socio-cultural context allowing for, and valuing, individualism. However, many traditional societies can be located more towards the 'dividual' end of Bloch's (1989) continuum between individual and dividual societies. The former denotes societies where individualistic action is possible, while the latter refers to societies where communal norms govern individual action more closely. It is unlikely that people would have been able to make decisions on the kind of body modifications performed on their bodies in traditional societies in isolation from communal socio-cultural and political ideas and norms. Further, some types of modifications known in the archaeological record can only be instigated in infancy, not allowing for individual choice by the subject. In the following, I will discuss headshaping as an example of a body modification that is inherently communal in that it can only be instigated by others, during a narrow time window in infancy, when the subject of the modification is hardly able to exercise agency over the modification performed.

Headshaping

Headshaping denotes the intentional modification of the human head form in infancy. Within a particular time window, between birth and the second year, the growth vectors of the infant cranium can be restricted by the use of a cradleboard, bandages (Figure 12.1), or other devices secured to the infant head. These kinds of devices have been extensively documented in the Americas (Allison *et al.*, 1981), but none have yet been found in the Near East or the Eastern Mediterranean, although depictions of cradleboards exist (Theodossiadou, 1991). The cranial bone element growing fastest during the first couple of years of life is the parietal, and thus this region shows usually the most notable changes. The restricting devices must be used consistently, over a considerable period of time, in order to have a lasting impact (information on timings can be gleaned from the modern Dynamic Orthotic Cranioplasty (DOC) practices, effectively a modern form of headshaping, see e.g. Ripley *et al.*,1994; Van Vlimmeren *et al.*, 2006). Headshaping is highly visual, and can be further accentuated by hairstyles and headdresses (Figure 12.1; Kiszely, 1978, Figure 41). While headshaping occurs worldwide (Dingwall, 1931; Kiszely, 1978; Ortner and Putschar, 1981), the oldest uncontested evidence for this cultural practice known to date derives from the Near East and Eastern Mediterranean (Meiklejohn *et al.*, 1992; Lorentz, 2003).

Headshaping, based on the plasticity of the human body, has been extensively documented through methods of physical anthropology (see e.g. Moss, 1958; McNeill and Newton, 1965; Ossenberg, 1970; El-Najjar and Dawson, 1977; Gottlieb, 1978; Schendel *et al.*, 1980; Brown, 1981; Cheverud and Midkiff, 1992; Cheverud *et al.*, 1992; 1993; Gerszten, 1993; Kohn *et al.*, 1993; Konigsberg *et al.*, 1993; Hoshower, *et al.*, 1995; O'Loughlin, 1996; White, 1996; Watson, 2000), especially in the Americas (see e.g. Tiesler Blos, 1998; Torres-Rouff, 2002), but also in the ancient Near East and Eastern Mediterranean (Figure 12.2; Özbek, 1974; 2001; Lorentz, 2003; 2004; 2006; 2008; in press). Very little work informed by body theory has been published to date (but see Lorentz, 2008), yet another example of the divisive nature of the split between disciplines studying the body. Headshaping as a bodily phenomenon cannot however be understood fully through the descriptive accounts in the physical anthropological literature focusing on typologies, effects of

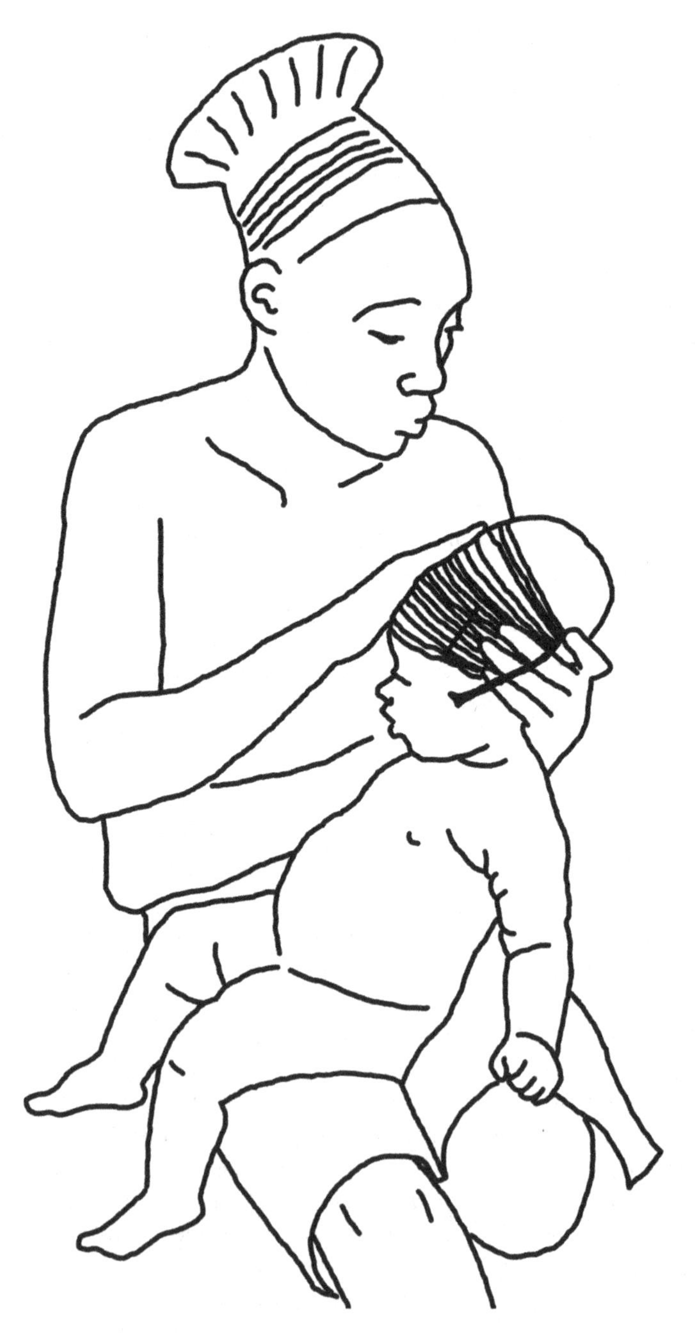

Figure 12.1 Mangbetu woman readjusting the headshaping bandages of an infant (redrawn after Cotlow, 1966).

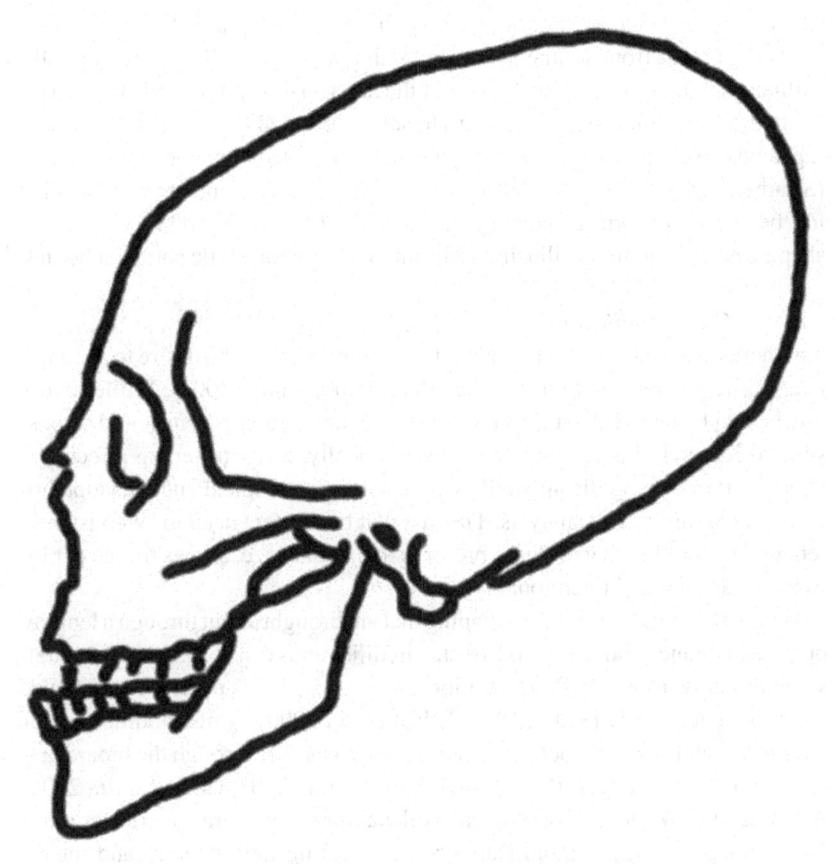

Figure 12.2 Cranium 117b from Byblos showing circumferential type headshaping (redrawn after Özbek, 1974).

sutural complexity, vessel patterns and other morphological changes. While these studies are important within a specific field of physical anthropology and skeletal biology, a biocultural approach is also needed. Theorising the body as between biology and culture allows integrated approaches to headshaping (Lorentz, 2003; 2004; 2008), taking into account both its cultural significance and morphological effects.

The high visibility and permanence of headshaping lends it to use as marker of various forms of cultural difference, including gender, status and socio-cultural group affiliation (ethnic or other), as documented by ethnographies (Dingwall, 1931). The patterning of headshaping in archaeological skeletal series does indeed point to its use in denoting gender, status and/or socio-cultural group affiliation in varying economic and political contexts (Özbek, 1974; Tiesler, 2002; Torres-Rouff, 2002). Headshaping indicates a culturally particular aesthetic for the human head and face, a point further supported by the Neolithic plastered crania from southern Levant, with the jaw modelled on the maxilla, further accentuating the visual effect of the anterior-posterior headshaping performed during life (Arensburg and Herschkovitz, 1989). As such, headshaping lends itself to analyses employing the concept of the 'becoming body' (Strathern, 1999). It can also be argued that, in particular contexts, headshaping has been used to form gendered physical capital (Lorentz, 2008). Thus headshaping as a phenomenon illustrates the mindful and non-static nature of bodies.

Other modifications

Another body modification that was clearly motivated by the desire to accumulate physical capital is foot-binding (Ping, 2000; Gates, 2001). While hardly touched on by physical anthropologists and bioarchaeologists, it should be possible to research this practice bioarchaeologically, as foot-binding affects the bones of the feet significantly, allowing osteoarchaeological and palaeopathological recognition and analysis. The fact that bound feet need to be constantly rebound in adulthood, in order to preserve their small size, allows for the exploration of agency and intentionality.

Unlike foot-binding and headshaping that are brought about through a lengthy process in infancy and childhood, dental modifications can be instigated through singular interventions of short duration. Dental evulsion is notoriously difficult to prove conclusively (Robb, 1997; Eshed *et al.*, 2006), while modification of the surface and shape of tooth crowns is more accessible through the bioarchaeological record (Tiesler, 2002; Williams and White, 2003; Cucina *et al.*, 2004; Arcini, 2005). While anthropological work has long concentrated on description and typology of dental modifications, investigating the visibility and meaning of these modifications requires theoretically informed bioarchaeological approaches integrating morphological and contextual analysis (Robb, 2002).

Soft tissues, such as skin, are rare in archaeological bodies, and the presence of such modifications as tattoos and piercings cannot frequently be ascertained (but see, e.g. Spindler, 2002, p. 155). However, some soft tissue modifications may lead to skeletal lesions, including labret use (Torres-Rouff, 2003), or scalping (Smith, 2003; Hogue, 2006). The latter leads to the final theme here, interpersonal violence.

Interpersonal violence

Many archaeological accounts of long-term changes in particular cultural regions evoke warfare as an explanation for increased fortification, relocation or discontinuity of settlements. Traditional accounts have tried to access warfare and violence through investigating distribution patterns of defensive settlements and fortifications, weaponry, iconography and warrior burials (Lambert, 2002, p. 207). However, the human skeleton constitutes the only firm evidence for interpersonal violence (Brothwell, 1999; Walker, 2001, p. 573; Larsen, 2002, p. 128), whether in the context of war, or smaller-scale conflict within a social unit.

Many physical anthropological accounts have focused on the recognition of different types of weapon-related trauma (see Boylston, 2000), while recent work attempts to investigate patterns of violent trauma against hypotheses of increased social unrest (Larsen, 2002, pp. 128–9; Smith, 2003; Andrushko *et al.*, 2005; Torres-Rouff and Junqueira, 2006). Quantifying types of lesions and their location, as well as patterns of healing according to gender, age, status, socio-cultural group affiliation and other forms of socio-cultural difference, allows a view of potentially differential involvement of individual members of populations as agents and victims of violence (Robb, 1998a; Roberts, 2000).

The presumed evidence for cannibalism in the past has been intensively debated (White, 1992; Dongoske *et al.*, 2000; Brantlinger, 2006), and while corroborative evidence for ingestion of human tissue has been derived from analyses of coprolites (Billman *et al.*, 2000; Lambert *et al.*, 2000; Marlar *et al.*, 2000), this still does not prove conclusively that human tissue was ingested for *nutritional* (as opposed to ritual) purposes, as some environmentally deterministic accounts claim, stating drought and lack of food as motivations (Billman *et al.*, 2000; Larsen, 2002, pp. 130–1). Understanding cannibalism requires theoretical insights into both food and violence, as well as into power, and disposal of the (dead) body. Body parts may not have been perceived as food for nutrition if/when ingested, but consumed to incorporate ancestors, or enemies, and their power (Pickering 1999; Obeyesekere, 2005; Arens, 1979; Goldman, 1999). The debates on cannibalism cannot be resolved without appreciation of the position of the human body between biology and culture.

Many current accounts of interpersonal violence in the past rely on simplistic understandings of power and the body, employing simple hypothesis about violence, social complexity and environment. In order to move forward towards more complex accounts, approximating the lived realities of violence inflicted on and by the bodies of past peoples, researchers not only need to increase the resolution of recording of weapon-related trauma (Boylston, 2000, pp. 375–6), as well as other types of lesions resulting from interpersonal violence (Roberts, 2000, p. 342; Jurmain, 2001; Andrushko *et al.*, 2005), but there is also a need to formulate more sophisticated understandings of power and the body, increasing social complexity and its relation (if any) to violence in particular cultural and socio-political settings. The enormous regional and temporal variation already evident in the bioarchaeological record of violence cannot be explained by universal theories implying uniformity of cause and effect (Thorpe, 2003, p. 145).

Theories of power (Foucault, 1977; Shilling, 1993; Price and Feinman 1995; Earle, 1997; Heyman, 2003), reformulated to take into account the specificities of bioarchaeological data (long-term temporal vision, cultural plurality, materiality), can inform the study of not only bodily violence, but also other, more indirect forms of power exercised through the body. These include, for example, the accumulation of physical capital, leading to socio-political advantage (Shilling, 1993). Not only can bioarchaeological approaches be informed by theories of body and power, but theoretically informed bioarchaeological research can be used to level a critique on the lack of time-depth, cross-cultural awareness, and materiality in some of the current theorising on the body and power. Shilling states that

> [i]n contrast to traditional societies, where power is exercised more directly by one embodied individual over another, the modern body has a far more complex role in the exercise of power and the reproduction of social inequalities. Different classes and class fractions tend to develop distinct orientations to their bodies, which result in the creation of various bodily forms
>
> (Shilling, 1993, p. 128)

Shilling's assertion is clearly untenable when considering the complexity of body modification practices in the past, lending themselves to exercise of power without inflicting violence on the body. It is not only the modern bodies that have a complex role in the exercise of power and reproduction of social inequalities.

Discussion

Bodies between biology and culture

The interesting results and interpretations already gained by the emerging bio-archaeological approaches bridging the cultural and biological aspects of the body (e.g. Grauer and Stuart-Macadam, 1998; Sofaer Derevenski, 2000a; Robb, 2002; Lorentz, 2008) are a proof in themselves of the utility of pursuing such a paradigm. The split between sciences and humanities is not unique to the archaeological study of the body and its remains, but is manifest also in other areas of archaeology (Jones, 2002; 2004; Knapp, 2000; 2002). Jones (2004) advocates focus on materiality in order to develop a framework that allows unifying the different facets of singular phenomena, currently studied through different, but potentially complementary, analytical procedures. Awareness of contemporary body theory can inform the ways in which human bones are studied, and allow for more nuanced reconstructions of past lives and bodies, as long as the theoretical frameworks formulated by bioarchaeologists take into account the specificities of bioarchaeological data: long-term temporal vision, cultural plurality and materiality.

Individualistic versus communal bodies

Human remains allow access to the individual like no other archaeological remains, while also allowing the investigation of communal bodies and social ideas and ideals of and for the body, through patterns of body modification, activity markers, nutrition, health, disease and trauma, investigated by sex, age, and other forms of difference. One of the strengths of bioarchaeology is that it allows a range of approaches at different scales of enquiry, including those concentrating on the biocultural life history of an individual, demographic approaches affording a communal view, and large-scale comparative studies in space as well as time.

Bioarchaeological and ethnographic studies demonstrate very poignantly that the Western concept of the individualistic body does not allow us to explore fully the physical form and socio-cultural meanings of bodies encountered in the skeletal record. When researching the bodies of past individuals, bioarchaeologists need to query any pre-existing assumptions on the possibility of individual/istic choice and intentionality (see Bloch's [1989] theory on individual and dividual societies discussed above).

Human bodies may be theorised as communal on another level also, in that bodies develop and grow in settings shaped by the presence and activities of other people. The body is not a static biological container to be filled with culture (Ingold, 1998, pp. 26–7), but physically embodies historical and cultural specificity (Keating and Miller, 1999; Toren, 1999; Dickens, 2001), created communally.

Conclusion

The study of the body is beginning to bridge the divide between science-based osteoarchaeology and interpretive archaeologies. The human body now needs to be addressed critically, using analytical methods and theoretical approaches developed to understand its materiality, its cultural as well as biological qualities. Some of the areas within which such complex approaches to the body are emerging include gender, movement and body modification, while potential in other areas is evident (interpersonal violence, food, health and disease).

Bioarchaeologists need to be wary of importing wholesale concepts and theories originating in disciplines with standpoints and agendas incompatible with the specificities of bioarchaeology. A uniquely long temporal vision, focus on the materiality of phenomena, and the plurality of societies, practices and processes unknown in the contemporary world afford the bioarchaeologist a viewpoint unique within current strands of academic enquiry. New, theoretically informed bioarchaeological research agendas strive to take full account of the cultural and biological realities of the human body.

References

Allison, M. J., Gerszten, E., Munizaga, J., Santoro, C. and Focacci, G. (1981). La practica de la deformacion craneana entre los pueblos andinos precolombinos. *Chungara Arica*, **7**, 238–45.

Ambrose, S., Buikstra, J. and Harold, W. (2003). Status and gender differences in diet at Mound 72, Cahokia, revealed by isotopic analysis of bone. *Journal of Archaeological Science*, **22**(3), 217–26.

Andrushko, V. A., Latham, K. A. S., Grady, D. L., Pastron, A. G. and Walker, P. L. (2005). Bioarchaeological evidence for trophy-taking in prehistoric central California. *American Journal of Physical Anthropology*, **127**(4), 375–84.

Appadurai, A. (1981). Gastropolitics in Hindu South Asia. *American Ethnologist*, **8**, 494–511.

Arcini, C. (2005). The Vikings bare their filed teeth. *American Journal of Physical Anthropology*, **128**(4), 727–33.

Arens, W. (1979). *The Man-eating Myth: Anthropology and Anthropophagy*. Oxford: Oxford University Press.

Arensburg, B. and Hershkovitz, I. (1989). Artificial skull 'treatment' in the PPNB period: Nahal Hemar. In *People and Culture Change: Proceedings of the Second Symposium on Upper Palaeolithic, Mesolithic and Neolithic Populations of Europe and the Mediterranean Basin*. BAR International Series, ed. I. Herschkovitz. Oxford: Archaeopress, pp. 115–31.

Armelagos, G. J. (1998). 'Introduction: Sex, gender and health status in prehistoric and contemporary populations. In *Sex and Gender in Palaeopathological Perspective*, eds. A. L. Grauer and P. Stuart-Macadam. Cambridge: Cambridge University Press. pp. 1–10.

Bachand, H., Joyce, R. A. and Hendon, J. A. (2003). Bodies moving in space: ancient Mesoamerican sculpture and embodiment. *Cambridge Archaeological Journal*, **13**(2), 238–47.

Barrett, J. (2000). A thesis on agency. In *Agency in Archaeology*, eds. M. Dobres and J. Robb. London: Routledge, pp. 60–8.

Bendelow, G. and Williams, S. (1995). Pain and the mind-body dualism: A sociological approach. *Body and Society*, **1**(2), 83–103.

Bienert, H. D. (1991). Skull cult in the prehistoric Near East. *Journal of Prehistoric Religion*, **5**, 9–23.

Billman, B. R., Lambert, P. M. and Leonard, B. L. (2000). Cannibalism, warfare, and drought in the Mesa Verde region during the twelfth century A.D. *American Antiquity*, **65**, 145–78.

Bloch, M. (1989). Death and the concept of the person. In *On the Meaning of Death*, eds. S. Cederroth, C. Corlin and J. Lindström. Cambridge: Cambridge University Press, pp. 11–29.

Bourdieu, P. (1977). *Outline of a Theory of Practice*. Cambridge: Cambridge University Press.

Bourdieu, P. (1990). *The Logic of Practice*. Cambridge: Polity Press.

Boylston, A. (2000). Evidence for weapon-related trauma in British Archaeological samples. In *Human Osteology in Archaeology and Forensic Science*, eds. M. Cox and S. Mays. London: Greenwich Medical Media Ltd, pp. 357–80.

Brantlinger, P. (2006). Missionaries and cannibals in nineteenth century Fiji. *History and Anthropology*, **17**(1), 21–38.

Bridges, P. S. (1989). Changes in activities with the shift to agriculture in the southeastern United States. *Current Anthropology*, **30**, 385–93.

Brothwell, D. (1999). Biosocial and bioarchaeological aspects of conflict and warfare. In *Ancient Warfare*, eds. J. Carman and A. Harding. Stroud: Sutton, pp. 25–38.

Brown, P. (1981). Artificial cranial deformation: a component in the variation in Pleistocene Australian Aboriginal crania. *Archaeologia Oceania*, **16**, 156–67.

Busby, C. (1997). Permeable and partible persons: a comparison of gender and body in south India and Melanesia. *Journal of the Royal Anthropological Institute*, **3**(2), 261–78.

Butler, J. (1990). *Gender Trouble: Feminism and the Subversion of Identity*. New York: Routledge.

Butler, J. (1993). *Bodies that Matter: On the Discursive Limits of 'Sex'*. London: Routledge.

Cauwe, N. (2001). Skeletons in motion, ancestors in action: early Mesolithic collective tombs in southern Belgium. *Cambridge Archaeological Journal*, **11**(2), 147–63.

Cheverud, J. and Midkiff, J. E. (1992). Effects of fronto-occipital cranial reshaping on mandibular form. *American Journal of Physical Anthropology*, **87**, 167–71.

Cheverud, J. M., Kohn, L. A. P., Konigsberg, L. W. and Leight, S. R. (1992). Effects of fronto-occipital artificial cranial vault modification on the cranial base and face. *American Journal of Physical Anthropology*, **88**(3), 323–45.

Cheverud, J. M., Kohn, L. A. P., Leight, S. R. and Jacobs, S. C. (1993). Effects of annular cranial vault modification on the cranial base and face. *American Journal of Physical Anthropology*, **90**, 147–68.

Conkey, M. W. and Gero, J. M. (1997). Programme to practice: gender and feminism in archaeology. *Annual Review of Anthropology*, **26**, 411–37.

Cotlow, L. (1966). *In Search of the Primitive*. Boston: Little Brown & Co.

Csordas, T. (ed.) (1994). *Embodiment and Experience: The Existential Ground of Culture and Self*. Cambridge: Cambridge University Press.

Cucina, A., Neff, H. and Tiesler, V. (2004). Provenance of African origin individuals from the colonial cemetery of Campeche (Mexico) by means of LA-ICP-MS. *American Journal of Physical Anthropology Supplement*, **38**, 80.

de Boeck, F. (1994). 'When hunger goes around the land': hunger and food among the aLuund of Zaire. *Man*, **29**(2), 257–82.

DeMarrais, E., Castillo, L. J. and Earle, T. (1996). Ideology, materialization and power ideologies. *Current Anthropology*, **37**, 15–31.

DeMarrais, E., Gosdon, C. and Renfrew, C. (eds.) (2005). *Rethinking Materiality: The Engagement of Mind with the Material World*. Cambridge: McDonald Institute.

Dickens, P. (2001). Linking the social and natural sciences: is capital modifying human biology in its own image? *Sociology*, **35**(1), 93–110.

Dietler, M. (1996). Feasts and commensal politics in the political economy: food, power and status in prehistoric Europe. In *Food and the Status Quest: An Interdisciplinary Perspective*, eds. P. Wiessner and W. Schiefenhövel. Oxford: Berghahn, pp. 87–125.

Dingwall, E. J. (1931). *Artificial Cranial Deformation: a Contribution to the Study of Ethnic Mutilations*. London: J. Bale, Sons & Danielsson.

Dobres, M. and Robb, J. (2000). Agency in archaeology: Paradigm or platitude? In *Agency in Archaeology*, eds. M. Dobres and J. Robb. London: Routledge, pp. 3–17.

Dongoske, K. E., Martin, D. L. and Ferguson, T. J. (2000). Critique of the claim of cannibalism at Cowboy Wash. *American Antiquity*, **65**, 179–90.

Douglas, M. (1966). *Purity and Danger: An Analysis of Concepts of Pollution and Taboo*. London: Routledge & Kegan Paul.

Earle, T. (1997). *How Chiefs Come to Power*. Stanford: Stanford University Press.

El-Najjar, M. Y. and Dawson, G. L. (1977). The effect of artificial cranial deformation on the incidence of wormian bones in the lambdoidal suture. *American Journal of Physical Anthropology*, **46**, 155–60.

Ervynck, A., Van Neer, W., Huster-Plogmann, H. and Schibler, J. (2003). Beyond affluence: The zooarchaeology of luxury. *World Archaeology*, **34**(3), 428–41.

Eshed, V., Gopher, A., Galili, E. and Hershkovitz, I. (2004). Musculoskeletal stress markers in Natufian hunter-gatherers and Neolithic farmers in the Levant: the upper limb. *American Journal of Physical Anthropology*, **123**(4), 303–15.

Eshed, V., Gopher, A. and Hershkovitz, I. (2006). Tooth wear and dental pathology at the advent of agriculture: New evidence from the Levant. *American Journal of Physical Anthropology*, **130**, 145–59.

Eves, R. (1996). Remembrance of things passed: memory, body and the politics of feasting in New Ireland, Papua New Guinea. *Oceania*, **66**, 266–77.

Falk, P. (1997). *The Consuming Body*. London: Sage Publications.

Favazza, A. R. (1996). *Bodies Under Siege: Self-mutilation and Body Modification in Culture and Psychiatry*. 2nd edn. London: The Johns Hopkins University Press.

Featherstone, M. (1982). The body in consumer culture. *Theory, Culture and Society*, **1**(2), 18–33.

Featherstone, M. (ed.) (2000). *Body Modification*. London: Sage Publications.

Featherstone, M., Hepworth, M. and Turner, B. S. (eds.) (1991). *The Body: Social Process and Cultural Theory*. London: Sage.

Foucault, M. (1977). *Discipline and Punish: The Birth of the Prison*. London: Allen Lane.

Garruto, R. M. (1995). Biological adaptability, plasticity and disease: patterns in modernizing societies. In *Human Variability and Plasticity*, eds. C. G. N. Mascie-Taylor and B. Bogin. Cambridge: Cambridge University Press, pp. 190–212.

Gates, H. (2001). Footloose in Fujian: Economic correlates of footbinding. *Comparative Studies of Society and History*, **43**(1), 130–48.

Geller, P. L. (2005). Skeletal analysis and theoretical complications. *World Archaeology*, **37**(4), 597–609.

Gerszten, P. C. (1993). Investigation into the practice of cranial deformation among the Pre-Columbian peoples of northern Chile. *International Journal of Osteoarchaeology*, **3**(2), 87–98.

Gibbs, L. (1987). Identifying gender representation in the archaeological record: A contextual study. In *The Archaeology of Contextual Meanings*, ed. I. Hodder. Cambridge: Cambridge University Press, pp. 79–89.

Giddens, A. (1979). *Central Problems in Social Theory: Action, Structure and Contradiction in Social Analysis*. Berkeley: University of California Press.

Giddens, A. (1984). *The Construction of Society: Outline of a Theory of Structuration*. Berkeley: University of California Press.

Giddens, A. (1991). *Modernity and Self Identity*. Cambridge: Polity Press.

Gilchrist, R. (1999). *Gender and Archaeology: Contesting the Past*. London: Routledge.

Goldman, L. (ed.) (1999). *The Anthropology of Cannibalism*. Westport: Bergin & Garvey.

Gottlieb, K. (1978). Artificial cranial deformation and the increased complexity of the lambdoid suture. *American Journal of Physical Anthropology*, **48**, 213–14.

Grauer, A. L. and Stuart-Macadam, P. (eds.) (1998). *Sex and Gender in Palaeopathological Perspective*. Cambridge: Cambridge University Press, pp. 1–10.

Grosz, E. (1994). *Volatile Bodies: Towards a Corporeal Feminism*. Bloomington: Indiana University Press.

Hamilakis, Y. (1999). Food technologies/technologies of the body: the social context of wine and oil production and consumption in Bronze Age Crete. *World Archaeology*, **31**(1), 38–54.

Hamilakis, Y. (2002). The past as oral history: towards an archaeology of the senses. In *Thinking Through the Body: Archaeologies of Corporeality*, eds. Y. Hamilakis, M. Pluciennik and S. Tarlow. New York: Kluwer Academic/ Plenum Publishers, pp. 121–36.

Hamilakis, Y., Pluciennik, M. and Tarlow, S. (eds.) (2002). *Thinking Through the Body: Archaeologies of Corporeality*. New York: Kluwer Academic/ Plenum Publishers.

Heyman, J. McC. (2003). The inverse power. *Anthropological Theory*, **3**(2), 139–56.

Hodder, I. (2000). Agency and individuals in long-term processes. In *Agency in Archaeology*, eds. M. Dobres and J. Robb. London: Routledge, pp. 21–33.

Hodder, I. (2005). Post-processual and interpretive archaeology. In *Archaeology: The Key Concepts*, eds. C. Renfrew and P. Bahn. London: Routledge, pp. 207–12.

Hogue, S. H. (2006). Determination of warfare and interpersonal conflict in the protohistoric period: A case study from Mississippi. *International Journal of Osteoarchaeology*, **16**(3), 236–48.

Honch, N. V., Higham, T. F. G., Chapman, J., Gaydarska, B. and Hedges, R. E. M. (2006). A palaeodietary investigation of carbon ($^{13}C/^{12}C$) and nitrogen ($^{15}N/^{14}N$) in human and faunal bones from the Copper Age cemeteries of Varna I and Durankulak, Bulgaria. *Journal of Archaeological Science*, **33**(11), 1493–646.

Hoshower, L. M., Buikstra, J. E., Goldstein, P. S. and Webster, A. D. (1995). Artificial cranial deformation at the Omo M10 site: A Tiwanaku complex from the Moquegua Valley, Peru. *Latin American Antiquity*, **6**(2), 145–64.

Huard, P. and Montagne, M. (1950). Le squelette humain et l'attitude accroupie. *Bulletin de la Société des Etudes Indochinoises*, **25**, 401–26.

Hulse, F. S. (1981). Habits, habitats and heredity: a brief history of studies in human plasticity. *American Journal of Physical Anthropology*, **56**(4), 495–501.

Ingold, T. (1998). From complimentary to obviation: on dissolving the boundaries between social and biological anthropology, archaeology and psychology. *Zeitschrift für Ethnologie*, **123**, 21–52.

James, A. (2000). Embodied being(s): understanding the self and the body in childhood. In *The Body, Childhood and Society*, ed. A. Prout. London: Macmillan Press Ltd, pp. 19–37.

Jones, A. (2002). *Archaeological Theory and Scientific Practice*. Cambridge: Cambridge University Press.

Jones, A. (2004). Archaeometry and materiality: Materials-based analysis in theory and practice. *Archaeometry*, **46**(3), 327–38.

Joyce, R. (1998). Performing the body in pre-Hispanic Central America. *Res*, **3**, 147–65.

Joyce, R. A. (2000a). Girling the girl and boying the boy: the production of adulthood in ancient Mesoamerica. *World Archaeology*, **31**(3), 473–83.

Joyce, R. A. (2000b). *Gender and Power in Prehispanic Mesoamerica*. Austin: University of Texas Press.

Joyce, R. A. (2002). Beauty, sexuality, body ornamentation and gender in ancient Meso-America. In *In Pursuit of Gender: Worldwide Archaeological Approaches*, eds. S. Milledge Nelson and M. Rosen-Ayalon. Walnut Creek: Alta Mira Press, pp. 81–91.

Joyce, R. A. (2003). Making something of herself: embodiment in life and death in Playa de los Muertos, Honduras. *Cambridge Archaeological Journal*, **13**(2), 248–61.

Jurmain, R. (2001). Paleoepidemiological patterns of trauma in a prehistoric population from central California. *American Journal of Physical Anthropology*, **115**(1), 13–23.

Karageorghis, J. (1977). *La Grande Déesse de Chypre et Son Culte: à Travers l'Iconographie, de l'Epoque Néolithique au VIéme s. a. C.* Lyon: Maison de l'Orient.

Karageorghis, J. (1999). *Coroplastic Art of Cyprus V: The Cypro-Archaic Period Small Female Figurines. B. Figurines Moulées*. Nicosia: Leventis Foundation.

Karageorghis, V. (1991). *Coroplastic Art of Cyprus I: Chalcolithic – Late Cypriot I*. Nicosia: Leventis Foundation.

Karageorghis, V. (1993a). *Coroplastic Art of Cyprus II: Late Cypriot II Cypro-Geometric III*. Nicosia: Leventis Foundation.

Karageorghis, V. (1993b). *Coroplastic Art of Cyprus III: The Cypro-Archaic Period Large and Medium Size Sculpture*. Nicosia: Leventis Foundation.

Karageorghis, V. (1995). *Coroplastic Art of Cyprus IV: The Cypro-Archaic Period Small Male Figurines*. Nicosia: Leventis Foundation.

Karageorghis, V. (1998). *Coroplastic Art of Cyprus V: The Cypro-Archaic Period Small Female Figurines. A. Handmade/Wheelmade Figurines*. Nicosia: Leventis Foundation.

Keating, D. and Miller, F. (1999). Individual pathways in competence and coping: from regulatory systems to habits of mind. In *Developmental Health and the Wealth of Nations*, eds. D. Keating and C. Hertzman. New York: Guildford, pp. 220–34.

Kiszely, I. (1978). *The Origins of Artificial Cranial Formation in Eurasia from the Sixth Millennium BC to the Seventh Century AD*. BAR International Series (Supplementary) 50. Oxford: BAR.

Knapp, B. (2000). Archaeology, science-based archaeology, and the Mediterranean Bronze Age metals trade. *European Journal of Archaeology*, **3**(1), 31–56.

Knapp, B. (2002). Disciplinary fault lines: science and social archaeology. *Mediterranean Archaeology and Archaeometry*, **2**(2), 37–44.

Knapp, B. and Meskell, L. (1997). Bodies of evidence on prehistoric Cyprus. *Cambridge Archaeological Journal*, **7**(2), 183–204.

Knüsel, C. (2000). Bone adaptation and its relationship to physical activity in the past. In *Human Osteology in Archaeology and Forensic Science*, eds. M. Cox and S. Mays. London: Greenwich Medical Media, pp. 381–402.

Kohn, L., Leigh, S., Jacobs, S. and Cheverud, J. (1993). Effects of annular cranial vault modification on the cranial base and face. *American Journal of Physical Anthropology*, **90**(2), 147–68.

Konigsberg, L. W., Kohn, L. and Cheverud, J. (1993). Cranial deformation and nonmetric trait variation. *American Journal of Physical Anthropology*, **90**(1), 35–48.

Kus, S. (1992). Toward an archaeology of body and soul. In *Representations in Archaeology*, eds. J.-C. Gardin and C. Peebles. Bloomington: Indiana University Press, pp. 168–77.

Lambek, M. and Strathern, A. (eds.) (1998). *Bodies and Persons: Comparative Studies from Africa and Melanesia*. Cambridge: Cambridge University Press.

Lambert, P. M. (2002). The archaeology of war: A North American perspective. *Journal of Archaeological Research*, **10**(3), 207–41.

Lambert, P. M., Leonard, B. L., Billman, B. M. *et al.* (2000). Response to critique of the claim of cannibalism at Cowboy Wash. *American Antiquity*, **65**, 397–406.

Laqueur, T. W. (1990). *Making Sex: Body and Gender from the Greeks to Freud*. Cambridge: Harvard University Press.

Larsen, C. S. (1995). Biological changes in human populations with agriculture. *Annual Review of Anthropology*, **24**, 185–213.

Larsen, C. S. (1997). *Bioarchaeology: Interpreting Human Behavior from the Human Skeleton*. Cambridge: Cambridge University Press.

Larsen, C. S. (2002). Bioarchaeology: The lives and lifestyles of past people. *Journal of Archaeological Research*, **10**(2), 119–66.

Lasker, G. (1969). Human biological adaptability. *Science*, **166**, 1480–6.

Lerner, R. M. (1984). *On the Nature of Human Plasticity*. Cambridge: Cambridge University Press.

Lorentz, K. O. (2003). Minding the body: The growing body in Cyprus from the Aceramic Neolithic to the Late Bronze Age. Unpublished Ph.D. thesis, University of Cambridge.

Lorentz, K. O. (2004). Age and gender in Eastern Mediterranean prehistory: Depictions, burials and skeletal evidence. *Ethnographisch-Archaologische Zeitschrift*, **45**, 297–315.

Lorentz, K. O. (2005). Late Bronze Age burial practices: age as a form of social difference. In *Cyprus: Religion and Society from the Late Bronze Age to the End of the Archaic Period*, ed. V. Karageorghis. Erlangen. Mohnesee–Wamel: Bibliopolis, pp. 41–55; Plates 5–8.

Lorentz, K. O. (2006). The malleable body: Headshaping in Greece and the surrounding regions. In *New Directions in the Skeletal Biology of Greece*, eds. S. C. Fox, C. Bourbou and L. Schepartz. Occasional Wiener Laboratory Series. Athens: Wiener Laboratory, the American School of Classical Studies in Athens.

Lorentz, K. O. (2008). From lifecourse to *longue durée*: headshaping as gendered capital? In *Gender Through Time in the Ancient Near East*, ed. D. Bolger. Lanham: AltaMira, pp. 281–312.

Lorentz, K. O. (in press). Ubaid headshaping: Negotiations of identity through physical appearance?. In *The Ubaid and Beyond: Exploring the Transmission of Culture in the Developed Prehistoric Societies of the Middle East*. eds. R. Carter and G. Philip.

Lösch, S., Grupe, G. and Peters, J. (2006). Stable isotopes and dietary adaptations in humans and animals at Pre-Pottery Neolithic Nevali Cori, Southeast Anatolia. *American Journal of Physical Anthropology*, **131**, 181–93.

Lovell, N. C. and Dublenko, A. (1999). Further aspects of fur trade life depicted in the skeleton. *International Journal of Osteoarchaeology*, **9**(4), 248–59.

Lucas, G. (1996). Of death and debt: a history of the body in Neolithic and Early Bronze Age Yorkshire. *Journal of European Archaeology*, **37**, 227–75.

Lyon, M. (1997). The material body, social processes and emotion: 'Techniques of the body' revisited. *Body and Society*, **3**(1), 83–101.

Marcus, M. (1993). Incorporating the body: adornment, gender, and social identity in ancient Iran. *Cambridge Archaeological Journal*, **39**(2), 157–78.

Marlar, R. A., Leonard, B. L., Billman, B. M., Lambert, P. M. and Marlar, J. E. (2000). Biochemical evidence of cannibalism at a prehistoric Puebloan site in southwestern Colorado. *Nature*, **407**, 74–8.

Mascie-Taylor, C. G. N. and Bogin, B. (eds.) (1995). *Human Variability and Plasticity*. Cambridge: Cambridge University Press.

Mauss, M. (1979) [1934]. Body techniques. In *Sociology and Psychology. Essays by Marcel Mauss, Part IV*, M. Mauss, trans. B. Brewster. London: Routledge & Kegan Paul, pp. 97–123.

Mays, S. (1999). A biomechanical study of activity patterns in a medieval human skeletal assemblage. *International Journal of Osteoarchaeology*, **9**(1), 68–73.

Mays, S. and Cox, M. (2000). Sex determination in skeletal remains. In *Human Osteology in Archaeology and Forensic Science*, eds. M. Cox and S. Mays. London: Greenwich Medical Media, pp. 117–30.

McNeill, R. W. and Newton, G. N. (1965). Cranial base morphology in association with intentional cranial vault deformation. *American Journal of Physical Anthropology*, **23**, 241–54.

Meiklejohn, C., Agelarakis, A., Akkermans, P. A., Smith, P. E. L. and Solecki, R. (1992). Artificial cranial deformation in the proto-Neolithic and Neolithic Near East and its possible origin: Evidence from four sites. *Paleorient*, **18**(2), 83–97.

Merbs, C. F. (1983). *Patterns of Activity-Induced Pathology in a Canadian Inuit Population*. National Museum of Man Mercury Series 119. Ottawa: Archaeological Survey of Canada.

Merleau-Ponty, M. (1989) [1962]. *Phenomenology of Perception*. London: Routledge.

Meskell, L. (1994). Dying young: the experience of death at Deir el Medina. *Archaeological Review from Cambridge*, **13**(2), 35–45.

Meskell, L. (1996). The somatization of archaeology: institutions, discourses, corporeality. *Norwegian Archaeological Review*, **29**(1), 1–16.

Meskell, L. (1998). The irresistible body and the seduction of archaeology. In *Changing Bodies, Changing Meanings: Studies of the Body in Antiquity*, ed. D. Montserrat, London: Routledge, pp. 139–61.

Meskell, L. (2001). Archaeologies of identity. In *Archaeological Theory Today*, ed. I. Hodder. Cambridge: Polity Press, pp. 187–213.

Meskell, L. (2002a). Cycles of life and death: narrative homology and archaeological realities. *World Archaeology*, **31**(3), 423–41.

Meskell, L. (2002b). Writing the body in archaeology. In *Reading the Body: Representations and Remains in the Archaeological Record*, ed. A. E. Rautman. Philadelphia: University of Pennsylvania Press, pp. 13–21.

Meskell, L. and Joyce, R. A. (2003). *Embodied Lives: Figuring Ancient Maya and Egyptian Experience*. London: Routledge.

Milner, G. R. and Larsen, C. S. (1991). Teeth as artefacts of human behaviour: Intentional mutilation and accidental modification. In *Advances in Dental Anthropology*, eds. M. A. Kelley and C. S. Larsen. New York: Wiley-Liss, pp. 357–78.

Molleson, T. (1989). Seed preparation in the Mesolithic: the osteological evidence. *Antiquity*, **63**, 356–62.

Molleson, T. (1994). The eloquent bones of Abu Hureyra. *Scientific American*, **271**, 70–5.

Molleson, T. and Cox, M. (1993). *The Spitalfields Project: Volume II – The Anthropology: The Middling Sort*. London: Council for British Archaeology Report 86.

Montserrat, D. (ed.) (1998). *Changing Bodies, Changing Meanings: Studies on the Human Body in Antiquity*. London: Routledge.

Moore, H. (1994). *A Passion for Difference: Essays in Anthropology and Gender*. Cambridge: Polity Press.

Moore, J. and Scott, E. (eds.) (1997). *Invisible People and Processes: Writing Gender and Childhood into European Archaeology*. London: Leicester University Press.

Morris, C. and Peatfield, A. (2002). Feeling through the body: gesture in Cretan Bronze Age Religion. In *Thinking through the Body: Archaeologies of Corporeality*, eds. Y. Hamilakis, M. Pluciennik and S. Tarlow. New York: Kluwer Academic/Plenum Publishers, pp. 105–120.

Moss, M. L. (1958). The pathogenesis of artificial cranial deformation. *American Journal of Physical Anthropology*, **16**(3), 269–86.

Nordbladh, J. and Yates, T. (1990). This perfect body, this virgin text: between sex and gender in archaeology. In *Archaeology after Structuralism*, eds. I. Bapty and T. Yates, London: Routledge, pp. 222–37.

O'Loughlin, V. D. (1996). Comparative endocranial vascular changes due to craniosynostosis and artificial cranial deformation. *American Journal of Physical Anthropology*, **101**, 369–85.

Obeyesekere, G. (2005). *Cannibal Talk: The Man-Eating Myth and Human Sacrifice in the South Pacific*. Berkeley: University of California Press.

Ortner, D. J. and Putschar, W. G. J. (1981). *Identification of Pathological Conditions in Human Skeletal Remains*. Washington: Smithsonian Institution Press.

Ossenberg, N. S. (1970). The influence of artificial cranial deformation on discontinuous morphological traits. *American Journal of Physical Anthropology*, **33**, 357–72.

Özbek, M. (1974). Etude de la deformation cranienne artificielle chez les chalcolithiques de Byblos (Liban). *Bulletin et Memoires de la Societé d'Anthropologie de Paris*, **1**, 455–81.

Özbek, M. (2001). Cranial deformation in a subadult sample from Degirmentepe (Chalcolithic, Turkey). *American Journal of Physical Anthropology*, **115**, 238–44.

Pearson, G. A. (1996). Of sex and gender. *Science*, **274**, 328–9.

Pearson, O. M. and Lieberman, D. E. (2004). The aging of Wolff's 'Law': Ontogeny and responses to mechanical loading in cortical bone. *Yearbook of Physical Anthropology*, **47**, 68–99.

Pickering, M. (1999). Consuming doubts: What some people ate? Or what some people swallowed?. In *The Anthropology of Cannibalism*, ed. L. Goldman. Westport: Bergin & Garvey.

Ping, W. (2000). *Aching for Beauty: Footbinding in China*. Minneapolis: University of Minnesota Press.

Pitts, V. (1999). Body modifications, self-mutilation and agency in Media accounts of a subculture. *Body and Society*, **5**(2–3), 291–303.

Pitts, V. (2003). *In the Flesh: The Cultural Politics of Body Modification*. New York: Palgrave Macmillan Press.

Prendergast, S. (2000). 'To become dizzy in our turning': girls, body-maps and gender as childhood ends. In *The Body, Childhood and Society*. ed. A. Prout, London: Macmillan, pp. 101–24.

Price, T. D. and Feinman, G. (eds.) (1995). *Foundations of Social Inequality*. New York: Plenum.

Privat, K. L., O'Connell, T. C. and Richards, M. (2002). Stable isotope analysis of human and faunal remains from the Anglo-Saxon cemetery at Berinsfield, Oxfordshire: dietary and social implications. *Journal of Archaeological Science*, **29**(7), 779–90.

Prout, A. (ed.) (2000). *The Body, Childhood and Society*. London: Macmillan Press.

Rautman, D. (ed.) (2000). *Reading the Body: Representations and Remains in the Archaeological Record*. Philadelphia: University of Pennsylvania Press.

Renfrew, C. (2005). Material engagement and materialisation. In *Archaeology: The Key Concepts*, eds. C. Renfrew and P. Bahn. London and New York: Routledge, pp. 159–63.

Ripley, C. E., Pomatto, J., Beals, S. P. *et al.* (1994). Treatment of positional plagiocephaly with dynamic orthotic cranioplasty. *Journal of Craniofacial Surgery*, **5**, 150–9.

Robb, J. (1997). Intentional tooth removal in Neolithic Italian women. *Antiquity*, **71**, 659–69.

Robb, J. (1998a). Violence and gender in early Italy. In *Violence and Warfare in the Past*, eds. D. W. Frayer and D. Martin, London: Routledge, pp. 111–44.

Robb, J. (1998b). The interpretation of skeletal muscle sites: a statistical approach. *International Journal of Osteoarchaeology*, **8**(5), 363–77.

Robb, J. (2002). Time and biography: Osteobiography of the Italian Neolithic lifespan. In *Thinking Through the Body: Archaeologies of Corporeality*, eds. Y. Hamilakis, M., Pluciennik and S. Tarlow. New York: Kluwer Academic/Plenum Publishers, pp. 153–71.

Robb, J. (2005). Agency. In *Archaeology: The Key Concepts*, eds. C. Renfrew and P. Bahn. London: Routledge, pp 3–7.

Roberts, C. (2000). Trauma in biocultural perspective: Past, present and future work in Britain. In *Human Osteology in Archaeology and Forensic Science*, eds. M. Cox and S. Mays. London: Greenwich Medical Media Ltd, pp. 337–56.

Roberts, D. F. (1995). The pervasiveness of plasticity. In *Human Variability and Plasticity*, eds. C. G. N. Mascie-Taylor and B. Bogin. Cambridge: Cambridge University Press, pp. 1–17.

Schell, L. M. (1995). Human biological adaptability with special emphasis on plasticity: history, development and problems for future research. In *Human Variability and Plasticity*, eds. C. G. N. Mascie-Taylor and B. Bogin. Cambridge: Cambridge University Press, pp. 213–37.

Schendel, S. A., Walker, G. and Kamisugi, A. (1980). Hawaiian craniofacial morphometrics: Average Mokapuan skull, artificial cranial deformation, and the 'rocker' mandible. *American Journal of Physical Anthropology*, **52**, 491–500.

Scheper-Hughes, N. and Lock, M. (1987). The mindful body: Prolegomenon to future work in medical anthropology. *Medical Anthropology Quarterly*, **1**, 6–41.

Scheuer, L. and Black, S. (2000). *Developmental Juvenile Osteology*. London: Academic Press.

Schulting, R. and Richards, M. (2001). Dating women and becoming farmers: new palaeodietary and AMS evidence from the Breton Mesolithic cemeteries of Téviec and Hoëdic. *Journal of Archaeological Science*, **20**(3), 314–44.

Schwartz, J. H. (1995). *Skeleton Keys: An Introduction to Human Skeletal Morphology, Development, and Analysis*. Oxford: Oxford University Press.

Scott, E. (1999). *The Archaeology of Infancy and Infant Death*. BAR International Series 819. Oxford: Archaeopress.

Sherratt, A. (1991). Palaeoethnobotany: from crops to cuisine. In *Paleoecologia e Arquelogia II*, eds. F. Queiroga and A. P. Dinis. Villa Nova de Famalicao: Centro de Estudos Arquelogicos Famalicences, pp. 221–36.

Sherratt, A. (1995). Alcohol and its alternatives: symbol and substance in pre-industrial cultures. In *Consuming Habits: Drugs in History and Anthropology*, eds. J. Goodman, P. E. Lovejoy and A. Sherratt. London: Routledge, pp. 11–46.

Shilling, C. (1993). *The Body and Social Theory*. London: Sage.

Smith, M. O. (2003). Beyond palisades: the nature and frequency of late prehistoric deliberate violent trauma in the Chickamauga Reservoir of East Tennessee. *American Journal of Physical Anthropology*, **121**(4), 303–18.

Sofaer, J. R. (2006). *The Body as Material Culture: A Theoretical Osteoarchaeology*. Cambridge: Cambridge University Press.

Sofaer Derevenski, J. (1998). Gender archaeology as contextual archaeology. Unpublished Ph.D. thesis, University of Cambridge.

Sofaer Derevenski, J. (2000a). Sex differences in activity-related osseous change in the spine and the gendered division of labour at Ensay and Wharram Percy, UK. *American Journal of Physical Anthropology*, **111**(3), 333–54.

Sofaer Derevenski, J. (2000b). *Children and Material Culture*. London: Routledge.

Sofaer Derevenski, J. (2000c). Rings of life: The role of early metalwork in mediating the gendered life course. *World Archaeology*, **31**(3), 389–406.

Sørensen, M. L. S. (1991). Gender construction through appearance. In *The Archaeology of Gender: proceedings of the 22nd Annual Chamcool Conference*, eds. D. Walde and N. D. Willows. Calgary: Archaeological Association of the University of Calgary, pp. 121–9.

Sørensen, M. L. S. (1997). Reading dress: the construction of social categories and identities in Bronze Age Europe. *Journal of European Archaeology*, **5**(1), 93–114.

Sørensen, M. L. S. (2000). *Gender Archaeology*. Cambridge: Polity Press.

Spindler, K. (2002). The Iceman's body – the 5000 year old glacial mummy from the Ötztal Alps. In *The Body*, eds. T. Sweeney and I. Hodder. Cambridge: Cambridge University Press, pp. 142–68.

Steele, J. (2000). Skeletal indicators of handedness. In *Human Osteology in Archaeology and Forensic Science*, eds. M. Cox and S. Mays. London: Greenwich Medical Media Ltd, pp. 307–23.

Strathern, A. J. (1999). *Body Thoughts*. Ann Arbor: The University of Michigan Press.

Strouhal, E. (1973). Five plastered skulls from Pre-Pottery Neolithic B Jericho – anthropological study. *Paleorient*, **1**, 231–47.

Sweeney, T. and Hodder, I. (eds.) (2002a). *The Body*. Cambridge: Cambridge University Press.

Sweeney, T. and Hodder, I. (2002b). Introduction. In *The Body*, eds. T. Sweeney and I. Hodder. Cambridge: Cambridge University Press, pp. 1–11.

Tarlow, S. (2000). Emotion in archaeology. *Current Anthropology*, **41**(5), 713–46.

Theodossiadou, M. (1991). Models of furniture in terracotta during Early and Middle Bronze Age in Cyprus. In *Cypriot Terracottas: Proceedings of the First International Conference of Cypriot Studies*. eds. F. Vandenabeele and R. Laffineur, Brussels-Liege: Buteneers s.p.r.l.

Thorpe, I. J. N. (2003). Anthropology, archaeology, and the origin of warfare. *World Archaeology*, **35**(1), 145–65.

Tiesler, V. (2002). New cases of an African tooth decoration from colonial Campeche, Mexico. *Homo*, **52**(3), 277–82.

Tiesler Blos, V. (1998). *La Costumbre de la Deformación Cefálica entre los Antiguos Mayas: Aspectos Morfológicos y Culturales*. México City: Instituto Nacional de Antropologia e Historia.

Toren, C. (1999). *Mind, Materiality and History: Explorations in Fijian Ethnography*. London: Routledge.

Torres-Rouff, C. (2002). Cranial vault modification and ethnicity in Middle Horizon San Pedro de Atacama, Chile. *Current Anthropology*, **43**(1), 163–71.

Torres-Rouff, C. (2003). Oral implications of labret use: a case from pre-Columbian Chile. *International Journal of Osteoarchaeology*, **13**(4), 247–51.

Torres-Rouff, C. and Junqueira, M. A. C. (2006). Interpersonal violence in prehistoric San Pedro de Atacama, Chile: Behavioral implications of environmental stress. *American Journal of Physical Anthropology*, **130**(1), 60–70.

Treherne, P. (1995). The warrior's beauty: The masculine body and self-identity in Bronze Age Europe. *Journal of European Archaeology*, **3**, 105–44.

Turner, B. (1984). *The Body and Society: Explorations in Social Theory*. London: Sage.

Valsiner, J. (2000). *Culture and Human Development: An Introduction*. London: Sage Publications.

Van der Veen, M. (2003). When is a food a luxury? *World Archaeology*, **34**(3), 405–27.

Van Vlimmeren, L. A., Helders, P. J. M., Van Adrichem, L. N. A. and Engelbert, R. H. H. (2006). Torticollis and plagiocephaly in infancy: Therapeutic strategies. *Pediatric Rehabilitation*, **9**(1), 40–6.

Walker, P. L. (2001). A bioarchaeological perspective on the history of violence. *Annual Review of Anthropology*, **30**, 573–96.

Walker, P. L. and Cook, D. C. (1998). Gender and sex: vive la difference. *American Journal of Physical Anthropology*, **106**(2), 255–9.

Watson, J. T. (2000). A quantitative study of artificial cranial deformation: bio-cultural behavior in Southwest prehistory. *American Journal of Physical Anthropology, 2000 Supplement AAPA Abstracts*, **111**(30, Suppl.), 314–5.

Weiss, G. and Haber, H. F. (eds.) (1999). *Perspectives on Embodiment: The Intersections of Nature and Culture*. New York: Routledge.

White, C. D. (1996). Sutural effects of fronto-occipital cranial modification. *American Journal of Physical Anthropology*, **100**, 397–410.

White, C. D., Healy, P. F. and Schwartz, H. P. (1993). Intensive agriculture, social status, and Maya diet at Pacbitun, Belize. *Journal of Anthropological Research*, **49**(375), 347.

White, T. D. (1992). *Prehistoric Cannibalism at Mancos 5MTUMR-2346*. Princeton: Princeton University Press.

Wiessner, P. and Schiefenhövel, W. (eds.) (1996). *Food and the Status Quest: An Interdisciplinary Perspective*. Oxford: Berghahn.

Williams, J. S. and White, C. D. (2003). Dental decoration during the Postclassic at Lamanai, Nelize: Sex and status differences. *American Journal of Physical Anthropology, Supplement*, **36**, 226.

Yates, T. (1993). Frameworks for the archaeology of the body. In *Interpretive Archaeology*, ed. C. Tilley, Oxford: Berg, pp. 31–72.

Index